T0309660

GEOLOGY OF CARBONATE RESERVOIRS

GEOLOGY OF CARBONATE RESERVOIRS
The Identification, Description, and Characterization of Hydrocarbon Reservoirs in Carbonate Rocks

WAYNE M. AHR
Texas A&M University

WILEY

A JOHN WILEY & SONS, INC., PUBLICATION

Copyright © 2008 by John Wiley & Sons, Inc. All rights reserved.

Published by John Wiley & Sons, Inc., Hoboken, New Jersey
Published simultaneously in Canada

No part of this publication may be reproduced, stored in a retrieval system, or transmitted in any form or by any means, electronic, mechanical, photocopying, recording, scanning, or otherwise, except as permitted under Section 107 or 108 of the 1976 United States Copyright Act, without either the prior written permission of the Publisher, or authorization through payment of the appropriate per-copy fee to the Copyright Clearance Center, Inc., 222 Rosewood Drive, Danvers, MA 01923, 978-750-8400, fax 978-750-4470, or on the web at www.copyright.com. Requests to the Publisher for permission should be addressed to the Permissions Department, John Wiley & Sons, Inc., 111 River Street, Hoboken, NJ 07030, 201-748-6011, fax 201-748-6008, or online at http://www.wiley.com/go/permiossion.

Limit of Liability/Disclaimer of Warranty: While the publisher and author have used their best efforts in preparing this book, they make no representations or warranties with respect to the accuracy or completeness of the contents of this book and specifically disclaim any implied warranties of merchantability or fitness for a particular purpose. No warranty may be created or extended by sales representatives or written sales materials. The advice and strategies contained herein may not be suitable for your situation. You should consult with a professional where appropriate. Neither the publisher nor author shall be liable for any loss of profit or any other commercial damages, including but not limited to special, incidental, consequential, or other damages.

For general information on our other products and services or for technical support, please contact our Customer Care Department within the United States at 877-762-2974, outside the United States at 317-572-3993 or fax 317-572-4002.

Wiley publishes in a variety of print and electronic formats and by print-on-demand. Some material included with standard print versions of this book may not be included in e-books or in print-on-demand. If this book refers to media such as a CD or DVD that is not included in the version you purchased, you may download this material at http://booksupport.wiley.com. For more information about Wiley products, visit www.wiley.com

Library of Congress Cataloging-in-Publication Data:

Ahr, Wayne M.
    Geology of carbonate reservoirs : the identification, description, and characterization of hydrocarbon reservoirs in carbonate rocks / Wayne M. Ahr.
        p. cm.
    Includes index.
    ISBN 978-0-470-16491-4 (cloth)
    1. Rocks, Carbonate. 2. Carbonate reservoirs–Geology. 3. Geology, Stratigraphic. I. Title.
    QE471.15.C3.A34 2008
    553.2'8—dc22

                                                                                2007051417

10 9 8 7 6 5 4 3

CONTENTS

PREFACE   xi

ABOUT THIS BOOK   xv

1   INTRODUCTION   1

     1.1   Definition of Carbonate Reservoirs / 2
         1.1.1   Carbonates / 2
         1.1.2   Reservoirs / 4
     1.2   Finding and Developing Carbonate Reservoirs / 6
         1.2.1   Sources of Data on Reservoirs / 7
     1.3   Unique Attributes of Carbonates / 9
         Suggestions for Further Reading / 12
         Review Questions / 12

2   CARBONATE RESERVOIR ROCK PROPERTIES   13

     2.1   Definitions / 13
     2.2   Fundamental Rock Properties / 14
         2.2.1   Texture / 15
         2.2.2   Fabric / 18
         2.2.3   Composition / 20
         2.2.4   Sedimentary Structures / 20
     2.3   Classification of Carbonate Rocks / 20
         2.3.1   Classification of Detrital Carbonates / 27
         2.3.2   Classification of Reef Rocks / 28
         2.3.3   Wright's Genetic Classification / 30

2.4   Dependent or Derived Rock Properties / 30

  2.4.1   Porosity / 31

  2.4.1.1   *Porosity Classifications* / 34

  2.4.1.2   *The Archie Classification* / 35

  2.4.1.3   *The Choquette–Pray Classification* / 36

  2.4.1.4   *The Lucia Classification* / 39

  2.4.2   A New Genetic Classification for Carbonate
  Porosity / 42

  2.4.3   Permeability / 44

2.5   Tertiary Rock Properties / 47

  2.5.1   Borehole Logs and Carbonate Reservoirs / 47

  2.5.2   Tertiary Rock Properties and the Seismograph / 53

  Suggestions for Further Reading / 54

  Review Questions / 54

3   PETROPHYSICAL PROPERTIES OF CARBONATE RESERVOIRS   56

3.1   Saturation, Wettability, and Capillarity / 56

  3.1.1   Saturation / 56

  3.1.2   Wettability / 62

  3.1.3   Capillarity / 63

3.2   Capillary Pressure and Reservoir Performance / 64

  3.2.1   Capillary Pressure, Pores, and Pore Throats / 66

  3.2.2   Converting Air–Mercury Capillary Pressures to
  Oil–Water Equivalents / 69

  3.2.3   Height of Oil Column Above Free-Water Level / 70

  3.2.4   Evaluating Seal Capacity / 70

3.3   Fluid Withdrawal Efficiency / 71

  Suggestions for Further Reading / 74

  Review Questions / 74

4   STRATIGRAPHIC PRINCIPLES   76

4.1   Carbonate Depositional Platforms / 77

  4.1.1   Rimmed and Open Shelves / 80

  4.1.2   Homoclinal and Distally Steepened Ramps / 82

4.2   Rock, Time, and Time–Rock Units / 83

  4.2.1   Rock Units / 83

  4.2.2   Time Units / 84

  4.2.3   Time–Rock Units / 86

4.3   Correlation / 86

4.4   Anatomy of Depositional Units / 88

   4.4.1   Facies, Successions, and Sequences / 91

   4.4.2   Environmental Subdivisions and Standard Depositional
            Successions / 93

4.5   Sequence Stratigraphy / 99

   4.5.1   Definitions and Scales of Observation / 99

   4.5.2   Sequence Stratigraphy in Carbonate Reservoirs / 102

   4.5.3   Sequence Stratigraphy in Exploration and Development / 102

   Suggestions for Further Reading / 104

   Review Questions / 105

5   DEPOSITIONAL CARBONATE RESERVOIRS   106

5.1   Depositional Porosity / 108

5.2   Depositional Environments and Processes / 109

   5.2.1   The Beach–Dune Environment / 110

   5.2.2   Depositional Rock Properties in Beach–Dune
            Successions / 112

   5.2.3   Tidal-Flat and Lagoon Environments / 117

   5.2.4   Depositional Rock Properties in Tidal Flat–Lagoon
            Successions / 119

   5.2.5   The Shallow Subtidal (Neritic) Environment / 121

   5.2.6   Depositional Rock Properties in Shallow Subtidal
            Successions / 123

   5.2.7   The Slope-Break Environment / 124

   5.2.8   Depositional Rock Properties in Slope-Break
            Successions / 125

   5.2.9   The Slope Environment / 126

   5.2.10 Depositional Rock Properties in the Slope and
            Slope-Toe Environments / 128

   5.2.11 Basinal Environments / 129

   5.2.12 Depositional Rock Properties in Basinal
            Environments / 130

   5.2.13 Ideal Depositional Successions Illustrated / 133

5.3   Paleotopography and Depositional Facies / 134

5.4   Diagnosis and Mapping of Depositional Reservoirs / 137

   Suggestions for Further Reading / 141

   Review Questions / 141

6   DIAGENETIC CARBONATE RESERVOIRS   144

6.1   Diagenesis and Diagenetic Processes / 144

   6.1.1   Definition of Diagenesis / 145

   6.1.2   Diagenetic Processes / 146

6.2   Diagenetic Porosity / 150

6.3   Diagenetic Environments and Facies / 153

    6.3.1   Diagenetic Facies / 155

6.4   Diagenetically Enhanced Porosity / 156

    6.4.1   Enhancement by Recrystallization / 158

    6.4.2   Enhancement by Solution Enlargement / 160

    6.4.3   Large-Scale Dissolution-Related Porosity / 161

    6.4.4   Porosity Enhancement by Replacement / 163

    6.4.5   Recognizing Enhanced Porosity / 163

6.5   Porosity Reduction by Diagenesis / 164

    6.5.1   Pore Reduction by Compaction / 165

    6.5.2   Pore Reduction by Recrystallization / 165

    6.5.3   Pore Reduction by Replacement / 166

    6.5.4   Pore Reduction by Cementation / 167

    6.5.5   Recognizing Diagenetically Reduced Porosity / 170

6.6   Diagnosing and Mapping Diagenetic Reservoirs / 171

Suggestions for Further Reading / 174

Review Questions / 175

7   FRACTURED RESERVOIRS   176

7.1   Fractures and Fractured Reservoirs / 176

    7.1.1   Definition of Fractures / 177

    7.1.2   Types of Fractures / 177

    7.1.3   Genetic Classification of Fractures / 178

    7.1.4   Fracture Morphology / 181

    7.1.5   Where Do Fractures Occur? / 184

7.2   Fracture Permeability, Porosity, and S_w / 186

    7.2.1   Fracture Permeability / 187

    7.2.2   Fracture Porosity / 188

    7.2.3 S_w in Fractured Reservoirs / 189

7.3   Classification of Fractured Reservoirs / 190

7.4   Detecting Fractured Reservoirs / 191

    7.4.1   Direct Observation of Fractures in the Borehole / 192

    7.4.2   Indirect Methods to Detect Fractures in the Borehole / 192

7.5   Predicting Reservoir Fracture Spacing and Intensity / 195

    7.5.1   Factors that Influence Fracture Spacing and Intensity / 195

7.6   Identifying and Developing Fractured Reservoirs / 195

Suggestions for Further Reading / 198

Review Questions / 198

8 SUMMARY: GEOLOGY OF CARBONATE RESERVOIRS   200

8.1 Rock Properties and Diagnostic Methods / 201
    8.1.1 Fundamental Rock Properties and Depositional
        Reservoirs / 202
    8.1.2 Reservoir Morphology / 203
    8.1.3 Derived Properties: Porosity and Permeability / 204
    8.1.4 Tertiary Properties and Petrophysical
        Characteristics / 204
8.2 Data Requirements / 206
    8.2.1 Regional Scale Investigations / 207
    8.2.2 Field Scale Studies / 207
    8.2.3 Quality Ranking of Flow Units / 208
    8.2.4 Pore Scale Features / 209
8.3 Depositional Reservoirs / 209
    8.3.1 Finding and Interpreting Depositional Reservoirs / 210
    8.3.2 Selected Examples of Depositional Reservoirs / 213
        8.3.2.1 North Haynesville Field / 214
        8.3.2.2 Conley Field / 219
8.4 Diagenetic Reservoirs / 224
    8.4.1 Finding and Interpreting Diagenetic Reservoirs / 224
    8.4.2 Field Examples of Diagenetic Reservoirs / 226
        8.4.2.1 Overton FIeld / 227
        8.4.2.2 Happy Field / 231
8.5 Fractured Reservoirs / 239
    8.5.1 Finding and Interpreting Fractured Reservoirs / 239
    8.5.2 Field Examples of Fractured Reservoirs / 240
        8.5.2.1 Quanah City Field / 241
        8.5.2.2 Dickinson Field / 244
8.6 Conclusions / 249
    Review Questions / 254

REFERENCES   **255**

INDEX   **269**

PREFACE

This is a book on the geology of hydrocarbon reservoirs in carbonate rocks. Although it is written for petroleum geologists, geophysicists, and engineers, it can be useful as a reference for hydrogeologists and environmental geologists because reservoirs and aquifers differ only in the fluids they contain. Environmental geoscientists interested in contaminant transport or hazardous waste disposal also need to know about porosity (capacity to store) and permeability (capacity to flow) of subsurface formations. The first two chapters focus on definitions and on rock properties that influence fluid movement. The third chapter focuses on reservoir properties—the interaction between rocks and fluids—and how rock properties influence saturation, wettability, capillarity, capillary pressure, and reservoir "quality." Although carbonate rocks differ in many ways from siliciclastic rocks, the laws of physics that govern fluid movement in terrigenous sandstones also govern fluid behavior in carbonates; therefore many of the principles discussed in this text are applicable to reservoirs and aquifers in any porous and permeable rock. There are fundamental differences between carbonates and siliciclastic rocks that will be emphasized thoughout, and knowing those differences can be used to advantage in exploration, development, and management of reservoirs and aquifers.

This book evolved from my graduate course on carbonate reservoirs at Texas A&M University. It is written as a textbook for geologists, engineers, and geophysicists in graduate and upper-level undergraduate courses. I hope it may also be useful for continuing education courses and as a reference book for industry professionals, especially for those who are not experts on carbonate rocks and reservoirs. It is not easy to write a survey of this subject in about 300 pages with a limited number of illustrations; consequently, this book emphasizes only fundamental principles. The vast literature on carbonate sedimentology, stratigraphy, geochemistry, and petrography makes it impractical if not impossible to include an extensive bibliography on all of those subjects. I did not include much material on borehole logging and seismology because they require lengthy explanations with examples that exceed the scope, purpose, and size limits of this book. I have tried to address these potential shortcomings by including suggestions for additional reading at the end of each chapter. This is a book for students—not for experts. Having taught university classes and continuing education courses for the past 38 years, I have learned that there are limits to what can be taught effectively in one university term or in a continuing education short course. I limited the material in the book to that which I believe can be taught in one university term or an intense, week-long short course. Clearly, I had to choose subjects and reference material carefully, focusing on the subjects I have found most helpful in understanding carbonate rocks and reservoirs. Other texts on carbonate reservoirs, including those by Chilingar et al. (1992), Lucia (1999), and Moore (2001), concentrate on engineering aspects of carbonate reservoirs (Chilingar and Lucia) or on sequence stratigraphy as it relates to carbonates (Moore), but they are not textbooks on the general geology of carbonate reservoirs. I have written this book to help university students and industry professionals learn more about how, when, why, and where carbonate reservoirs form, and about how to recognize, analyze, and map the end-member reservoir types in carbonates—reservoirs with depositional, diagenetic, or fracture porosity systems. Special emphasis is given to relationships between genetic pore types and carbonate reservoir properties. To that end, a new classification of carbonate porosity that focuses on the genetic pore types is presented. Two themes are repeated throughout the book: (1) it is not possible to understand carbonate reservoirs without looking at the rocks; and (2) one cannot accurately predict the spatial distribution of rock and reservoir properties that are linked by cause–effect mechanisms without using a genetic classification of carbonate porosity.

Development geologists and engineers will find the book useful, as will exploration geophysicists and geologists. Development geologists and engineers will find the book helpful because it emphasizes the relationships between rock and reservoir characteristics. Explorationists should find the distinction between genetic pore types in carbonate reservoirs helpful because exploration strategies need to be built around geological concepts that are in turn based on knowledge of how and where porosity and permeability may occur together in depositional and diagenetic facies or in fractured rocks.

There is a tradition among petroleum geologists to search for analogs or "look-alikes" for exploration or production prospects. This noncritical application of geological form over critically analyzed substance presumes that reservoir models can be exported from one geological age and setting to another with little concern about possible differences in reservoir characteristics. All too often, geologists find themselves having to explain why the "look-alike" failed to predict depositional or diagenetic porosity loss, or why structural and stratigraphic models for exploration prospects did not turn out to be realistic after the drill reached target depth. Analogs offer comfortable "sameness" but they provide no help to explain the unexpected. They lack information to find hydrocarbon reservoirs in the wide variety of geological situations that typify carbonates and they lack information needed to develop carbonate reservoirs in the most efficient and profitable ways.

This book emphasizes ways to formulate geological concepts rather than "look-alikes" to predict the spatial distribution of porosity and permeability. Optimum combinations of porosity and permeability and least resistance to fluid flow are called flow units (the origins of this term are discussed later). When flow units are identified and ranked on their rock and reservoir properties, accurate maps, volumetric calculations, and economic forecasts can be made. Primary recovery methods have produced only about one-third of the world's original oil in place, leaving an estimated 891 billion barrels or more (Ahlbrandt et al., 2005). If unconventional sources of oil and natural gas are included, the figure will be even larger. If reservoir flow units could be mapped with a higher degree of precision than was available previously, then a significant percentage of those 1 trillion barrels of remaining oil could be within reach with novel methods of improved recovery. Knowing the size, shape, and connectivity of flow units, secondary and tertiary recovery methods are economically attractive, especially at current oil prices. This rings especially true when one considers the extreme cost of deep water drilling and production, the risk of geopolitical conflicts, and the risk of drilling dry holes as compared to extracting bypassed hydrocarbons from proven fields. Also importantly, if hydrogeologists have accurate maps of aquifer connectivity, their models for groundwater flow or contaminant transport pathways will be greatly improved. If flow barriers were more accurately mapped, site evaluation for dangerous waste disposal could be improved significantly. These are only a few of the exciting reasons to learn more about carbonate reservoirs and aquifers.

I would not have embarked on this project without the encouragement of the graduate students who have taken my course on carbonate reservoirs over past years and who have continually asked me to write a book for the course. Old friend Robert Stanton read some of the early chapters and offered helpful comments. Rick Major and P. M. (Mitch) Harris read early versions of the entire manuscript and gave encouragement, guidance that kept me on track, and criticisms that greatly improved the book.

WAYNE M. AHR

College Station, Texas
December 2007

ABOUT THIS BOOK

To understand carbonate rocks at reservoir scale, one first has to understand them at pore scale. Carbonate reservoirs are porous and permeable rocks that contain hydrocarbons. Carbonate porosity includes three end-member genetic categories: purely depositional pores, purely diagenetic pores, and purely fracture pores. Intermediate types exist, of course, but the point is that there are three main types of carbonate porosity that represent distinctly different geological processes. Before one can fully appreciate these differences and be proficient at distinguishing between the varieties of carbonate reservoir types, one must understand what carbonates are, how and where they form, and how they become reservoirs. One must understand the differences between reservoirs, traps, and seals and learn to appreciate that reservoir characterization is the study of rocks *plus* the fluids they contain. The operative word is rocks. Carbonate rocks consist of component particles and maybe some lime mud matrix and cement. The skeletal and nonskeletal particles, along with mud and cement, hold an enormous amount of information about the depositional and diagenetic environments that produced the reservoir rock. This book begins with definitions, with discussions about how, where, and why carbonates are formed and about how fundamental rock properties are used to create a language for communicating information about the rocks—carbonate rock classifications. Reservoir porosity and permeability are variables that depend on fundamental rock properties. The book explores how rock classifications do or do not correspond with conventional porosity classifications. Reservoirs contain fluids; therefore we explore reservoir properties such as saturation, wettability, capillarity, and capillary pressure.

Geophysical (borehole) logs are briefly mentioned because they provide information about third-order rock properties. Logs provide important information to develop static and dynamic reservoir models, to calculate fluid properties such as saturation and movable oil volumes, to make stratigraphic correlations, and to interpret lithological characteristics in boreholes where no rock samples are available.

Logs are only briefly mentioned because an extensive literature on logs and log interpretation already exists. Today's digital technology and sophisticated computer software have expanded the need for petrophysicists who specialize in computer-assisted log interpretation. Even with the modern computer-assisted log evaluation software available in almost every company and university laboratories, the working geoscientists still must be familiar with the types of logs that are useful in studying carbonate reservoirs.

Seismic methods for exploration and development are mentioned only briefly because a satisfactory treatment of seismological methods in exploration and reservoir analysis is beyond the scope of this book. Selected references are given at the end of each chapter to help the reader find more information.

Following the discussions on the hierarchical order of rock properties and the different reservoir characteristics, basic sedimentological and stratigraphic principles are reviewed to explain carbonate platform characteristics, stratigraphic relationships, and depositional facies. This background is intended to guide the reader into depositional models and greatly simplified, "standard depositional successions" that characterize different platform types. The standard depositional successions will become models for depositional reservoirs—reservoir rock bodies with depositional porosity. Following the discussions of depositional models and depositional reservoir types, diagenetic environments and diagenetic processes are introduced to illustrate how carbonate reservoir porosity is enhanced, reduced, or created by the chemical and mechanical processes that typify each diagenetic environment. Finally, fractured reservoirs are reviewed after the reader has a thorough grasp of rock and reservoir properties, along with of depositional and diagenetic processes and attributes. Checklists for the diagnosis and interpretation of depositional, diagenetic, and fractured reservoirs are given at the end of each of the respective chapters. A summary of the topics covered in the book and selected field examples of depositional, diagenetic, and fractured reservoirs round out the final chapter.

W. M. A.

CHAPTER ONE

INTRODUCTION

The goal of this book is to explain in plain language for the nonspecialist how and where carbonate rocks form, how they do, or do not, become reservoirs, how to explore for carbonate reservoirs or aquifers in the subsurface, and how to develop them once they have been found. The book is organized around a genetic classification of carbonate porosity and ways it can be employed in exploration and development. The genetic categories include three end members—depositional pores, diagenetic pores, and fractures. Genetic pore categories are linked with geological processes that created, reduced, or enlarged pores during lithification and burial. In the end, a chronology of pore origin and evolution is developed to put in the larger stratigraphic context for identification of reservoir flow units, baffles, and barriers. Connectivity can be evaluated by determining the range of porosity and permeability values for the different pore categories within reservoirs. Connected pore systems can be correlated stratigraphically to identify reservoir zones that have the highest combined porosity and permeability and the least resistance to the passage of fluids. Such zones are defined in this book as reservoir *flow units* somewhat similar to the definition of Ebanks (1987; Ebanks et al., 1992) but different in that rock units that impede flow are defined as baffles and units that prevent flow are defined as barriers. Each end-member reservoir type generally has characteristic pore-scale features (porosity and permeability) that correspond to petrologic and stratigraphic properties (borehole-scale features). When the zones with good, fair, and poor connectivity are identified, the characteristic petrologic and stratigraphic features that correspond with them can become *proxies for connectivity*. The larger scale features, or proxies, are generally easier to identify in borehole cores, on wireline log traces, and in some sequence stratigraphic "stacking patterns." When mode and time of origin of the proxies are known, geological concepts can be formulated to predict the

Geology of Carbonate Reservoirs: The Identification, Description, and Characterization of Hydrocarbon Reservoirs in Carbonate Rocks
By Wayne M. Ahr   Copyright © 2008 John Wiley & Sons, Inc.

spatial distribution of reservoir flow units at field scale. In other words, the fundamental rock properties that correspond to good, fair, and poor combined values of porosity and permeability can be identified and put in larger stratigraphic context, or "scaled-up." Then the temporal and genetic characteristics of the large-scale petrologic and stratigraphic properties (proxies) are used for reservoir prediction and flow unit mapping.

Carbonate reservoir porosity usually represents the combined effects of more than one geological process. Sometimes it reflects multiple episodes of change during burial history; therefore particular care must be given to identification of the sequence of events that led to the final array of rock properties and pore characteristics. Usually it is possible to identify *cross-cutting relationships* between rock properties so that their relative times of origin are distinguishable. Reservoir porosity governed only by depositional rock properties, a rather uncommon occurrence, will not exhibit cross-cutting relationships because rock texture, fabric, porosity, and permeability share a single mode and time of origin. In that case, reservoir architecture and spatial distribution conform to depositional facies boundaries. These reservoirs are referred to as *stratabound*, and porosity is *facies-selective, fabric-selective*, or both. Diagenesis and fracturing do not always follow depositional unit boundaries. Although carbonate reservoirs exist in which diagenetic porosity corresponds with depositional rock properties (fabric-selective or facies-selective diagenesis), in many instances it does not. In the latter case, it is especially important to identify the type of alteration, how it was formed, when it was formed, and what cross-cutting relationships it shares with other diagenetic and fracture attributes. Fractures cut across most rock boundaries but there are some fundamental rock properties that dictate how and where fractures will form. Fractures happen as a result of brittle failure under differential stress, usually in conjunction with faulting or folding. Fault and fold geometry can be determined; therefore it follows that associated fracture patterns can also be determined. In short, there are many rock and petrophysical characteristics in carbonates that expose a wealth of information about the origin and architecture of carbonate reservoirs.

1.1 DEFINITION OF CARBONATE RESERVOIRS

1.1.1 Carbonates

Carbonates are anionic complexes of $(CO_3)^{2-}$ and divalent metallic cations such as Ca, Mg, Fe, Mn, Zn, Ba, Sr, and Cu, along with a few less common others. The bond between the metallic cation and the carbonate group is not as strong as the internal bonds in the CO_3 structure, which in turn are not as strong as the covalent bond in carbon dioxide (CO_2). In the presence of hydrogen ions, the carbonate group breaks down to produce CO_2 and water. This breakdown reaction, commonly experienced when acid is placed on limestone, is the chemical basis for the fizz test that distinguishes carbonates from noncarbonates. It is also used to distinguish dolostones, which fizz slowly, from limestones, which fizz rapidly. Carbonates occur naturally as sediments and reefs in modern tropical and temperate oceans, as ancient rocks, and as economically important mineral deposits. The common carbonates are grouped into families on the basis of their crystal lattice structure, or the internal arrangement of atoms. The families are known by the crystal systems in which they form, namely, the hexagonal, orthorhombic, and monoclinic crystallographic systems. The most common carbonate minerals are in the hexagonal system, notably calcite ($CaCO_3$) and dolomite ($Ca,Mg(CO_3)_2$) (Figures 1.1 and 1.2). Aragonite has the same composition as calcite, $CaCO_3$, but it crystallizes in the orthorhombic system. The monoclinic system is characterized by the beautiful blue and green copper carbonates—azurite and malachite, respectively. Calcite and aragonite are polymorphs of calcium carbonate because they share the same composition but have different crystal structures. Dolomite, like calcite, crystallizes in the hexagonal

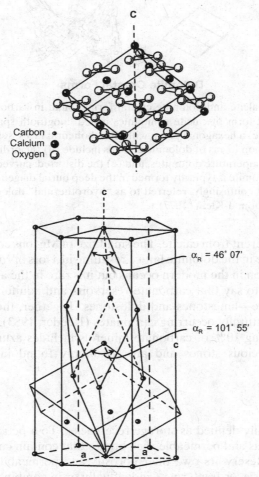

Carbon •
Calcium ●
Oxygen ○

$\alpha_R = 46° 07'$

$\alpha_R = 101° 55'$

Figure 1.1 Internal atomic (lattice) structure of calcite. The ball-and-stick model at the top of the figure shows the position and orientation of calcium and carbonate ions in layers, or sheets, within the lattice. Note that the orientation of the triangular carbonate ions changes in alternate layers from top to bottom. The bottom drawing shows the hexagonal crystal structure of calcite, the scalenohedral calcite unit cell, and the position of cleavage rhombs with respect to the c crystallographic axes. (Adapted from illustrations in Hurlbut and Klein (1977).)

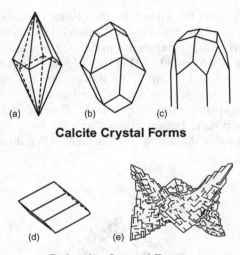

Calcite Crystal Forms

Dolomite Crystal Forms

Figure 1.2   Typical calcite and dolomite crystal forms found in carbonate reservoir rocks: (a) the scalenohedral form of calcite sometimes called "dogtooth spar"; (b) a compound rhombohedral form; (c) a hexagonal prism with rhombohedral faces, sometimes called "nail-head spar"; two common forms of dolomite crystals include (d) the ordinary rhombohedron, typical of most low-temperature dolomites, and (e) the distorted, curved form called "saddle dolomite." Saddle dolomite is typically formed in the deep burial diagenetic environment and is sometimes, perhaps confusingly, referred to as "hydrothermal" dolomite. (Adapted from illustrations in Hurlbut and Klein (1977).)

system, but it is different from calcite. The small size of Mg ions compared to calcium ions causes a change in the dolomite lattice resulting in a loss of rotational symmetry. Aragonite is common in the modern oceans but it is rare in the ancient rock record; therefore it is safe to say that carbonate reservoirs and aquifers are composed of calcite and dolomite—limestones and dolostones. Together, those rocks make up about 90% of all naturally occurring carbonates (Reeder, 1983). Only a small fraction of the remaining 10% of carbonate minerals includes azurite and malachite, which are semiprecious stones and are commonly found in jewelry or other ornaments.

1.1.2   Reservoirs

Reservoirs are usually defined as storage receptacles. To a petroleum geoscientist, reservoirs are porous and permeable rock bodies that contain commercial amounts of hydrocarbons. Reservoirs owe their porosity and permeability to processes of deposition, diagenesis, or fracturing—individually or in combination. Although we will focus on hydrocarbon reservoirs in carbonate rocks, many porous and permeable carbonates are groundwater aquifers. Reservoirs are three-dimensional bodies composed of rock matrix and networks of interconnected pores. If the three-dimensional geometry (size and shape) of a connected pore system is known, it is possible to (1) determine drilling locations in exploration or development prospects, (2) estimate the volume of the resource in the reservoir or aquifer, (3) achieve optimum extraction of the resource, (4) determine the practicality of drilling additional (infill) wells to achieve the optimum spacing between field wells during development, and (5) predict the path that will be taken by injected fluids as they "sweep" remaining hydrocarbons during secondary and enhanced recovery. In the broad sense, reservoir studies include reservoir geology, reservoir characterization, and reservoir engineering. To avoid confusion in terminology about carbonate reservoirs, some common terms are discussed in the following paragraphs.

Reservoir geology deals with the origin, spatial distribution, and petrological characteristics of reservoirs. The reservoir geologist utilizes information from sedimentology, stratigraphy, structural geology, sedimentary petrology, petrography, and geochemistry to prepare *reservoir descriptions*. Those descriptions are based on both the fundamental properties of the reservoir rocks and the sequence of geological events that formed the pore network. Data for these descriptions comes from direct examination of rock samples such as borehole cores and drill cuttings. Borehole logs and other geophysical devices provide useful information, but they are indirect measurements of derived and tertiary rock properties. They are not direct observations. Direct observations of depositional textures, constituent composition, principal and accessory minerals, sedimentary structures, diagenetic alterations, and pore characteristics provide the foundation for reservoir descriptions. The geological history of reservoir formation can be traced by interpreting depositional, diagenetic, and tectonic attributes. The goal of such interpretations is to formulate *geological concepts* to guide in predicting reservoir size, shape, and performance characteristics. In the absence of direct lithological data from wells, as in the case of frontier exploration and wildcat drilling, geologists commonly study nearby outcrops of the same age and geological formation as the expected reservoir. A measure of care is given to interpreting reservoir geology from distant outcrops because depositional and diagenetic characteristics may vary significantly from place to place and from outcrops that have been altered by surface weathering to subsurface reservoirs that have never been exposed to weathering.

Reservoir characterization, like reservoir geology, deals with physical characteristics of the reservoir. It differs from geological description in that data on petrophysics and fluid properties are included. In addition to data from direct examination of reservoir rocks, reservoir characterization involves interpretation of borehole logs, porosity–permeability measurements, capillary pressure measurements, reservoir fluid saturations, and reservoir drive mechanisms.

Reservoir engineering deals with field development after discovery. The main goal of the reservoir engineer is to optimize hydrocarbon recovery as part of an overall economic policy. Reservoirs are studied throughout their economic lives to derive the information required for optimal production. In addition to geological data and borehole log characteristics, reservoir engineering deals with reservoir pressures, oil–water saturation, and gas–oil ratio in order to provide estimates of in-place hydrocarbon volumes, recoverable reserves, and production potential for each well in a field (Cossé, 1993).

Petroleum geoscientists not only study reservoirs, but they also study *traps, seals*, and *source rocks* that make up most of the petroleum system described by Magoon and Dow (1994). *Traps* are bodies of rock where hydrocarbons accumulate after migrating from their source and are restricted from further movement. It is convenient to think of traps as large-scale geometrical features that form boundaries around porous and permeable reservoir rocks. Traps are created by structural, strati-graphic, hydrodynamic, or diagenetic processes. It is important to recognize that the geometry of the reservoir–trap system may or may not correspond with present-day structural configurations. Subsurface structures may form and then be deformed by later episodes of tectonism. Ancient or paleo-highs may become present-day lows or saddles. Likewise, paleo-lows may be tectonically elevated to exhibit present-day structural closure and be "high and dry." This is called structural inversion and it is especially characteristic of basins with mobile salt or shale in the subsurface and in some structural settings where multiple episodes of tectonism have changed older structures.

Seals are the physical mechanisms that restrict fluids from flow out of the trap and are usually described in terms of capillary pressures. Seals may extend along the top, side, or bottom of the trap. Later we will define seals on the basis of the high capillary pressure exhibited by the seal rock as compared to the reservoir rock. These differences usually correspond to changes in rock type such as a change from sandstone to siltstone or shale in the case of siliciclastics, or porous grainstone to mudstone in carbonates. Most seals are not completely impermeable and will allow some leakage of hydrocarbons. Less commonly, seals may consist of totally imper-meable barriers to flow such as evaporite deposits.

Source rocks are rich in kerogen, the parent organic matter that produces petro-leum hydrocarbons when it reaches a threshold temperature during burial and thermal maturation. Source rocks usually consist of shales or lime mudstones that were deposited in oxygen-deficient environments where lipid-rich organic matter was preserved and converted to kerogen on further burial.

An integrated petroleum exploration program includes geophysical and geologi-cal studies of basin stratigraphy and structure to isolate the regions where reservoir rocks are most likely to be found, where structural, stratigraphic, or diagenetic pro-cesses have formed traps and seals, and where the basin contains an ample thickness of source rocks buried to a depth at which the temperature would have been high enough to liberate hydrocarbons from kerogen. In the initial phases of exploration, knowledge of how and where reservoir rocks form is critical; however, until a well is drilled the reservoir remains a hypothetical entity. Trap configurations may be identified as structural and stratigraphic anomalies, but without a hydrocarbon-filled reservoir, they only beckon explorationists to drill dry holes. After a successful well is drilled, the discovery is evaluated to predict the size and shape of the reservoir, to estimate its economic value, and to formulate a development program. At this stage, knowledge of reservoir characteristics is obviously the most important consideration.

1.2   FINDING AND DEVELOPING CARBONATE RESERVOIRS

The main reasons to study carbonate reservoirs and aquifers are to learn more about how to find, extract, and manage the oil, gas, usable water, or other resources they contain. Carbonates hold about half of the world's oil and gas, much of its ground-water, and extensive deposits of metallic ores, yet of the relatively few texts on res-ervoir geology, only a handful deal with carbonates. Carbonate reservoirs occur in the subsurface so most of the data used to study them comes from borehole cores, drill cuttings, logs, well tests, production data, and geophysical measurements. Modern reservoir geological studies are much more than geological descriptions of reservoir rocks. They include detailed research on the origin and spatial distribution of effective porosity, how it is connected through pore throats, and how rock and pore properties influence fluid storage and transmission.

In the following chapters, depositional, diagenetic, and fractured reservoirs are described and interpreted to illustrate methods of critical analysis of reservoirs rather than offering case histories that may be applied uncritically as analogs, or "look-alikes," for exploration and development programs. Petrophysical as well as petrological and structural characteristics of carbonate reservoirs are described, borrowing from both reservoir characterization and reservoir engineering. The study of *flow units* that make up all or part of a reservoir is emphasized throughout the book. Flow units are the rock bodies that transmit fluids, to a greater or lesser degree, through connected porosity. Identification and mapping of flow units requires knowledge of reservoirs in three or four dimensions (porosity in three dimensions plus a fourth dimension that may include connectivity or fluid saturation, for example). Such knowledge involves understanding the timing and mode of origin of the pore systems. Mapping flow units also requires a basic knowledge of subsurface mapping techniques and methods for evaluating reservoir connectivity. In the end, accurate and realistic reservoir models to provide the basis for finding new reservoirs and developing existing ones can be constructed from flow unit maps. I hope these discussions will stimulate analytical and synthetic thinking that will enable you to evaluate each reservoir on its petrological and petrophysical characteristics—not on its similarity to a "look-alike." Finding and developing carbonate reservoirs requires critical analysis and interpretation of essential geological data. Some of the kinds of data used in reservoir studies are mentioned in the following section.

1.2.1   Sources of Data on Reservoirs

Carbonate reservoirs may have mixed-origin pore systems, but an understanding of the end-member categories, how and when they formed, and how they are connected is necessary before a geological concept for exploration or development can be created. The most important data for reservoir studies comes from direct examination of rock samples. Basic rock properties studied in cores and cuttings include texture, mineral and grain composition, fossil content (taxonomic diversity), and sedimentary structures. These properties provide data on depositional characteristics. Petrographic study provides data about the diagenetic history of both the rock matrix and the pore system. Microscopic descriptions of borehole cores also provide data on the presence or absence of through-going, natural fractures. Porosity and permeability are measured from cores and porosity can be calculated from appropriate borehole logs. The statistical relationship between porosity and permeability is commonly used to distinguish between pore systems of different geological origins. Fractured reservoirs may exhibit high permeability at comparatively low porosity, for example. Another important kind of data is capillary pressure measurements. These measurements on cores provide data on the behavior of fluids in the reservoir pore system. Plots of capillary pressure at different fluid saturations within the reservoir rock sample, or capillary pressure curves, are related to pore throat size distribution, pore throat characteristics, and height of the hydrocarbon column above free water in a reservoir. Capillary pressure data are also used in estimating connectivity, fluid recovery efficiency, reservoir quality, thickness of the hydrocarbon column in reservoirs, and in evaluating seal capacity.

Data from indirect measurements made with wireline logs are the "bread and butter" of everyday work in subsurface geology. Some logs provide much more than others, however. Laboratory measurements of nuclear magnetic resonance (NMR) responses to different pore characteristics in carbonates are helping us develop better methods for interpreting reservoir quality from new-generation NMR log data. Of course, the traditional wireline log data are analyzed to extract useful information about reservoirs, traps, and seals. Of these logs, gamma ray, acoustic, resistivity, photoelectric effect, neutron, and density logs are typically included in modern logging runs. Even ordinary mud logs with drilling times, mud characteristics, and basic "lith log" descriptions of borehole cuttings can offer valuable information to the well analyst. Other indirect measurements of reservoir characteristics include reflection seismic traces that can be interpreted to reveal trap geometry and, in some cases, reservoir rock attributes. Borehole testing such as pressure transient tests helps confirm the presence or absence of fracture permeability. When used with cores or imaging logs, fractured reservoirs can be classified according to the relative contribution of fractures to overall reservoir performance.

Gravity and magnetic measurements are less commonly used in exploring for carbonate reservoirs than for terrigenous sandstone reservoirs, probably because gravity anomalies associated with shallow salt domes are comparatively easy to identify and are relatively common in sand–shale geological provinces such as the Gulf of Mexico onshore areas. Magnetic anomalies may sometimes be useful in identifying paleostructural trends associated with depositional carbonate reservoir trends that overlie magnetically susceptible basement rocks.

Reflection seismology is a widely used and fundamental tool in exploration for carbonate reservoirs. In some cases, particularly with today's high-resolution, three-dimensional (3D) seismic data, seismic attribute analyses might be used in field development where the analyses can help discriminate between reservoir and non-reservoir zones. High porosity zones, high fracture intensity, or strong contrasts in fluid content might be detectable as a seismic wave characteristic, or attribute. Seismic data usually provide the essential information to identify structural anomalies and can sometimes pinpoint stratigraphic anomalies if the anomalies are large enough. A rule of thumb is that the target reservoir zone must be thicker than one-quarter of the seismic wavelength. Also, the impedance contrast between reservoir and nonreservoir horizons must be strong enough for the difference between them to be detected. Seismic records are commonly helpful in detecting fractured reservoirs and a substantial literature exists on this subject. Suggestions for additional reading are presented at the end of each chapter of this book.

Analyses of these different kinds of data help to determine the size and shape of the reservoir body, the spatial distribution of the pore types within it, and how the pore system interacts with reservoir fluids. Evaluation of depositional characteristics draws from carbonate sedimentology to utilize depositional sequences and lithofacies in establishing links between depositional setting, sedimentary processes, and pore types. *Paleostructure* can have a strong influence on depositional facies, as is easily seen by examining depositional facies maps overlain on interval isopach maps. Diagenesis can be a pore-forming and a pore-destroying process. Diagenetic patterns are related to environments at the surface, in the shallow subsurface, and in the deep burial environment. Shallow diagenetic environments are commonly linked to depositional or paleostructural trends and these relationships can be seen with map overlays. Deep-burial diagenetic trends record the burial and thermal history of basin evolution, although the odd tectonic or geothermal event may also influence late diagenesis. Finally, natural fracture trends can be detected with a combination of special core analyses and acoustic or electronically generated borehole images. These combined types of data provide information on the size, intensity, and orientation of natural fractures. Fracture data plotted on structure maps can sometimes be correlated with reservoir geometry, because in most cases fracture trends correspond with geometrical orientation and position of the borehole with respect to the geometry of subsurface structures such as faults and folds. Cavernous and connected-vug pore systems could be mistaken for fractures, especially during drilling; therefore borehole log responses, digital imaging logs, drilling history characteristics, microscopic examination of drill cuttings, and the behavior of the drilling mud system are commonly used together to test for the presence of natural fractures. Pressure transient tests provide very important information to help confirm fracture behavior.

1.3 UNIQUE ATTRIBUTES OF CARBONATES

Texture is defined as the size, shape, and arrangement of detrital grains in a sedimentary rock. In siliciclastic rocks, it is strongly influenced by parent rock type, weathering, and transportation history. Most sandstones are classified on the basis of how much quartz, feldspar, rock fragments, and matrix they contain. Fossils are generally ignored. Sandstone diagenesis is usually treated as a process that alters depositional texture, fabric, and porosity only after extensive burial. Sandstone porosity and permeability are nearly always described as facies-specific; that is, the rock properties of the depositional facies determine reservoir characteristics. Fractured reservoirs in terrigenous clastics are less commonly reported in the literature than are fractured reservoirs in carbonates, although fractures are certainly not exclusive to one rock type or the other.

In contrast, carbonates have unique attributes that distinguish them from siliciclastics and that require different methods of study. Some of these attributes were recognized decades ago by Ham and Pray (1962). First, carbonates form within the basin of deposition by biological, chemical, and detrital processes. They do not owe their mineralogical composition to weathered, parent rocks and their textures do not result from transport down streams and rivers. Carbonates are largely made up of skeletal remains and other biological constituents that include fecal pellets, lime mud (skeletal), and microbially mediated cements and lime muds. Chemical constituents, including coated grains such as ooids and pisoids, cements, and lime mud, are common in carbonates but are absent in most siliciclastics. Clastic grains exist in carbonates, as they do in siliciclastics. In carbonates, however, these grains are mainly clasts of intraformational, lithified sediment (intraclasts) or of reworked, older rock (lithoclasts). The second major difference between carbonates and siliciclastics is that carbonates depend greatly on biological activity. They are composed mainly of biogenic constituents, component grains may have undergone size and shape changes as they were eaten by organisms, and the stratification of carbonate rock bodies is extensively modified by burrowing and boring organisms. The third major difference is that carbonates are susceptible to rapid and extensive diagenetic change. Carbonate minerals are susceptible to rapid dissolution, cementation, recrystallization, and replacement at ambient conditions in a variety of diagenetic environments. Finally, although not stressed by Ham and Pray (1962), fractured reservoirs are probably more common in carbonate rocks than in siliciclastics (as indicated in Table 1.1), but work by Laubach (1988, 1997) and Laubach et al. (2002) suggests that fractures are more common in siliciclastic reservoirs than was previously recognized. In short, porosity and permeability in carbonate reservoirs depend on a broad array of rock properties, on diagenetic episodes that may continue from just after deposition through deep burial, and on fracture patterns related more to the geometry of stress fields than to rock type. Choquette and Pray (1970) highlighted some of the differences between carbonate reservoirs and those in siliciclastics. A summary of their findings is given in Table 1.1.

Three other significant differences between carbonate and terrigenous sandstone reservoirs are: (1) electrofacies maps from gamma ray and resistivity log data *do not* indicate depositional facies in carbonates as they can do with terrigenous sandstones; (2) Focke and Munn (1987) demonstrated that a strong relationship exists between pore type and petrophysical characteristics in carbonate reservoirs such that saturation calculations using the Archie equation will vary greatly depending on the chosen "*m*" exponent and its dependence on the proportion of vuggy and moldic pores compared to interparticle pores; and (3) carbonates form in temperate as well as tropical environments. Because temperate carbonates have decidedly different mineralogical and component grain type compositions from tropical carbonates, their reservoir characteristics could also be different than expected.

Differences between sandstone and carbonate reservoirs influence the way we study them. Sandstone porosity is mainly interparticle; therefore it is related geometrically to depositional texture and fabric. Because permeability usually correlates rather well with interparticle porosity in sandstones, it can be related to depositional texture and fabric, as illustrated in a study of pore geometry in sphere packs and in terrigenous sandstones (Berg, 1970). Assuming that porosity and permeability are closely related, laboratory measurements made on small core plugs of terrigenous sandstones may be assumed to be representative of large rock volumes. That is, small samples are representative of large populations if the populations are homogeneous. Carbonates do not always exhibit interparticle porosity; they may have a variety of pore sizes, shapes, and origins, and measured porosity values do not always correspond closely with permeability. In short, carbonate pore systems are not usually homogeneous. While a 1-inch perm-plug will provide reliable data on sandstone porosity and permeability, entire core segments 4 inches in diameter and 1 foot long may be required for reliable measurements on carbonates. Relatively simple porosity classification schemes are useful for siliciclastics but a compound scheme of genetic classification augmented by measurements of pore geometry is needed for carbonates. Carbonate porosity classifications are discussed in Chapter 2.

TABLE 1.1   A Comparison of Terrigenous Sandstone and Carbonate Reservoir Characteristics

Reservoir Characteristic	Terrigenous Sandstones	Carbonates
Amount of primary porosity	25–40%	40–70%
Amount of ultimate porosity	Half or more of primary porosity, commonly 15–30%	Small fraction of original porosity, commonly 5–15%
Type of primary porosity	Almost exclusively interparticle	May be interparticle, intraparticle, intercrystalline, moldic, vuggy, cavernous, fenestral, or "constructed void"
Type of ultimate porosity	Almost exclusively primary interparticle	Highly variable owing to different origins or pore types
Typical pore size	Diameter and throat sizes related to depositional texture	Diameter and throat size may not be related to depositional texture
Typical pore shape	Dependent on particle shape; typically a "negative" of particles	Varies from strongly related to particle shape to totally unrelated
Uniformity of pore size and shape distribution	May be relatively uniform in homogeneous sand bodies	Variable from fairly uniform to extremely heterogeneous—even within a body made up of a single rock type
Influence of diagenesis	Minor—usually reduction by compaction or cementation	Major—can create, obliterate, or totally modify porosity; cementation and solution important
Influence of fracturing	Not of major importance in reservoir properties	Major importance in reservoir properties if present
Visual estimation of porosity and permeability	Semiquantitative estimates may be relatively easy	Semiquantitative estimates may be easy or impossible; instrumental measurements commonly required
Adequacy of core analysis for reservoir evaluation	Core plugs of 1-inch diameter may be adequate to determine matrix porosity	Core plugs commonly inadequate; may require whole core analyses (~4-inch by 1-foot segments) for large pore sizes
Porosity–permeability relationships	Relatively consistent; may be dependent on particle texture	Highly varied; may be independent of particle texture
Reliability of log characteristics as indicators of depositional facies (electrofacies mapping)	Standard practice that may provide reliable proxies for depositional facies	Not reliable because logs cannot generally detect differences in carbonate grain types or textures

Source: Adapted from Choquette and Pray (1970).

SUGGESTIONS FOR FURTHER READING

A good general reference on carbonate sedimentology is by M. E. Tucker and V. P. Wright (1990), *Carbonate Sedimentology*; and an encyclopedic review of petroleum geology is by F. K. North (1985), *Petroleum Geology*.

REVIEW QUESTIONS

1.1. What is the difference between a reservoir and a trap?

1.2. What units of measure would you use to describe traps? To describe reservoirs?

1.3. What is the difference between a trap and a seal?

1.4. What measurements are used to determine the "efficiency," or capacity, of a seal?

1.5. What differentiates reservoir characterization from reservoir description? From reservoir engineering?

1.6. What are at least three differences between carbonate and terrigenous sandstone reservoirs?

1.7. What type of data would explorationists use to identify subsurface structural anomalies in untested areas?

1.8. What, according to Focke and Munn (1987), is a characteristic of carbonate reservoir porosity that must be taken into consideration when making fluid saturation calculations with the Archie equation?

1.9. Which wireline logs would you use to create depositional facies maps in carbonate reservoirs?

1.10. What is the main difference between carbonate reservoirs and carbonate aquifers?

CHAPTER TWO

CARBONATE RESERVOIR ROCK PROPERTIES

2.1 DEFINITIONS

This chapter focuses on the definitions of fundamental rock properties, how the properties are used to classify both rocks and porosity, and how fundamental rock properties are related to reservoir properties. It is traditional in geology to use purely descriptive terms for rock properties because objective descriptions are less likely to contain subjective interpretations or biases. This philosophy has merit in most cases, but in the end, the task of the reservoir geologist is to formulate interpretive models for use in exploration and development. In carbonate rocks, reservoir porosity and permeability can be formed by a variety of processes. These processes create the rock properties we describe with rigidly objective terms. Some of the formative processes may have affected reservoir rocks more than once; therefore an accurate reservoir description should incorporate terminology that classifies the altered properties, the processes that created them, and at least an estimate of the number of times the rock properties underwent change. To produce such a classification requires the use of genetic terms along with descriptive ones. For example, subjective interpretations are required to determine which processes caused diagenetic changes at which times during the burial history of the rocks. Diagenetic porosity can be classified in a totally objective manner, but without interpretations of how, when, and where different diagenetic events changed preexisting pore characteristics it is hardly possible to predict the spatial distribution of the ultimate porosity.

Traditional geological literature includes the terms primary and secondary to describe rock properties. This is not helpful because those terms are ambiguous. Primary may be used in a temporal sense to indicate the depositional origin of rock properties such as grain size, grain composition, or skeletal morphology in the case

---

Geology of Carbonate Reservoirs: The Identification, Description, and Characterization of Hydrocarbon Reservoirs in Carbonate Rocks
By Wayne M. Ahr   Copyright © 2008 John Wiley & Sons, Inc.

of calcified organisms. Secondary properties traditionally refer to features produced by diagenesis. This oversimplified language creates more problems than it solves because diagenesis is a continuous process that may affect a reservoir rock many times in its burial history. If diagenetic and fracture porosity are "secondary," then how do we distinguish between multiple episodes of diagenesis or fracturing that had major impacts on reservoir characteristics? How do we develop accurate reservoir models that take into account the different times and modes of change? Some texts imply that primary rock properties such as texture and fabric are depositional only. However, texture and fabric can also be used to describe diagenetic properties of carbonate reservoir rocks. Because diagenetic characteristics represent multicycle changes, not just one-time, secondary changes, it can be confusing to classify rock and reservoir properties simply as primary and secondary. It is important to distinguish between timing and mode of origin of the various diagenetic events to determine the history of porosity development. Similarly, it is necessary to distinguish between timing and mode of origin of different fracture sets in fractured reservoirs. The problem of dealing with descriptive terms is made less troublesome if carbonate rock properties and pore categories are classified genetically as products of depositional, diagenetic, and fracture-related processes. Those are end-member processes. Further subdivision of the classifications can be made to identify detrital, chemical, and biogenic deposition. Diagenetic attributes can be linked to time and mechanism of change. Mechanically produced fractures, cataclastic textures, mylonites (gouge), ductile folding, and plastic deformation can be related to different times, stress conditions, and material properties. Hybrid properties such as those produced by depositional facies-selective (texture plus-or-minus fabric) diagenesis, or strain recrystallization, or stylolitization, can also be described in terms of time and mode of origin. Time and mode of origin of depositional, diagenetic, and fracture rock properties are, as we will demonstrate throughout this book, critical to understanding the architecture of carbonate reservoirs.

Depositional, diagenetic, and tectonic rock properties, although they are genetic, still represent basic descriptive characteristics of carbonate reservoirs. They can even be thought of as *fundamental properties* in the sense that other reservoir properties are dependent on them. Properties such as porosity, permeability, and bulk density are dependent or derived properties. Yet another set of properties is encountered in the study of carbonate reservoirs: third order or *tertiary properties*. Tertiary properties include electrical resistivity and conductivity, acoustic transmissivity, natural radioactivity, and the various attributes measured by most wireline logs, gravity meters, magnetometers, and the seismograph. Those characteristics depend on porosity, fluid content, radioactive element content, rock density, magnetic susceptibility, and acoustic characteristics, none of which are measures of fundamental rock properties. In describing rock properties, the words primary, secondary, and tertiary have no time significance and they may not be related to deposition, diagenesis, or mechanical fracturing. It is preferable to describe rock properties as fundamental, dependent, and tertiary.

2.2 FUNDAMENTAL ROCK PROPERTIES

Fundamental properties of carbonate rocks include texture, fabric, grain type, mineralogical composition, and sedimentary structures. Note that texture and fabric are not interchangeable terms. *Texture* is defined as the size, shape, and arrangement of the grains in a sedimentary rock (Pettijohn, 1975). Among carbonate sedimentologists, texture is sometimes thought of in the context of *depositional texture*, which forms the basis for several carbonate rock classification systems. *Fabric* refers to the spatial arrangement and orientation of the grains in sedimentary rocks. It can also refer to the array geometry or mosaic pattern of crystals in crystalline carbonates and the growth form (macroscale) and skeletal microstructure (microscale) of reef organisms. *Mineralogical composition* refers to original mineralogy. Original mineralogical composition has great significance in the study of carbonate diagenesis and it provides important clues about the chemical evolution of the earth. It is not, however, a reliable clue to the origin and distribution of reservoir flow units because carbonates in a wide variety of depositional settings may consist of calcite, aragonite, or dolomite, individually or in mixtures. It is more practical for the reservoir geoscientist to substitute constituent grain type, such as skeletal grains, peloids, clasts, or ooids, among others, for composition. *Sedimentary structures* are preserved bedforms created by fluid processes acting on the sediment interface, by desiccation, slope failure, thixotropy, compaction, fluid expulsion, and *bioturbation* by burrowing and boring organisms. These definitive rock properties are discussed in more detail in the following sections.

2.2.1 Texture

There are many textural terms in the literature on sedimentary rocks, but most geologists today describe grain sizes according to the Wentworth (1922) scale in millimeters, or in "phi units," which are logarithmic transformations to the base 2 of the size (in millimeters). It is rarely possible to disaggregate lithified limestones into component grains; consequently, direct size measurements by sieve, pipette, or hydrometer are limited to unconsolidated sediments. Estimates of grain size can be made from thin sections of lithified carbonates, although the method requires statistical manipulation of grain size measurements to compensate for the fact that two-dimensional microscope measurements do not provide the true three-dimensional grain size. Tucker (1988) and Tucker and Wright (1990) discuss the problem of determining grain sizes from thin section measurements in more detail.

The Wentworth scale (Figure 2.1) classifies all grains with average diameters greater than 2 mm as *gravel*, those with average diameters between 2 mm and $\frac{1}{16}$ mm (62 μm) as *sand*, and those finer than 62 μm as *mud*. In this context, sand denotes texture rather than composition. Other terms for gravel, sand, and mud include the Greek derivatives psephite, psammite, and pelite, but they are rarely used in modern literature. The Latin terms rudite, arenite, and lutite appear in the comprehensive but unwieldy sedimentary rock classification scheme of Grabau (1960). The terms appear in modern literature as *calcirudite*, *calcarenite*, and *calcilutite*, indicating carbonate gravel, sand, and mud, respectively. Embry and Klovan (1971) blended rudite with Dunham's (1962) carbonate rock classification terminology to create *rudstone* in their classification of reef carbonates. Lithified lime mud that exhibits a mosaic of calcite crystals 1–4 μm in diameter became known as *micrite*, a contraction of *micro*crystalline and cal*cite*, coined by Folk (1959). Some workers now classify all carbonate mud, regardless of its size and mineralogical composition, as micrite, even though that is inconsistent with the original definition.

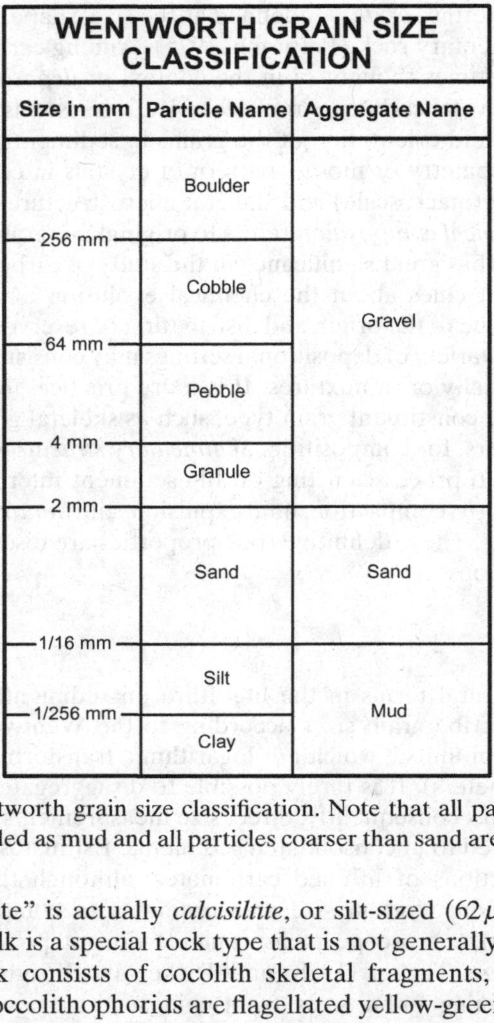

WENTWORTH GRAIN SIZE CLASSIFICATION		
Size in mm	Particle Name	Aggregate Name
	Boulder	
— 256 mm —		
	Cobble	
		Gravel
— 64 mm —		
	Pebble	
— 4 mm —		
	Granule	
— 2 mm —		
	Sand	Sand
—1/16 mm —		
	Silt	
— 1/256 mm —		Mud
	Clay	

Figure 2.1 The Wentworth grain size classification. Note that all particles finer than sand ($\frac{1}{16}$–2.0 mm) are included as mud and all particles coarser than sand are included as gravel.

Much of this "micrite" is actually *calcisiltite*, or silt-sized (62 μm to 3.90 μm) sediment. Note that chalk is a special rock type that is not generally classified as micrite or mud. True chalk consists of cocolith skeletal fragments, usually in a grain-supported fabric. Coccolithophorids are flagellated yellow-green algae that produce a spheroidal mass of platelets that become disarticulated after death and rain down to the sea floor as disk-shaped particles 2–20 μm in diameter (Milliman, 1974). Electron micrographs of chalk show grain-supported depositional textures without a matrix of aragonite or calcite crystals finer than the cocoliths; therefore chalk is not strictly a mud or micrite in the sense of the detrital micrites described earlier. Of course, there are "gray" areas. Calcisiltites (lime muds) may contain some cocoliths, but they are not proper chalks.

Grain size is not generally as useful for interpreting ancient hydrologic regimes in carbonate depositional environments as it is with terrigenous sandstones nor is grain size consistently related to carbonate reservoir porosity or permeability. Carbonates consist mainly of biogenic particles that owe their size and shape to skeletal growth rather than to a history of mechanical transport, deposition, and arrangement. Most carbonate grains originate in the marine environment where waves and currents fragment, winnow, and sort sediment, primarily along strand plains and on slope changes (usually associated with bathymetric highs) that occur above fair-weather wave base. However, carbonate grains are produced in a wide variety of environments; consequently, it is risky to use grain size and shape alone as indicators of the hydrologic regime. It is better to consider grain size, sorting, grain shape, amount and type of grain fragmentation, and mechanical durability of the grains as clues for interpreting depositional environments. Some grains are produced by bioerosion—boring, rasping, grinding, and digesting carbonate constituents by organisms. The biological reworking makes grain size and shape virtually useless as indicators of the hydrologic regime. Perhaps the most effective villain in altering carbonate grain size and shape is diagenesis. Micrite can be produced from sand and gravel, grains can be enlarged by cement overgrowths, and shape can be changed by cementation or dissolution. Fortunately, diagenetic changes are not difficult to identify in thin sections. Sorting and grain size (Figure 2.2) are textural attributes that can be useful in studying carbonate rocks because they influence depositional porosity and permeability. Porosity is independent of grain size where grains are ideal spheres, but permeability varies with particle size because small grains have small intergranular pores with small pore throats. Sorting and grain packing are also strongly related to permeability because sorting and packing influence the geometrical relationship between pores and pore throats. A high correlation exists between permeability and pore throat dimensions, as we will see later in our discussions on capillary pressure, permeability, and reservoir quality, but pore geometry alone is not strongly related to permeability. It is the pore–pore throat relationship that is so important.

Mechanical abrasion, along with hydraulic size and shape sorting, are important processes in beach, dune, and shallow, slope-break environments where mud-free

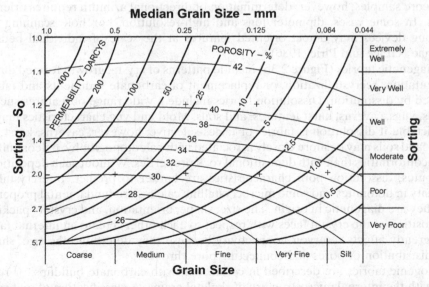

Figure 2.2 A plot showing the relationships between grain size, sorting, and porosity in unconsolidated sands. Based on data in Beard and Weyl (1973) and adapted from unpublished notes with permission from R. M. Sneider (1988). Note that porosity does not vary with grain size but does vary with sorting. Permeability varies with both grain size and sorting.

deposits are produced. The removal of mud by winnowing is important because the leading carbonate rock classifications are based on the presence or absence of mud. Rocks with high mud content usually represent sediment "sinks," or areas where water movement has been slow and mud has settled out of suspension. If fractures, dissolution diagenesis, or alteration to microporous crystalline fabrics are absent, muddy carbonates are not good reservoir rocks. They may be good source rocks instead if they contain enough sapropelic (lipid-rich) organic matter. Carbonate sands and gravels with little or no mud represent either effective winnowing or a lack of mud production. Well-sorted and mud-free carbonate rocks have high depositional porosity and permeability.

2.2.2 Fabric

Depositional, diagenetic, or biogenic processes create carbonate rock fabrics. Tectonic processes such as fracturing and cataclasis are not part of the depositional and lithification processes but may impart a definite pattern and orientation to reservoir permeability. Fractured reservoirs are discussed in Chapter 7.

Depositional fabric (Figure 2.3a) is the spatial orientation and alignment of grains in a detrital rock. Elongate grains can be aligned and oriented by paleocurrents. Flat pebbles in conglomerates and breccias may be imbricated by unidirectional current flow. These fabrics affect reservoir porosity and can impart directional permeability, ultimately affecting reservoir performance characteristics. Elongate skeletal fragments such as echinoid spines, crinoid columnals, spicules, some foraminifera, and elongate bivalve and high-spired gastropod shells are common in carbonate reservoirs. Presence or absence of depositional fabric is easily determined with core samples; however, determination of directional azimuth requires oriented cores. In some cases, dipmeter logs and high-resolution, borehole scanning and imaging devices may detect oriented features at the scale of individual beds or laminae (Grace and Pirie, 1986).

Diagenetic fabrics (Figure 2.3b) include patterns of crystal growth formed during cementation, recrystallization, or replacement of carbonate sediments and fabrics formed by dissolution. Dissolution fabrics include a wide range of features such as molds, vugs, caverns, karst features, and soils. Mold and vug characteristics may be predictable if dissolution is fabric- or facies-selective; however, caverns, karst features, and soils may be more closely associated with paleotopography, paleoaquifers, or unconformities than with depositional rock properties. Without such depositional attributes, dissolution pore characteristics are harder to predict. Intercrystalline porosity in dolomites and some microcrystalline calcites are fundamental properties but they are diagenetic in origin. The size, shape, orientation, and crystal "packing" (disposition of the crystal faces with respect to each other) create an internal fabric that greatly affects reservoir connectivity because they determine the size, shape, and distribution of pores and connecting pore throats.

Biogenic fabrics are described in connection with carbonate buildups, or reefs, and with the internal microstructure of skeletal grains. A classification of reef rocks was conceived to cope with variability in reservoir characteristics within a single reef complex (Embry and Klovan, 1971). They described three end-member biogenic fabrics, including (1) skeletal frameworks in which interframe spaces are filled

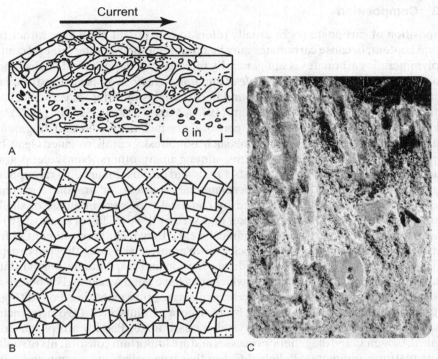

Figure 2.3 (a) Depositional fabric in detrital rocks. Grain orientation and alignment produced by currents at time of deposition. The larger grains at the top of the figure are imbricated and those at the bottom of the figure are simply oriented with long axes parallel to the direction of flow. Permeability is highest in the direction of grain alignment. (b) Diagenetic fabric. Complete replacement of limestone by dolomite creates a diagenetic fabric totally unrelated to depositional rock properties. In this case the dolomite rhombohedra occur in an "open" fabric with a great amount of intercrystalline porosity. (c) Biogenic (skeletal growth) fabric. The great variety of internal growth fabrics created by reef-building organisms creates depositional fabrics dramatically different from those in detrital (particulate) rocks. The photo illustrates a Pleistocene patch reef from Windley Key, Florida. Porosity and permeability in the coral framestone are influenced by biologically constructed, skeletal structures, not by granular or crystalline fabrics.

with detrital sediments, (2) skeletal elements such as branches or leaves that acted as "baffles" that were subsequently buried in the sediment they helped to trap, and (3) closely bound fabrics generated by encrusting organisms. The skeletal microstructure of many organisms is porous and may provide intraskeletal porosity, even in nonreef deposits. The pores within sponge, coral, bryozoan, stromatoporoid, or rudist skeletons, for example, are intraparticle pores, although the individual skeletons are part of larger reef structures. All three fabric categories are closely related to reservoir properties because fabric influences pore to pore throat geometry and may influence directional permeability. An example of combined biogenic and detrital fabric is illustrated in a Pleistocene coral framestone reef with detrital interbeds (Figure 2.3c).

2.2.3 Composition

Composition of carbonate rocks usually refers to constituent grain type rather than mineral content, because carbonates may be monomineralic and the mineral content of polymineralic carbonates is not generally indicative of depositional environment. Carbonate grains are classified as *skeletal* and *nonskeletal*. Extensive, illustrated discussions of constituents commonly found in carbonates of different geological ages are found in Bathurst (1975), Milliman (1974), Purser (1980), Scoffin (1987), and Tucker and Wright (1990). Skeletal constituents include whole and fragmented remains of calcareous plants and animals such as mollusks, corals, calcified algae, brachiopods, arthropods, and echinoderms, among many others. Nonskeletal grains include ooids, pisoids, peloids, and clasts. Ooids and pisoids (Figure 2.4a) are spheroidal grains that exhibit concentric microlaminae of calcite or aragonite around a nucleus. The marine variety is formed by chemical processes in agitated, shallow water, usually less than 2 m deep (Tucker and Wright, 1990). Clasts (Figure 2.4b) are particles produced by detrition (mechanical wear); they include resedimented fragments of contemporaneous or older rock known as intraclasts and lithoclasts, respectively, following Folk (1959). Clasts indicate erosion and resedimentation of lithified or partly lithified carbonates, some of which may have been weakened by bioerosion (rock boring and grinding by specialized organisms) or by weathering. Peloid (Figure 2.4c) is an all-inclusive term coined by McKee and Gutschick (1969) to include rounded, aggregate grains of microcrystalline carbonate. Peloids are produced by chemical, biogenic, and diagenetic processes and are important constituents of shallow marine platform sediments. Pellets differ in that true pellets are compacted bits of fecal matter that have distinctive shapes or internal structures (Figure 2.4d). Pellets can be useful in determining the environment of deposition (Moore, 1939). Peloids that were probably formed as fecal pellets are prominent constituents of Wilson's (1975) "standard microfacies 8" in the "restricted platform" environment.

2.2.4 Sedimentary Structures

Sedimentary structures are useful aids for interpreting ancient depositional environments. They may affect reservoir characteristics because their internal fabrics are usually oriented and there may be regular patterns of grain size change within them. A complete discussion of sedimentary structures and their hydrodynamic significance is beyond the scope of this book. Instead, representative categories of sedimentary structures are grouped in Table 2.1 according to origin. Brief descriptions are included on the characteristics that distinguish the types of sedimentary structures, their environmental significance, and their potential influence on reservoir performance. Some common sedimentary structures are illustrated in Figure 2.5. Extensive discussions and illustrations of sedimentary structures can be found in Allen (1985), Purser (1980), Reading (1996), Reineck and Singh (1973), and Tucker and Wright (1990).

2.3 CLASSIFICATION OF CARBONATE ROCKS

There are many classification schemes for carbonate rocks. In 1904 Grabau devised one of the most comprehensive, but it is cumbersome and has never been popular

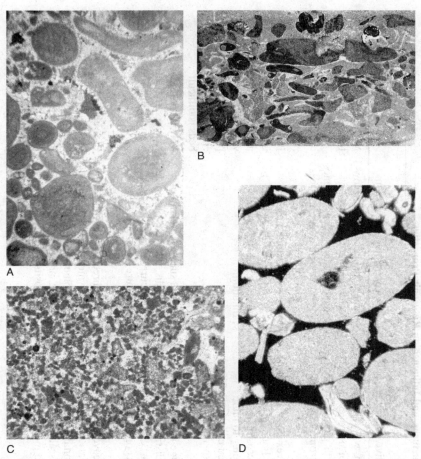

Figure 2.4 (a) Photomicrograph of a lime grainstone with ooids, intraclasts, and pisoids. Note the "dogtooth spar" isopachous rim cement on grain surfaces. Porosity in this rock is mainly intergranular but the grains have been altered and intragranular microporosity is also present. The horizontal width of the photo is 2.5 mm. (b) Photomicrograph of an intraclastic (nonskeletal grains) conglomerate from the Cambrian of Central Texas. Note that the clasts are aligned in a fabric created by currents at the time of deposition. The width of the photo is 4 mm. (c) Photomicrograph of an intraclastic, peloidal grainstone from the Cambrian of Central Texas. The width of the photo is 2.5 mm. The most common peloids are probably microbial in origin and, of those, most are found as cavity-fillings in reefs and mounds. (d) Photomicrograph of ovoid, polychaete worm fecal pellets from the Pleistocene Campeche Calcilutite, Yucatan ramp, Mexico. The long axis of each pellet is about 2 mm.

(Grabau, 1960). Popular, modern classifications for detrital carbonates were developed by Folk (1959, 1962) (Figure 2.6) and Dunham (1962) (Figure 2.7). Classifications for reef rocks were developed by Embry and Klovan (1971) (Figure 2.8) and Riding (2002). A scheme to include depositional, diagenetic, and biological aspects of carbonates in one classification system was proposed by Wright (1992). There are two main purposes for classification systems: (1) to make descriptions of

TABLE 2.1 Physical Processes, Their Sedimentary Structures, and Influence on Reservoir Properties

Formative Processes	Sedimentary Structures	Descriptive Characteristics	Environmental Association	Influence on Reservoir Performance
Deposition only—Ordinary bedding	Ordinary bedding "planes" with variations due to surface irregularities, bioturbation, or diagenesis	May be even, wavy, or nodular; continuous or not over m-scale distances; beds parallel or not; thicknesses from millimeter to meter scale	Sediment "sinks"—areas of low-velocity fluid motion; may be protected shallows, deeper zones below wave agitation, and areas free from current scour	May confine permeability to horizontal direction
Deposition only—Cross bedding	Commonly called "spillover beds" or "simple cross stratification"	Curved, convex-upward bedding where lower boundaries of bed sets are nonerosional; scale varies from centimeter to multi-meter size; bed curvature difficult to detect in small-diameter cores	Eolian dunes, slope-break calcarenites ("spillover lobes"), sandwaves with high rate of sedimentation	Pronounced bedding inclination, repetitive sets, and common textural variability influence both porosity and permeability; correlation of individual beds is difficult
Erosion only	Truncated surfaces, channels, "sharp" contacts between lithological units; unconformities and disconformities	Abrupt change in bedding or lithology at distinct surface; may be curvilinear; clasts of eroded bed may be incorporated in overlying beds	Exposure surfaces on land or paleo-highs, tidal flats and inlets, subtidal current scour zones, slope failure scars	May be permeability barriers; may affect porosity distribution
Erosion with deposition	Migrating sets of sinusoidal ripple marks that create trough or festoon crossbeds; bed aspect differs in bedding plane and cross-sectional views	Curved, concave-upward bedding; lower boundaries of bedsets always erosional—scour-and-fill mechanism; scale varies from centimeter to meter in size	Any unidirectional current in wind or water; typical in sand waves, eolianites, middle shoreface, and turbidite/density flow deposits	Pronounced bedding, inclination, repetitive sets, and textural variations influence both porosity and permeability

Biological growth patterns	Constructed voids, shelter voids, skeletal growth fabrics, and patterns of organic lamination (e.g., algal laminae); includes *Stromatactis* cavities	Voids (pores) of irregular size and shape formed by organic growth in reefs, voids conforming to size and shape of skeletal component, as sheltered beneath a bivalve shell, or algal–microbial laminae	Reefs, algal and microbial mats in intertidal or subtidal zones, and in detrital carbonates with high content of large bioclasts forming shelters	Large pores may have major impact on reservoir porosity, but connectivity may be poor; fabric similar to "vuggy" porosity formed by dissolution diagenesis; recovery efficiency may be poor
Desiccation	Mud cracks, desiccation polygons (related to "teepee" structures), desiccation wrinkles in algal laminae, and fenestral or bird's-eye fabric	Broken laminae in lime mudrocks and algal laminated carbonates; meter-scale polygons with mudcrack boundaries; irregular, wrinkled laminae in algal and microbial laminites; irregular, lenticular pores parallel to bedding (fenestrae)	Subaerial exposure surfaces as in supratidal zone, perimeter of water bodies, intermittent lagoons, playa lakes, etc.; commonly associated with evaporites in arid climates; and with flat-pebble conglomerates in tidal flats	Shrinkage and fenestral pores may have major impact on total porosity but permeability may depend more on "matrix" pore network connectivity; recovery efficiency may be poor
Slope failure and fluidization	Contorted bedding, fluid escape structures, and detached bodies of "exotic" sediment	Bedding appears "folded" and contorted; laminations indicate plastic (soft sediment) deformation; exotic blocks; upward-penetrative features indicating fluid escape under high pressure	Base of slope at shelf edge or paleo-high margin; may be in basinal surroundings indicating debris flow and slope margin failure	May transport porous and permeable rock masses into basinal setting to produce exotic reservoir in unlikely setting; may have small volume; difficult to detect
Compaction	Stylolites, diagenetic enhancement of bedding irregularities, and closure of intergranular pores	Irregular, spiked traces in subparallel arrangement with bedding, exaggerated bed irregularities, especially "nodular" bedding, and "overcompaction" of intergranular pores	A consequence of burial; greater burial depth or application of tectonic stresses exaggerates the features	Stylolites are barriers to flow in many Middle Eastern reservoirs; compaction is probably the major cause of porosity reduction in granular carbonates

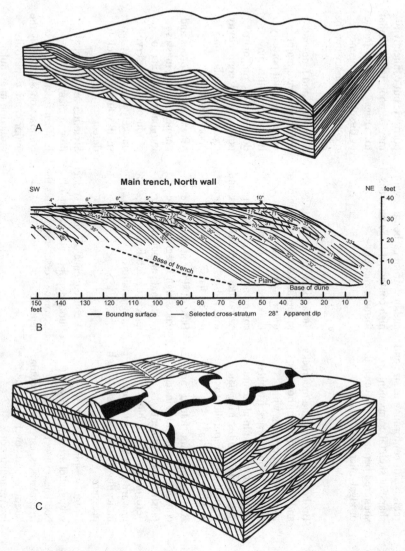

Figure 2.5 Typical sedimentary structures found in carbonate rocks. (a) Block diagram showing a three-dimensional view of asymmetrical oscillation (wave) ripples. The scale of these ripples varies from a few millimeters to tens of centimeters in crest-to-trough height and a few centimeters to a meter or more in crest-to-crest wavelength depending on the size and period of the waves that formed them. (b) Illustration of the large-scale and common "spillover"-type bedding common in eolian dunes. (c) Block diagram of asymmetrical ripples formed by unidirectional currents. Note that these ripples produce trough crossbeds visible in the lower right face of the diagram. Current direction is from left to right. Note the sinusoidal shape of the advancing ripple crests and the scoured depressions in front of them. The scale of these features varies from a few millimeters to as much as a meter in ripple height, depending on the depth and power of the flow. (Parts (a) and (c) adapted from illustrations in Reineck and Singh (1973); part (b) adapted from an illustration in McKee (1966).)

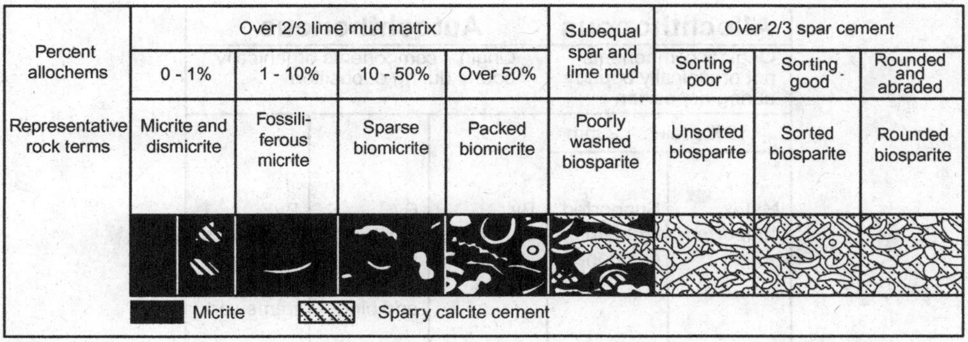

Percent allochems	Over 2/3 lime mud matrix				Subequal spar and lime mud	Over 2/3 spar cement		
	0 - 1%	1 - 10%	10 - 50%	Over 50%		Sorting poor	Sorting good	Rounded and abraded
Representative rock terms	Micrite and dismicrite	Fossili-ferous micrite	Sparse biomicrite	Packed biomicrite	Poorly washed biosparite	Unsorted biosparite	Sorted biosparite	Rounded biosparite

▓ Micrite	▨ Sparry calcite cement	

Figure 2.6 The classification scheme for detrital carbonate rocks developed by R. L. Folk first published in 1959 and later in 1962. Folk classified biogenic (reef) carbonates as bio-lithites. He did not include a term for completely recrystallized or replaced rocks. (Adapted from the illustration in Folk (1962).)

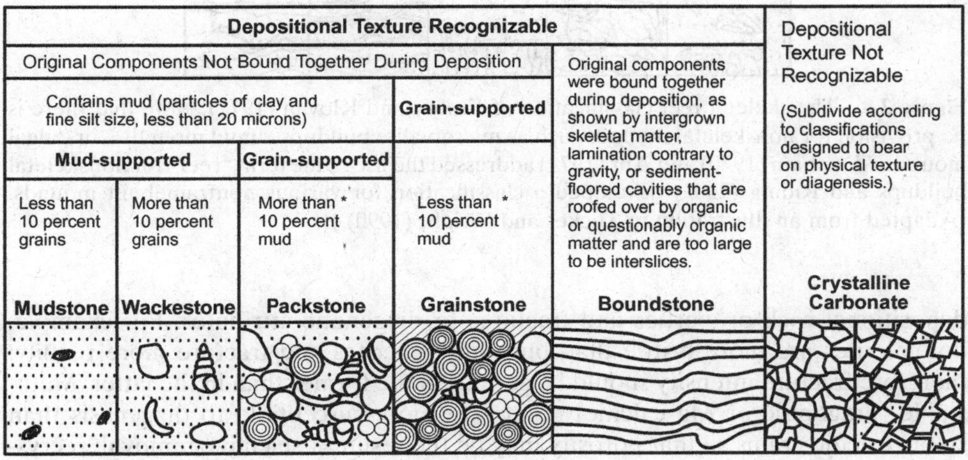

Depositional Texture Recognizable						Depositional Texture Not Recognizable
Original Components Not Bound Together During Deposition					Original components were bound together during deposition, as shown by intergrown skeletal matter, lamination contrary to gravity, or sediment-floored cavities that are roofed over by organic or questionably organic matter and are too large to be interslices.	(Subdivide according to classifications designed to bear on physical texture or diagenesis.)
Contains mud (particles of clay and fine silt size, less than 20 microns)				Grain-supported		
Mud-supported		Grain-supported				
Less than 10 percent grains	More than 10 percent grains	More than * 10 percent mud	Less than * 10 percent mud			
Mudstone	Wackestone	Packstone	Grainstone		Boundstone	Crystalline Carbonate

* Modification of original Dunham classification by changing percent mud from 1 to 10%

Figure 2.7 R. J. Dunham's classification of carbonate rocks. Note that the classification includes detrital carbonates as mudstones through grainstones, biogenic (reef) carbonates as boundstones, and diagenetically altered carbonates as crystalline carbonate. (Adapted from the classification scheme illustrated in Dunham (1962).)

fundamental rock properties systematic and reproducible, and (2) to facilitate information transfer. Reservoir geologists are concerned with both aspects, but they are even more concerned with relating rock classifications to reservoir properties. In order for that to be possible, reservoir and rock properties must have characteristics in common. That is, rock classifications correspond closely with reservoir properties only if reservoir properties depend on parameters used in the rock classifications. For example, a classification of detrital carbonates based on depositional texture is independent of pore types formed by diagenesis, by fracturing, or by biological growth patterns in reef rocks. But some indirect relationships may exist between

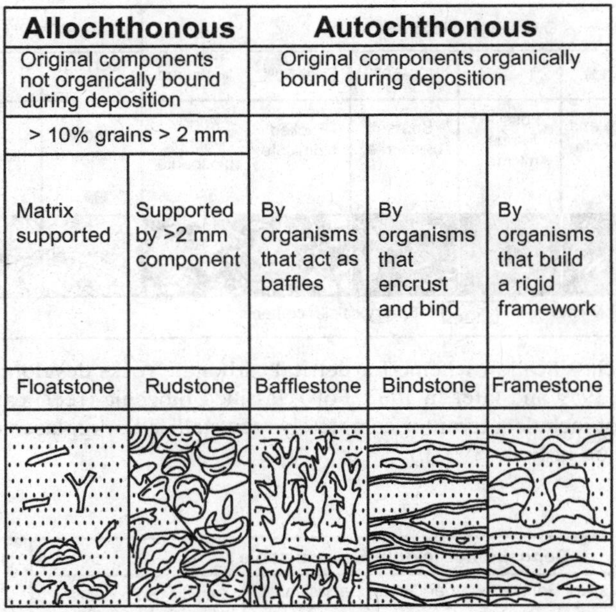

Allochthonous		Autochthonous		
Original components not organically bound during deposition		Original components organically bound during deposition		
> 10% grains > 2 mm				
Matrix supported	Supported by >2 mm component	By organisms that act as baffles	By organisms that encrust and bind	By organisms that build a rigid framework
Floatstone	Rudstone	Bafflestone	Bindstone	Framestone

Figure 2.8   The skeletal reef classification of Embry and Klovan (1971). Note that there is no provision for nonskeletal mounds such as microbialite buildups, "mud mounds," or "algal mounds." Dunham (1970) and Ahr (1971) addressed the use of the term "reef" for nonskeletal buildups and Riding (2002) developed a classification for various nonframebuilt mounds. (Adapted from an illustration in Tucker and Wright (1990).)

depositional rock properties and fractures or diagenetic attributes. For example, dolomicrites are more brittle than pure lime micrites and fracture more readily; therefore fracture intensity should be greater in the former than in the latter. Metastable (aragonitic or Mg calcitic) grains are more susceptible to diagenesis than stable calcitic grains so that porosity may be the result of selective removal, recrystallization, or replacement of original minerals. We will see that several carbonate porosity classifications do not include the mode of origin of rocks; consequently, it is difficult to use those classifications to distinguish pore types that formed as results of depositional processes from those that were modified or created by postdepositional diagenesis or fracturing. It is equally difficult to distinguish flow units by their geological origin, without which it is difficult if not impossible to predict their spatial distribution at stratigraphic scale. This book presents a genetic classification of porosity linked to the complete geological history of reservoir rocks as an alternative. It is based on the idea that there are three end-member pore types in carbonate reservoirs: depositional, diagenetic, and fracture pores. These different processes impart distinctive characteristics to both rock matrix and pores. Because the distinctive characteristics were imparted to pores and rocks at the same time and by the same processes, key rock properties may act as "markers" or proxies for pore types that can be identified and traced at stratigraphic scale. To the extent that the proxies are identifiable and mappable, so will be the accompanying pore types and, of capital importance, their petrophysical attributes.

2.3.1 Classification of Detrital Carbonates

The Folk (1959, 1962) and Dunham (1962) classifications work well for detrital carbonates but they are not as useful to describe reef rocks or diagenetic textures and fabrics. Folk and Dunham coined words to describe reef rocks. Folk chose *biolithite* and Dunham chose *boundstone*, but those terms treat all reefs alike and as if the entire reef mass were homogeneous. Porosity and permeability vary greatly in reefs depending on the type of reef organisms present, the reef growth forms, the ratio of skeletal framework to loose detritus, and the internal microstructure of the reef-building organisms. Embry and Klovan (1971) and Riding (2002) devised reef classifications that include more detailed systems for describing the variability found in reef reservoirs. Diagenetic properties are not included in the Folk and Dunham schemes either, except that Dunham included the term *crystalline carbonate* for a rock in which depositional texture is unrecognizable because it was obliterated by diagenesis.

The Folk and Dunham classifications share a common theme. They are based on the mud-to-grain ratios in carbonate rocks and on the packing arrangement of the framework grains. These similarities exist because Dunham and Folk shared discussions on carbonate classifications (Folk, 1962) and how the concept of textural maturity used for terrigenous sandstones could be applied to carbonates. Textural maturity in terrigenous sandstones refers to the amount of matrix (clay or mud) that has been removed by winnowing and the extent to which sorting and rounding are visible in framework grains. When applied to carbonates, rocks with high lime mud content (more than 90% lime mud) are classed as *mudstones* by Dunham and as *micrites* by Folk. Rocks with only grains and no mud are classed as *sparites* by Folk and *grainstones* by Dunham. The term sparite implies that sparry cement occupies intergranular pores. Unconsolidated sediments were excluded by Folk because his scheme was devised for lithified limestones. Between high and low mud content are the carbonates with variable proportions of mud and grains. Folk chose the terms "sparse" and "packed" to modify micrites with 10%–50% and over 50% grains, respectively. By doing so, he set the requirement that grain percentage determines the rock name. Dunham chose a different approach, probably because he saw that less than 50% of irregularly shaped grains could create a self-supporting fabric; hence the origin of the term *packstone*. He resorted to grain percentage as the determining factor for naming muddy rocks with grains but without a self-supporting grain fabric. These muddy, grainy mixtures in which the grains are widely dispersed ("floating") in the mud matrix are termed *wackestones*. Most industry professionals use the Dunham classification today because the terms are shorter, and easier to log when working on large quantities of rock, they do not require tedious counting or percentage estimates, and they seem to evoke mental images of rock properties that can be related to reservoir properties.

For the reservoir analyst, detrital rock classifications based on depositional texture are the most practical and easiest to use. First, the muddy rocks can be assumed to have formed in environments where winnowing was insignificant and rocks with high grain content represent environments with extensive winnowing, the "high-energy" environments. Second, excluding diagenesis, fracturing, and special forms of intragranular porosity, mud content is inversely related to intergranular porosity. Grainstones and packstones have the highest percentage of depositional porosity, they usually have comparatively simple intergranular pore systems, and porosity is predictably related to facies geometry. However, because they may have high intergranular porosity, they are susceptible to early cementation and compaction that reduce pore and pore throat size. As rocks with high grain content commonly occur near the tops of shallowing-upward cycles, they are relatively easy to locate in repetitive sequences of these cycles. Some cycles terminate in evaporite "capping facies" with pores in the underlying grainstones and packstones plugged with gypsum, anhydrite, or halite. In those cases, diagenesis sometimes compensates for pore plugging at the cycle tops, because dolomitization commonly accompanies evaporite formation and it may be linked with enhanced porosity in midcycle wackestone and packstone facies.

2.3.2 Classification of Reef Rocks

The word "reef" still prompts animated discussion and disagreement among geologists. Much of the older terminology on reefs centers on the academic issue of whether reefs are "ecological" or "stratigraphic," as described by Dunham (1970). An ecological reef is built by constructor organisms that have the "ecological potential" to form wave-resistant frameworks. That is, they must be made of sturdy skeletal structures that grew presumably in the midst of breaking waves. Many reefs throughout time grew in environments that were not exposed to breaking waves and many biogenic buildups lack sturdy skeletal frameworks, especially buildups constructed of micrite, or carbonate cement, or microbial thrombolites and stromatolites. Terminology is not a major issue for reservoir studies or for carbonate sedimentologists who follow the more modern style of classifying all sturdy skeletal buildups as *frame-built*, or *skeletal reefs*, and all of those buildups without sturdy skeletal frameworks as *reef mounds* (Tucker and Wright, 1990). However, additional detail is needed in reef classification schemes to describe the fundamental rock properties of the reef as they relate to reservoir porosity, permeability, and connectivity.

Reservoir characteristics in reefs vary with the type of constructor organisms, with the relationship between constructor organisms and associated reef detritus, and with growth patterns of reef complexes in response to prevailing hydrologic conditions. Reefs built by calcified microbes, for example, have high proportions of lime mud and cement but few skeletal framebuilders. Skeletal framework reefs, reefs built up as repeated layers of pavement-like organic encrustations, and reefs formed by the current-baffling and sediment-trapping action of benthic organisms such as sea grasses and algae present unique rock fabrics and pore characteristics. Dense encrustations by calcareous algae exhibit internal microstructures that differ from those of porous sponge or coral skeletons. Patterns of reef growth vary in response to the depth of the photic zone, to oxygenation and nutrient content, to turbidity, and to water agitation by waves and currents. For example, modern corals grow in sheet-like or dome-like fashion in deeper water because they need light for their photosynthesizing symbionts, the zooxanthellae. Stromatoporoids, major constructor organisms in Silurian and Devonian reefs, took on specific growth forms in response to higher or lower levels of wave and current activity.

Facies patterns associated with reefs vary as a function of the hydrologic regime. Shallow-water (in most modern oceans this is less than about 10-m depth) reefs have distinct seaward and leeward sides because they are shaped or "polarized" by the prevailing wind–wave direction. Windward sides of reefs are characterized by massive and encrusted organic growth with boulder-to-gravel-sized particles as rudstones. Leeward sides of reefs are characterized by higher percentages of detrital sediments such as floatstones, grainstones, and more delicate growth forms of reef organisms. Shallow-water patch reefs and shelf-edge reefs tend to be streamlined in plan view with buttress-like structures and more massive skeletal frameworks on the windward side. In modern coral–algal reefs, structures called *spur and groove* or *buttress and chute* develop on the windward sides of reefs (Shinn, 1963; James, 1983). Deep-water buildups, or those that grew in protected shallows, do not exhibit windward and leeward sides, streamlining, or polarized facies geometry. In short, reservoir characteristics of reef rocks are related to variables different from those that form the basis for detrital rock classifications. Embry and Klovan (1971) found it hard to map reef reservoir porosity zones with nothing more definitive than Dunham's boundstone or Folk's biolithite terminology. They cited papers from the 1960s focusing on, if not lamenting, the problem. As an improvement, they adapted Dunham's classification to reef reservoirs and developed a more detailed scheme to account for different organic growth forms and for the associated detrital carbonates that surround and fill-in open spaces within reefs.

The Embry–Klovan terms *framestone*, *bindstone*, and *bafflestone* refer to growth patterns of reef organisms. That is, a reef constructed of stout coral skeletons in a girder-like frame arrangement is called a framestone. One in which the reef is constructed by pavement-builders or encrusters is known as a bindstone, and reefs that exhibit detrital carbonate accumulations in the midst of organic thickets such as seagrass beds are known as bafflestones. There is some controversy over whether reefs can truly be formed by baffling action of sessile benthonic organisms because sessile benthonic animals are mainly filter feeders that would be smothered by a rain of carbonate sediment. If the bafflers were plants, true grasses can be eliminated for much of geological history, as they have existed only since the Mesozoic Era. Bafflestones tend to have high proportions of lime mud, and there is evidence, especially in mud mounds of Lower Carboniferous age, that muds are formed in place by biological or biochemical processes rather than having been trapped by organisms. The term bafflestone is considered by Tucker and Wright (1990) to be "rather subjective." Those authors also point out, that reef rocks are subject to extensive diagenesis and bioerosion that may dramatically alter the original rock fabric. Diagenetic micritization of reef rocks is common and may account for the loss of 20–70% of the original reef framework (Tucker and Wright, 1990). Finally, the Embry–Klovan terms *rudstone* and *floatstone* refer to detrital rocks associated with reefs. Rudstones are the reef-derived, gravel equivalent of grainstones and packstones; floatstones are the gravel and sand equivalent of wackestones.

Riding (2002) developed an alternative classification for reef rocks. He defines reef as "in-place calcareous deposits with topographic relief, created by sessile organisms." The different types of reef rocks are classified on the basis of whether there is matrix (mud) support (the *carbonate mound* category), skeletal support (the *frame reef* category), or cement support (the *cement reef* category). Terms such as sparse and dense are used to describe the three-dimensional fabric of skeletal elements in matrix supported reefs, and open, tight, and solid describe the architecture of the constructor assemblages in frame reefs. These terms refer to spatial patterns and evoke mental images of framework/detritus ratio, which translates into type and spatial distribution of pore categories, assuming that diagenesis has not radically altered them. It is reasonable to infer that depositional porosity and permeability are highest in frame reefs and lowest in micrite mounds and cement reefs. Reservoirs exist in those reef categories but usually owe their existence to enhanced porosity and permeability formed by diagenesis or fracturing. Predicting reservoir connectivity is difficult in all reef categories but is especially hard to predict in diagenetically altered, complex mixtures of frame and detritus. Diagenetic porosity may be strongly bimodal in size. Microporosity is common in lime mud portions of reef rocks, for example.

2.3.3   Wright's Genetic Classification

The Folk and Dunham classifications for detrital carbonates were introduced nearly a half-century ago when our understanding of diagenetic processes and their products was in its infancy and before much effort was made to develop classifications for reef rocks. Recognizing these deficiencies in descriptive terminology, Wright (1992) proposed an integrated scheme that links the depositional classification of Dunham (1962) and the biological classification of Embry and Klovan (1971) with a new classification for diagenetic rocks. In concept, this classification is logical and more utilitarian for the reservoir geoscientist than any existing single classification. This is a *genetic* classification system in which carbonate rocks are grouped by mode of origin—depositional, biological, and diagenetic. Each category has subheadings to distinguish the various rock properties that typify each mode. The terms introduced for diagenetic carbonates draw attention to whether or not the diagenetic process has obliterated the original texture and fabric. This distinction requires examination and interpretation of thin sections under the polarizing microscope, however. In addition, the terms for compacted rocks with microstylolitic grain contacts should include packstone along with grainstone. It is difficult to have a perfect classification for all applications, but this genetic scheme represents an advance. Later in this chapter we will discuss classifications of porosity and we will see that purely descriptive classifications of porosity, like nonintegrated classifications for carbonate rocks, may be less useful in analyzing carbonate reservoirs than integrated, genetic classifications.

2.4   DEPENDENT OR DERIVED ROCK PROPERTIES

Porosity, permeability, and bulk density depend on fundamental properties such as texture, mineralogical composition, and fabric. Dependent properties, especially porosity and permeability, are among the most important variables that determine reservoir quality. While rocks are classified according to their fundamental properties and inferences are made from rock classifications about depositional environments, porosity is classified according to physical attributes that may not be related to mode of origin. But unless mode of origin is included in porosity classifications, it is not possible to deduce the environment in which the porosity was formed, when it was modified, and which genetic pore types correspond to highest permeability. Comprehensive reservoir description depends on identification and description of correspondence between rock matrix and pore characteristics, how they are genetically and temporally related, and how they influence petrophysical attributes. Porosity is measured directly from core samples and indirectly with some types of borehole logs. Permeability is measured as the coefficient of proportionality in Darcy's equation for fluid flow through porous media. It is measured directly from core samples and it is the yardstick by which many quality rankings are assigned to reservoirs. Special wireline testers and pressure buildup tests can measure flow rates and provide meaningful estimates of permeability and petrophysical experts argue that permeability can be estimated from wireline log data. Not everyone agrees, especially those who work on carbonate reservoirs. Bulk density is a measure of the solid/void ratio in reservoir rocks and is measured directly in core analyses or indirectly with wireline logs. Bulk density values can be used to aid in estimating porosity.

2.4.1   Porosity

Reservoir rocks consist of solid material and interstitial pore spaces that may or may not be connected, such that

V_p = Pore volume
V_s = Solid volume
V_t = Total rock sample volume = $V_p + V_s$

Porosity is usually designated by the symbol ϕ and is expressed as a percentage

$$\phi = (V_p/V_t) \times 100$$

Reservoir specialists are primarily concerned with the fraction of total porosity that transmits fluids, that is, the interconnected or *effective porosity*, ϕ_e. Effective porosity is the ratio of the interconnected pore volume to the total rock volume. Direct measurements of V_p in the laboratory are measurements of effective porosity. Not all pores are interconnected, however. Unconnected porosity is called *residual porosity*, ϕ_r, so that total porosity is the sum of $\phi_e + \phi_r$. Total porosity is the quantity derived from borehole measurements made with the various "porosity logging" devices (Monicard, 1980). Porosity varies with texture, fabric, and fracture geometry in the reservoir rock. Grain shape, sorting, and packing are the main variables that affect porosity in *detrital rocks*, growth fabric and skeletal microstructure affect inter- and intraparticle porosity in *biogenic rocks*, and porosity in *fractured rocks* is determined by fracture width, fracture spacing, and presence/absence of mineralization. Diagenesis may plug pores with cement, close pores with compaction, open pores with dissolution, or create new pores by recrystallization or replacement.

Berg (1970) illustrated the geometrical relationship between pore size and grain size with identical spheres in a packing arrangement with about 30% porosity. Although the example is unrealistic in terms of "real world" reservoir rocks, it is a useful demonstration of the relationships between fundamental rock properties, texture in this case, and pore characteristics. Imagine that Berg's idealized example is an oolite grainstone with well-sorted spherical grains and unaltered depositional porosity. Depositional pore size is a function of grain size, packing, and sorting.

According to Berg's calculations, pore diameter in a sample with about 30% porosity and well-sorted spherical grains varies from about $0.4D$ to $0.06D$ (D = grain diameter), depending on packing: cubic packing has the largest pore diameter, while rhombohedral has the smallest. High porosity grainstones with cubic packing and near-spherical grains will have pore sizes in the 0.06–0.4-mm range for 1-mm diameter ooids. Porosity does not vary with grain diameter, but it does vary with packing and sorting. Cubic and rhombohedral packing arrangements for idealized, spherical grains have 47.6% and 25.96% porosity, respectively (Figure 2.9). If a second size of spherical grains is introduced, that is, if sorting is poor (Figure 2.10), the porosity of a cubic-packed sample is reduced from about 48% to about 13%. Grain shape also affects porosity. A comparatively small number of irregularly shaped grains,

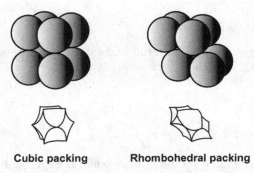

Cubic packing   **Rhombohedral packing**

Figure 2.9   The influence of grain packing on porosity. Cubic packing of spherical grains has 47.6% porosity regardless of grain size, but rhombohedral packing of the same grains reduces porosity to 26.9%. Pores, shown below the grain packs, are "negative images" of the grains. In the cubic packing example it is clear that each pore is connected to others by three pore throats. Later this will be known as a pore system with a coordination number of 3.

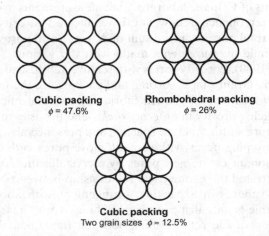

Cubic packing   **Rhombohedral packing**
$\phi = 47.6\%$ $\phi = 26\%$

Cubic packing
Two grain sizes $\phi \approx 12.5\%$

Figure 2.10   Cubic and rhombohedral packing and the importance of sorting. When a second, smaller but uniform grain size is introduced, the original 47.6% porosity is reduced to 12.5%.

usually skeletal in origin, can support high-porosity packing arrangements as Dunham (1962) illustrated in photographs. As we have seen earlier in this chapter, reef growth form and skeletal microstructure are types of biogenic rock fabric that have a major effect on effective porosity. In some reef rocks porosity values may be high, but effective porosity may be low because some intraparticle pores are open to fluid movement in only one or two directions. Some intraparticle pores are, in fact, totally disconnected. They make up part of the residual porosity in a reef reservoir. Conversely, if large interskeletal and intraskeletal pores are present and connected, a reef reservoir may have very high effective porosity.

Porosity in modern carbonate sediments ranges from about 40% to 70% but is about 5%–15% in ancient rocks (Choquette and Pray, 1970). Porosity reduction is complex and can involve cementation, compaction, or combinations of the two. Some studies show that porosity in carbonate reservoirs is reduced by a factor of 2 during burial to a depth of 1740 m and that burial depth has a greater effect on porosity reduction than the amount of time during burial (Schmoker and Halley, 1982). They found that porosity in South Florida carbonates decreased exponentially with depth from over 40% at the surface to less than 10% at 5486 m (see Figure 5.7 for a variety of porosity vs depth curves). They also found that porosity in dolostones was lower than that of limestones near the surface, but greater than limestones at depths greater than 1700 m, and that the rate of decrease in dolostone porosity was less than for limestones with increasing burial depth. Budd's (2001) study of shallow Cenozoic carbonates in Florida revealed that permeability is lost more quickly during burial than is porosity in the same rocks. He also found that low permeability, lime-muddy rocks with median permeability values of 35 md (millidarcies) or less did not show a clear trend of permeability change with depth, but limestones with median permeability of 69 to over 400 md showed a clear trend of decreasing permeability with increasing depth of burial. He concluded that the best limestone reservoir rocks in his study were those with grain-supported textures and higher permeability before burial. He found that depth-related permeability loss was due mainly to mechanical compaction in shallower depths and to chemical plus mechanical compaction at greater depths. For limestone reservoirs in general, porosity and permeability loss due to cementation is probably an early diagenetic phenomenon. More pronounced porosity and permeability loss with depth is caused by mechanical and chemical compaction. Amthor et al. (1994) found that if burial depth is not considered, limestones and dolomitic limestones have higher porosity and permeability than dolostones in the Devonian LeDuc Formation in Canada, but they noted that porosity and permeability decreased with increasing depth. At depths of up to about 2000 m, limestones and dolomitized limestones had nearly equal values of porosity and permeability, but at depths greater than 2000 m, dolostones had significantly greater porosity and permeability than limestones. They concluded that dolostones undergo less porosity and permeability loss with depth than limestones because dolostones are more resistant to chemical and mechanical compaction than limestones.

Most carbonate reservoirs have porosity of about 5–15%, as compared with terrigenous sandstone reservoirs, which have porosities of 15–30% (see Table 1.1). The percentage of sample surface area covered by visible porosity can be used to obtain a qualitative estimate of the "quality" of reservoir porosity, following Archie (1952):

Porosity Range	Qualitative Description
5% or less	Poor
10%	Fair
15%	Good
20% or more	Excellent

Note that porosity together with permeability determine reservoir quality. Porosity is a measure of capacity to store fluids; permeability is a measure of capacity to transmit fluids. Pores are connected by *pore throats*, which strongly influence permeability and which are related to pore size, shape, and arrangement. If one rock is composed of spherical, well-sorted grains of a given size, intergranular pores will be only a fraction of the grain size and the attendant pore throats will be even smaller but generally uniform in size. Rocks with poorly sorted grains will have poorly sorted pore and pore throat sizes. Sorting and distribution of pore throat sizes in rocks determines pore–pore throat size ratio and *accessibility* of fluids to the pore throat system in rocks. Imagine molecules of oil migrating through pores and pore throats. If pore throat sizes are uniformly distributed, the oil molecules will encounter more-or-less the same size pathway for migration and the threshold force (pressure) required to move through the rock will be more-or-less uniform for any given pore throat size. If the pore throats are poorly size-sorted, the migrating oil molecules will encounter different sized pathways with different pressure thresholds to pass through. In most terrigenous sandstones and in some carbonate grainstones, intergranular pore sizes have a strong statistical correlation to grain size. In such cases, pore throat sizes may correlate well with pore sizes, as illustrated by a plot of the log of permeability as a function of porosity. In the ideal case, data points on this "Φ–k" plot will form a straight line cluster. In most carbonates, pore and pore throat geometries are not as predictably regular as this ideal case; therefore it is helpful to be able to identify the pore types that have the most influence on permeability. In order to understand pore characteristics and how pore types are related to other readily identifiable rock properties, a system for classifying carbonate porosity is necessary.

2.4.1.1 Porosity Classifications

Classifications are schemes for sorting or grouping things. They vary from simple sorting of items into groups with common characteristics to tiered schemes in which characteristics are grouped in sets, or which take on modifiers to aid in identifying complex or obscure relationships. For example, a simple classification for porosity might designate three categories of pores as interparticle, vug, and fracture. A tiered classification scheme might group according to average pore size, pore shape, petrophysical characteristics, and mode of origin. Classifications are ways of organizing material into useful forms. Some schemes aid in interpreting the origin of pore types. However, a weakness in much of our traditional approach is that we treat rocks and reservoirs as separate entities. The links between rock classifications and pore classifications are poorly represented in most geological literature. Even less well represented are the relationships between rock and pore classifications and reservoir petrophysical characteristics. The concept of "rock typing"—identifying reservoir

rock categories on the basis of their porosity, permeability, and pore throat sizes—
is widely discussed in the petroleum engineering and log analysts' literature,
however.

A simple method for grouping pore characteristics is all that is required for a
basic reservoir rock description. The results can be combined with other data to
compare pore characteristics with rock classifications, capillary pressures, saturation
values, and borehole log signatures. Ideally, porosity classifications should compare
closely with rock classifications in order to reveal which kinds of rock data correspond most closely with porosity data in order to identify rock properties that can
act as proxies for porosity. Ultimately, these rock–pore data sets could be compared
with petrophysical characteristics in order to achieve the goal of identifying and
mapping reservoir flow units (a version of petrophysical rock typing). Traditional
classifications of carbonate porosity were not designed for that purpose, but the
well-known schemes by Archie (1952), Choquette and Pray (1970), and Lucia (1983)
are useful to illustrate the evolution in thinking about carbonate reservoir pore
systems. Finally, a new classification based on end-member genetic categories—
depositional, diagenetic, and fracture porosity—is presented as a more useful
alternative.

2.4.1.2 The Archie Classification

One of the first, if not *the* first, carbonate porosity classifications was developed by
G. E. (Gus) Archie (1952), who pioneered the study of electrical resistivity in rocks,
developed the principles that led to the Archie saturation equation, and investigated
methods to integrate geological data with laboratory petrophysical data and borehole log signatures. His objective was to illustrate relationships between rock and
petrophysical properties in reservoirs.

The Archie porosity classification is based on textural descriptions of reservoir
rock along with the "character" of any visible porosity. Three textural categories are
termed Type I, II, and III, and four classes for visible porosity are identified as classes
A through D. Class A has no visible porosity at 10 magnifications, class B has visible
pores between 1 and 10 μm, and class C has visible pores larger than 10 μm but
smaller than rotary cuttings (roughly, about 2.0 mm). Class D includes large visible
pores such as solution vugs larger than cuttings samples.

Archie described Type I carbonates as "crystalline, hard, dense, with sharp edges
and smooth faces on breaking." Under the binocular microscope, these rocks have
a matrix made of tightly interlocking crystals that do not exhibit visible intercrystalline porosity. For practical purposes, these rocks correspond to today's mudstones
and dolomudstones. The Solenhofen Limestone is a good example of this type of
rock. Type II rocks are described as "earthy" or "chalky" with grains not larger than
about 50 μm in diameter (just below the finest silt size), and they are composed of
"fine granules or sea organisms." These rocks correspond to today's true chalk, or
mudstones and wackestones that have probably undergone diagenetic alteration to
attain the chalky appearance. Type III carbonates are "granular or saccharoidal."
Saccharoidal is a somewhat arcane term derived from the Greek, σακκηαρου,
meaning "sugar." Many fine grained carbonates, especially dolostones, exhibit small
crystalline mosaics that reflect light like so many sugar crystals. Granular carbonates
include today's grainstones and packstones.

Archie's classification is the pioneering effort to relate rock properties and reservoir petrophysics but because Archie was an engineer rather than a geologist, his classification does not deal with pore origin or modification. The major contribution of the Archie classification is that it relates petrophysical properties, including capillary pressures, electrical properties, and saturation characteristics, to different rock types. For the first time, groups of capillary pressure curve shapes could be compared with their corresponding groups of carbonate pore types. But by today's standards these pore types do not have an obvious relationship to depositional textures, diagenetic fabrics, or pore types by geological origin; consequently, the Archie classification is more interesting from a historical standpoint than it is from a practical one.

Archie's goal was to develop a classification that emphasized pore structure, fluid flow characteristics, fluid distribution, and the flow of electricity. He avoided terms that denote rock composition or that suggest a geological origin for the porosity. His visionary work resulted in a combined petrological–petrophysical classification system for carbonate porosity. He did not consider it important to identify the geological thread that binds pore origin to rock origin, although many carbonate pore types are altered or created long after the host rock was deposited. Unfortunately, when the mode of origin—the genesis—of carbonate pore types is not part of the porosity classification, the job of establishing correspondence between pore types and rock matrix properties is all the more difficult. Even with the present technology, it is difficult to correlate individual porous zones from well to well, and always difficult to map the spatial distribution of flow units at field scale. Performing these tasks requires data on how porosity relates to other rock properties that serve as *proxies* or *markers* for effective porosity. The Archie classification does not provide that kind of information.

2.4.1.3 The Choquette–Pray Classification

Choquette and Pray (1970) recognized the need to incorporate time and mode of origin in their classification of carbonate porosity. Their classification (Figure 2.11) is practical. They recognized 15 basic pore types and organized them into three classes depending on whether they are *fabric selective, not fabric selective*, and *fabric selective or not*. The origin of fabric-selective pores could be depositional, diagenetic, or both, although that point is not emphasized in the classification. Intergranular pores in an oolite grainstone, intercrystalline pores in a crystalline dolomite, or grain-moldic pores in a skeletal packstone are examples of fabric-selective pores that have different origins. Non-fabric-selective porosity includes fractures or dissolution cavities of various sizes that cut across rock fabric. Fabric selective or not is a category that includes mainly penetrative features such as animal or plant borings and burrows and desiccation cracks. Breccias may be conglomeratic residue from solution collapse, the products of clast-producing erosion followed by resedimentation, or the result of tectonism.

Time of origin and direction of pore-altering, diagenetic processes are incorporated as "genetic modifiers." Paralleling Archie (1952), terms for pore size and abundance also modify root names for pore types. Times of origin are designated as primary and secondary. Primary origin includes pores formed by depositional and early postdepositional processes. Secondary porosity is interpreted to be later

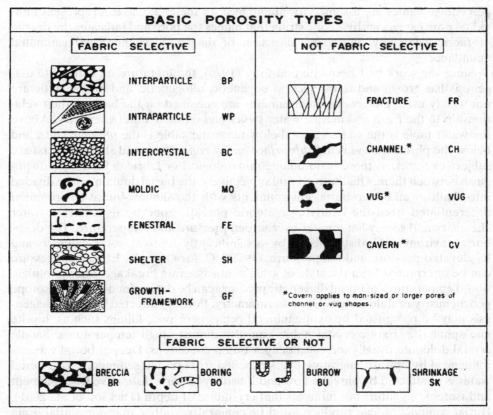

Figure 2.11 The essential elements of the Choquette–Pray (1970) porosity classification for carbonate rocks. The essence of the classification is whether or not porosity conforms to rock "fabric." Note that rock fabric in this sense includes both depositional and diagenetic fabrics. (Adapted from an illustration by Choquette and Pray (1970) in Scholle (1978).)

diagenetic in origin and to have occurred in the "eogenetic," "telogenetic," or "meso-genetic" domains. Eogenetic changes are those affecting newly buried rocks still within the influence of a depositional surface or an intraformational erosional surface. Telogenetic changes affect long buried rocks that later became "connected" to an erosional surface. Mesogenetic changes are those unaffected by surface conditions or processes. The direction of change in pore characteristics is indicated by the terms enlarged, reduced, and filled. These changes result from solution, cementation, or internal sediment (infilling).

Pore abundance in the Choquette–Pray classification is given as a percentage, expressed as a ratio of pore categories, or as a ratio of pore types plus a percent figure. Pore size categories are termed "megapore, mesopore, and micropore," with size limits closely paralleling the Wentworth (1922) grain size scale. Megapore sizes are equivalent to gravel (4–256 mm), mesopore sizes are equivalent to sand-plus-granules (4 to 1/16 mm), and micropores are equivalent to mud (<1/16 mm). Letter codes are assigned to pore types, size modifiers, and genetic modifiers, and percentage values are assigned for abundance. In the end, one can represent individual pore categories by a code string that names the pore and indicates the degree of fabric selectivity, pore size, direction of diagenetic change, and estimated abundance.

Since the work by Choquette and Pray (1970), the literature on carbonate diagenesis has grown and terms such as eogenetic, telogenetic, and mesogenetic are not widely used. Diagenetic environments are classified on the basis of their relationship to the fresh and marine water tables and on their water chemistry. Above the water table is the *vadose zone*, below the water table is the *phreatic zone*, and below the phreatic zone is the *subsurface burial zone*. Deep and shallow burial are subjective terms, as there is no unique mineralogical or fabric definition to distinguish between them. Qualitatively and subjectively, the burial domain can be divided into shallow- and deep-burial environments with the shallow-burial environment differentiated from the overlying meteoric phreatic zone by its different water chemistry and somewhat greater overburden pressure and temperature. The deep-burial environment is distinguished by its significantly different water chemistry and its elevated pressure and temperature: 60–200 °C, for example. Effects of pressure can be interpreted from the style of grain contacts, grain breakage, and stylolitization. Temperatures of burial diagenetic products can be determined by stable isotope geochemistry or fluid inclusion geothermometry. Porosity affected by burial diagenesis may be recognized by exotic mineral cements or pore-fillings such as fluorite and sphalerite. Exotic crystal habits can also indicate high temperatures. Saddle crystal dolomite usually indicates deeper burial conditions. Deeper burial water is influenced by upward migrating fluids expelled from buried sediments. This burial water is unaffected by phreatic flow and it usually imparts distinctive trace element and isotopic signatures to minerals that crystallize at depth (Dickson et al., 2001). Burial compaction can produce fitted or penetrative rather than tangential grain contacts, along with other pressure-solution fabrics such as stylolites. Some of these diagenetic characteristics are treated in the Choquette–Pray classification. Finally, fracture porosity is classified as non-fabric-selective. Normally, fractures cut across depositional and diagenetic fabrics, but in the case of some dolomite–limestone rock combinations, dolomite may fracture selectively because it behaves as a more brittle material than limestone.

In sum, the Choquette–Pray classification is a useful method to describe carbonate porosity but it was not designed to aid in determining the spatial distribution of different pore types. For example, fabric-selective pores in crystalline dolostones are treated identically as intergranular pores in an oolite grainstone, yet the origin of the two pore types—diagenetic and depositional, respectively—is significantly different and requires different strategies for correlating the pore types at stratigraphic scale. The non-fabric-selectivity criterion is also insufficient to differentiate between mechanical fractures and large-scale diagenetic features such as caverns and connected vugs, all of which exhibit extreme petrophysical behavior and which may require different methods for correlation between wells within a field.

The Choquette–Pray classification has limited usefulness in determining relationships between rock and petrophysical properties because it focuses on fabric selectivity, drawing one to relate reservoir porosity to rock fabric. As discussed earlier, rock fabric may represent mechanical sedimentation, biological growth processes, or diagenetically produced crystallinity. Fabric selectivity that is also equivalent to depositional facies selectivity transforms those facies into proxies for porosity. Many of the penetrative features classified as fabric selective or not (burrows, borings, and desiccation cracks) may in fact be facies selective. As such, they can appear at certain positions in stratigraphic cycles where the occurrences of facies-selective porosity are more predictable. Fabric-selective porosity may not be mappable at reservoir scale, especially if it does not conform to facies boundaries as is commonly the case with diagenetic porosity. Diagenetic rock and pore properties must be identified, linked to the type of diagenesis, tied to its time of occurrence, and ultimately placed in the stratigraphic framework.

2.4.1.4 The Lucia Classification

Lucia's (1983) classification of carbonate porosity (Figure 2.12) evolved from work done at Shell Oil and Shell Development Companies during the 1960s. Although the influence of Archie's work is evident in Lucia's classification, Lucia's division of carbonate pore types into vuggy and interparticle categories distinguishes it. Like Archie, the objective of Lucia's classification is to provide a practical field and laboratory method for visual description of porosity in carbonate rock samples. For rocks with interparticle and separate vug porosity, Archie's m factor can be estimated if the particle size, amount of separate vug porosity, and total porosity are known. Lucia's emphasis on the relationship between porosity, permeability, inferred capillary displacement pressure, and particle size, some of which relationships were also recognized by Craze (1950) and Bagrintseva (1977), is an early method for "rock typing" or ranking reservoir zones on the basis of petrophysical characteristics.

A particularly important attribute of Lucia's classification is its emphasis on the petrophysical significance of separate and touching vugs. Vugs are pores larger than surrounding framework grains. They may have originally been moldic, interparticle,

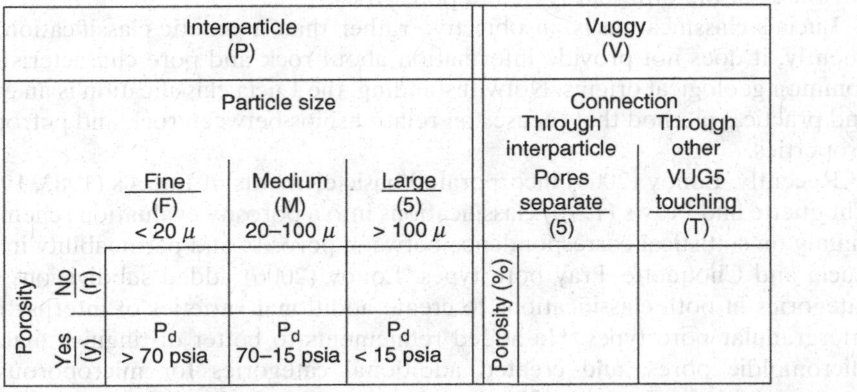

Interparticle (P)			Vuggy (V)	
Particle size			Connection	
			Through interparticle	Through other
Fine (F) < 20 μ	Medium (M) 20–100 μ	Large (5) > 100 μ	Pores separate (5)	VUG5 touching (T)
Porosity — No (n)			Porosity (%)	
Porosity — Yes (y) P_d > 70 psia	P_d 70–15 psia	P_d < 15 psia		

Figure 2.12 Classification of carbonate porosity by Lucia (1983). This scheme is especially important because it emphasizes that interparticle (grains or crystals) porosity and separate or touching vuggy porosity have profound effects on such reservoir petrophysical characteristics as Archie's m (cementation exponent), porosity–permeability relationships, and capillary pressure behavior. The latter influence is reflected by the "P_d" values in psia, which indicate the mercury displacement pressure required to enter the pore systems corresponding to small, intermediate, and large particle sizes.

intraparticle, or intercrystalline pores, but they were enlarged by dissolution to become vugs. Dissolution does not follow a predictable pattern in most cases; consequently, the size, shape, and spatial distribution of vugs may be quite irregular. They may begin as fabric-selective dissolution or non-fabric-selective enlargement of fractures by leaching. Flow between separate vugs has to pass through matrix porosity and permeability to drain the vugs; therefore the contribution of separate vugs to total reservoir porosity and permeability can be estimated if matrix characteristics and total porosity are known. Clearly, the only way to obtain that kind of information is by direct observation of rock samples—such as cores—that are large enough to display vugs that may be centimeter scale in size. Fluid flow through touching vugs is much less affected by matrix permeability and behaves more like flow through open fractures. Because most vugs, particularly touching vugs, are larger than rotary drill cuttings, they may be overlooked during sample examination, which again emphasizes the importance of examining full-diameter cores when working with carbonate reservoirs.

Non-vuggy or interparticle pores are classified by Lucia as visible or not visible in cuttings. Visible pores are grouped according to particle size as fine ($<20\,\mu$m), medium (20-$100\,\mu$m), and large ($>100\,\mu$m). There are no genetic modifiers or categories for time and direction of pore alteration, as in the Choquette and Pray (1970) scheme, but Lucia provides a basis for estimating the displacement pressure (mercury injection capillary pressure) for each particle size range in interparticle porosity. This is important because it offers clues about the ease with which fluids can move through rocks of different particle (grains or crystals) sizes. The relationships between porosity, permeability, and particle characteristics were further investigated by Lucia (1995) and expanded upon in his 1999 book *Carbonate Reservoir Characterization*. His 1995 paper is an extension of the original classification and presents discussions on pore characteristics in limestones, dolostones, and grain-dominated and mud-dominated fabrics. Especially useful are the discussions on petrophysical attributes of the different rock and pore types.

Lucia's classification is an objective rather than a genetic classification; consequently, it does not provide information about rock and pore characteristics with common geological origins. Notwithstanding, the Lucia classification is an excellent and practical method that focuses on relationships between rock and petrophysical properties.

Recently, Lonoy (2006) incorporated basic elements of Lucia's (1983, 1995) and Choquette and Pray's (1970) classifications into a porosity evaluation scheme based mainly on statistical correspondence between porosity and permeability in various Lucia and Choquette–Pray pore types. Lonoy (2006) added subdivisions of pore categories in both classifications to create additional varieties of interparticle and intergranular pore types. He added refinements to better distinguish macro- and micromoldic pores, and created additional categories for microporous, mud-supported fabrics. Lonoy's (2006) study provides valuable information about correspondence between porosity and permeability in different carbonate pore types and it should provide useful information for refining pore volume estimates used in volumetric calculations. For its merits, it does not provide information on how to improve our ability to correlate flow units, baffles, and barriers in stratigraphic space and it does not focus on combined petrophysical and conventional geological rock typing techniques.

All of the pore classifications discussed so far, though useful and informative, lack information necessary to group pore types with common geological origins except for purely depositional varieties. The value of a porosity classification depends largely on how reliably the pore types can be placed in correlatable stratigraphic space at field scale. In other words, it depends on how easy it is to map flow units, baffles, and barriers that are defined by the classified pore types. High correspondence between porosity and permeability alone does not offer clues for ways to correlate pore types at reservoir scale. Although the Lonoy (2006) scheme is useful, it does not provide new information with which to predict the spatial distribution of flow units, baffles, and barriers. That information must come from rock properties that co-vary with porosity. They co-vary because the rock properties and pore types were formed at the same time by the same geological processes. Usually there are distinctive rock properties—signatures of a sort—left by the geological processes that created the final set of rock and pore properties. It is those "signatures" that can serve as "tags," or proxies, for the pores that were formed concurrently with the signature rock properties. It is those signatures—the rock or stratigraphic characteristics—that can be correlated at reservoir scale.

Porosity classifications should be simple to use, they should illuminate the genetic relationship between rock, pore, and petrophysical attributes, and they should serve as aids in predicting the spatial distribution of reservoir flow units. That is a big order. It requires a classification based on integrated data from rocks, pores, petrophysical attributes, and on ways that pore types can be correlated at stratigraphic scale. Classification data must include time and mode of origin of rocks and pores, as well as percentage abundance of pores along with attributes such as average pore size and sorting. Rock texture, fabric, and mode of pore origin are obtained from petrographic study of samples, usually from cores. Because cores reveal fundamental rock properties of larger scale than cuttings, cored intervals can be correlated directly with borehole logs and with the stratigraphic column. Cores provide a greater volume of rock for more representative measurements of petrophysical properties in reservoirs that have widely varying pore categories and pore sizes. Without rock samples, pore characteristics remain unseen, and their origin, geometrical properties, and distribution within the rock remain unknown. Petrophysical attributes of different pore types are not visible in thin sections or on sample surfaces but can be determined from capillary pressure measurements that provide information about pore throat size distribution and aperture size sorting. Oriented pore fabrics, especially fractures, vugs, and linear or planar features, can be identified visually, and comparison of vertical and horizontal permeability measurements from core analyses provides additional clues about directionality.

Information about the origin of both rocks and pores can be incorporated in a genetic classification. If pores and rock matrix formed contemporaneously, as in reservoirs with purely depositional porosity, the common and synchronous origin of rocks and pores allows depositional facies to become proxies for porosity. Pores not formed contemporaneously with deposition, as in diagenetic and fracture porosity, must be interpreted differently. Methods for correlating flow units based on nondepositional pore types may be significantly different from methods that simply map facies and expect the maps to correspond with flow unit boundaries. Times and modes of origin for diagenetic and fracture pores are important in a genetic classification of carbonate porosity in order to compare pores formed by different types of diagenesis or fracturing with other rock properties that reflect pore-forming geological events—events that may leave traces other than pore characteristics in the rock record. Those traces are geological clues that can help in correlating genetic pore types from borehole to borehole. Modes of origin are defined explicitly from sample examination, usually by thin section petrographic study. Relative timing of pore origins is established by interpreting cross-cutting diagenetic features or fractures. Once the distribution of genetic pore types is known, the different categories can be compared with permeability, capillary pressures, borehole log data, well test data, or production characteristics to establish links between pore types, petrophysical "rock types," and their locations in the stratigraphic column. Before flow units can be defined, ranked by quality, and correlated from well to well, pore types must be classified according to their mode and time of origin.

2.4.2   A New Genetic Classification for Carbonate Porosity

Carbonate porosity is created or altered by (1) depositional processes, (2) diagenetic processes, and (3) mechanical fracture. It is intuitive to plot those processes as end members on a triangular diagram (Figure 2.13) with notations about hybrid pore types along its sides (Ahr and Hammel, 1999; Ahr et al., 2005). Pores associated with mechanically sedimented detrital deposits will conform to original grain texture and fabric such that depositional facies maps are proxies for reservoir porosity maps. In such cases, porosity is facies selective much as some of Choquette and Pray's (1970) fabric-selective pores may also be facies selective. Depositional pores related to reef fabric and skeletal or microbial textures have properties that reflect biological processes rather than mechanical sedimentation of loose grains. Clearly, pore geometry in reefs and mounds can be significantly different from that in detrital carbonates.

Depositional porosity that has been altered by diagenesis is classified as one of three hybrid pore types: hybrids of depositional and diagenetic processes, hybrids of diagenetic and fracture processes, and hybrids of depositional and fracture processes. If depositional characteristics are dominant but somewhat modified by diagenesis, depositional facies remain reliable proxies for porosity. If more than about half of pores visible in thin section are determined to have been altered by diagenesis (diagenetic attributes dominate), the types of diagenesis that created the hybrid pores must be determined because depositional facies are less reliable as proxies for porosity and facies maps will not be reliable guides to the spatial distribution of porosity at reservoir scale. In this case, it is necessary to determine the types of diagenesis that caused the alteration and at what times diagenesis was active during the burial history of the rocks. This information can be used to identify rock properties that may be new proxies for porosity when depositional facies boundaries are no longer reliable indicators of flow unit dimensions.

Diagenesis alters depositional porosity by dissolution, cementation, compaction and pressure solution, recrystallization, and replacement. It may enhance or reduce original porosity or it may create totally new pore types. Purely diagenetic porosity cannot be identified from depositional attributes when alteration is pervasive and dissects depositional textures and fabrics, rendering them useless as guides to flow unit boundaries. Alteration such as cementation, compaction, replacement, or

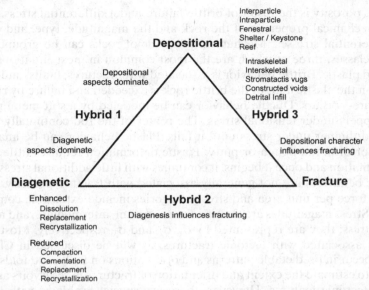

Figure 2.13 A genetic classification for porosity in carbonate rocks by the author. Pores are created by three end-member processes that include depositional, diagenetic, and fracture mechanisms. The end-member processes are independent but hybrid pore types exist between them because more than one mechanism can affect the formation of a given pore system at different times during its genetic history. For example, depositional porosity altered by diagenesis, but with depositional texture, fabric, or bedforms still recognizable, is classified as a hybrid in which depositional attributes are dominant. Fracturing, especially fracture intensity, is influenced by mineralogy and crystal size (diagenetic influence) and by bed thickness and grain size (depositional attributes). Complete porosity analysis must include the total amount (percent) of porosity present, and ideally the amounts of separate versus touching vugs. Pore geometry is important too, because some pore types can be identified by their size and shape with NMR measurements (Genty et al., 2007). A version of this classification appears in Ahr et al. (2005).

recrystallization may have reduced porosity to the extent that previous flow units become baffles or barriers. Alternatively, diagenesis may enhance or create new porosity by dissolution or by some types of replacement or, rarely, by recrystallization. For example, dissolution creates caves, connected vugs, and karst features including solution-collapse breccias that do not correspond to depositional facies boundaries. Instead, they may correspond to positions of ancient water tables, positions on antecedent structure, or to the locations of ancient mixing zones. Diagenetic-fracture hybrids are those fractures that form preferentially in rocks with diagenetically altered mineralogy and texture. Dolostones behave as more brittle material than limestones such that fracture intensity is higher in dolostones than in limestones with the same crystal size and bed thickness. As we will see in later chapters, fracture porosity and permeability are computed differently than porosity in unfractured rock matrix; therefore it is important to recognize the geological factors other than stress concentration and stress orientation that influence the degree and type of fracturing.

Fracture porosity is the result of brittle failure under differential stress. It varies with the mechanical properties of the rock and the magnitude, type, and direction of the differential stresses. Mechanical behavior of rocks can be grouped into a variety of classes, three of which are the most common in most situations: brittle, ductile, and plastic. Brittle behavior is associated with fractures, faults, and joints. It occurs when the elastic limit of the brittle rock is exceeded and failure by rupture—brittle failure—occurs. Ductile behavior can be modeled by a soft metal rod (e.g., lead or copper) under tensional stress. The center of the rod continually becomes thinner and thinner under stress until it fails. Plastic behavior can be imagined as the behavior of bread dough or putty. Plastic deformation requires little stress to start deformation and once it begins, it continues with little additional stress. Ductile and plastic behaviors are not generally associated with fracture porosity. Stress is defined as force per unit area and stress categories include extension, compaction, and shear. Stress magnitudes are classified as maximum, intermediate, and minimum principal stress; they are represented by σ_1, σ_2, and σ_3, respectively. Most fracture porosity is associated with tectonic fractures, as will be discussed in Chapter 7. Fractures occur in predictable patterns and orientations on faults and folds, making it possible to estimate the extent and orientation of fractures in reservoirs associated with such tectonic features. However, there are special problems with fractured reservoirs that will be discussed later. Fracture intensity varies with depositional bed thickness and depositional texture to the extent that a hybrid category of fractures influenced by depositional attributes can be useful. Characteristics of depositional, diagenetic, and fracture porosity are discussed in more detail in Chapters 5, 6, and 7, respectively.

2.4.3 Permeability

Nineteenth century engineers Henri Darcy and Charles Ritter conducted experiments to establish the laws that govern the flow of water through sand. Their purpose was to explain these principles as aids to planning and managing water distribution for Dijon and other cities in France (Darcy, 1856). They filled a cylinder with different mixtures of sand and gravel, packed the mixture, and passed water through the column to determine flow rate. The experiments involved pure water and atmospheric pressure such that the principal variables were sand and gravel textural characteristics. Flow rates and pressure differences were small in the original Darcy–Ritter experiments as compared to those in hydrocarbon reservoirs. "Darcy flow" is defined as *laminar* flow.

Today the Darcy–Ritter expression is written with different letter designations for parameters and measurements than in their 1856 paper but the method and the outcome are unchanged. Discharge (Q) through a known cross-sectional area (A) and length (L) of a cylindrical, sand-packed cylinder is proportional to the hydraulic gradient ($h_1 - h_2$)/L along the cylinder, and *Darcy's law* is given by

$$\frac{Q}{A} = k\left(\frac{h_1 - h_2}{L}\right)$$

The Darcy–Ritter expression has to be modified for application to hydrocarbon reservoirs because fluids other than water are involved and there are wide variations

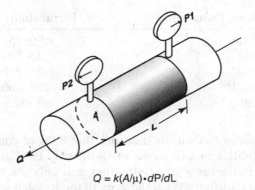

$$Q = k(A/\mu) \cdot dP/dL$$

Figure 2.14 A modified Darcy apparatus to measure permeability, where fluid viscosity and pressure can be varied to be more representative of subsurface reservoir conditions. The original Darcy apparatus, a glass cylinder with mercury manometers and flow valves, was designed to operate at atmospheric pressure with water only.

in pressures. Consider a laboratory apparatus (Figure 2.14) to measure permeability of reservoir rock samples. In this system a rock sample of length dL and cross-sectional area A is saturated with a fluid of dynamic viscosity μ, which flows through dL at a rate Q. Under steady-state conditions, the upstream pressure is P, and the downstream pressure is $(P - dP)$. There is no flow through the sides of the sample and there is no reaction between fluid and rock. Under these conditions, the modified Darcy's law is

$$\frac{Q}{A} = \frac{k}{\mu} \cdot \frac{dP}{dL}$$

In this expression, k is the permeability coefficient and represents rock properties. It is known as *absolute* or *specific permeability* and has the dimensions of an area. The permeability expression is written

$$Q\frac{cm^3}{s} = k(darcies) \cdot \frac{A}{\mu}\frac{cm^2}{centipoise} \cdot \frac{dP}{dL}\frac{atm}{cm}$$

Or, 1 *darcy of permeability is defined as when a fluid with unit viscosity flows at a rate of 1 cm³/s from a rock sample with a cross-sectional area of 1 cm² under a differential pressure of 1 atm/cm.* It would be nice if reservoirs delivered 1 cm³ of fluid for each cm² of surface area; however, most do not; consequently, the practical working unit of permeability in the petroleum industry is the *millidarcy* (md), which is equal to $0.0987 \times 10^{-15}\,m^2$. In ordinary applications, $1\,md = 10^{-15}\,m^2$ and 1 darcy $= 1\,\mu m^2$. Permeability varies greatly in carbonate reservoirs from values of less than 0.1 md in tight, crystalline mosaics in mudstones to over 10 darcies in fracture, cavern, or connected vug systems. Qualitatively, reservoir permeability values can be graded in the following manner (North, 1985):

Qualitative Description	Permeability (md)
Poor to fair	<1.0–15
Moderate	15–50
Good	50–250
Very good	250–1000
Excellent	>1000

Research by Russian geoscientists indicates that 80% of common sedimentary rocks have permeabilities in the range of $0–10^{-3}$ md, 13% are in the range of $10^{-3}–1.0$ md, 5% are in the range of 1.0–1000 md, and only 2% have permeabilities of more than 1000 md (North, 1985). If the work of the Russian scientists is generally applicable, it indicates that over 90% of all sedimentary rocks are either seals or "tight gas sands" that produce no more than 5 bbls of oil per day (North, 1985). Evaporites are the least permeable rocks, being impermeable to water. Shales are permeable to water but not generally permeable to oil. Very high permeability through connected vugs and fractures is relatively common in carbonate rocks, notably in limestones rather than dolostones. The El Abra Formation of the Poza Rica and Golden Lane trends in Mexico (Cretaceous) and some of the Permian carbonates in West Texas and New Mexico are known for high permeability, dissolution-diagenetic porosity. Individual wells in vuggy-fractured Asmari Limestone (Oligocene–Miocene) of Iran have produced over 100,000,000 barrels of oil (North, 1985).

Permeability is expressed as (1) *specific permeability*, (2) *effective permeability*, and (3) *relative permeability*. Specific permeability, described in the previous paragraphs, is the permeability of a reservoir rock to a single fluid. It is measured on core samples, commonly by commercial laboratories. Effective permeability is a measure of the permeability to another fluid when the reservoir is already saturated, that is, the effective permeability to oil of a reservoir rock already saturated with water. The presence of a wetting fluid impedes the entry of a nonwetting fluid; therefore effective permeability is lower than specific or absolute permeability. In other words, the sum of $k_o + k_w + k_g$ is less than absolute permeability because the mutually interfering presence of oil, water, or gas retards flow. Relative permeability is the ratio of effective permeability at a given saturation to absolute permeability at 100% saturation, or $k_r = k_e/k$. Next to basic lithology, effective porosity and specific permeability are the most important variables used to describe reservoir rocks. Absolute permeability, or simply permeability, may vary directly with interparticle porosity in *detrital reservoir rocks* such that $\varphi = a + b \log k$.

Permeability, like porosity, is sensitive to variations in texture and fabric in reservoir rocks, but for each 1% change of porosity, at least in siliciclastic sandstone reservoirs, the change in permeability is greater by a factor of 7–10 (North, 1985). Unlike porosity, permeability varies with grain size, as well as packing, sorting, and fabric. Fine-grained detrital rocks with comparatively high intergranular porosity have low permeability. In ideal reservoirs with intergranular porosity and uniform grain size, permeability varies approximately as the fourth power of the average pore radius, or approximately as the square of the grain diameter (North, 1985). Most reservoirs, especially carbonate reservoirs, are not represented by this ideal model. The wide variety of genetic pore types and attendant varieties of pore–pore throat geometries affect permeability greatly. Permeability depends on the geometry of pore throats rather than on the largest pore dimensions, although in some cases, pore dimensions may vary predictably with pore throat dimensions. Reservoir analysts must determine which geological characteristics of the reservoir rock serve as "markers" for the pore types that consistently have the largest pore throats. This task requires information from petrographic studies on reservoir rocks. When the data are collected they are compared with other petrologic and petrophysical data to formulate the *geological concept* that enables recognition and mapping of three-dimensional *flow units* at field scale.

2.5 TERTIARY ROCK PROPERTIES

Tertiary properties are measured indirectly by geophysical tools such as borehole logging devices and to an extent by the seismograph. Neither borehole logs nor the seismograph can make direct measurements of fundamental rock properties such as depositional texture, sedimentary structures, mineralogical composition, or rock fabric. In the case of borehole logs, the geoscientist or engineer makes inferences about fundamental properties from log characteristics or from calculations based on log data. Tertiary properties are twice removed from fundamental rock properties and once removed from dependent or secondary properties such as porosity, permeability, and bulk density. Some of the more widely used measurements of tertiary properties include electrical conductivity, electrical resistivity, acoustic transmissivity (including seismic reflection characteristics), natural radioactivity, nuclear magnetic resonance relaxation time, and photoelectric effect. One log that does make direct measurements of borehole properties is the caliper log. It records the diameter of the borehole. Otherwise, tertiary properties depend on porosity, pore fluid composition and saturation, presence or absence of radiogenic materials, and reaction of rocks and fluids in the borehole to external energy sources. Because measurements of some tertiary properties require external energy sources for their measurement (e.g., the photoelectric effect and acoustic logs), those properties are sometimes called *latent* properties.

2.5.1 Borehole Logs and Carbonate Reservoirs

Logging is done by lowering a device called a sonde down the borehole and raising it at a prescribed rate. As they are raised, the tools transmit data through cables to recording devices at the surface. Some tools are designed for open-hole logging; others are used in cased holes. Responses of logging tools vary with an array of parameters such as size of the borehole, mud properties, speed of tool movement up the borehole, and temperature; consequently, the novice interpreter should not assume that log measurements always provide accurate and representative values. Uncorrected logs usually require corrections before they are interpreted or compared with direct measurements such as those made during core analyses.

Older, analog recording devices produce the familiar strip logs that have been used for decades, but with today's technology analog records can be digitized easily and quickly. Digital data are acquired by modern logging tools and either processed by computers at the wellsite or transmitted directly to a distant office or laboratory for further analysis. Detailed descriptions of the wide variety of logging devices, logging principles, and methods of interpretation are beyond the scope and purpose of this book, but a list of typical logs in use today, along with a brief list of their applications to the study of carbonate reservoirs, are shown in Table 2.2.

Distinctive patterns on analog wiggle traces and trends in numerical values that can be read from log traces are commonly compared with lithological descriptions to establish log signatures that correspond to certain rock types. This correlation technique is especially useful in fields where only a few cores exist in widely separate locations. Most texts on reservoir engineering, for example, Cossé (1993), include

TABLE 2.2   Logging Tools for Carbonate Reservoirs

Device	Application	Limitations
Open-Hole Logging Tools		
Spontaneous potential	Correlation and lithology	Unusable in oil-based mud; R_{mf} and R_w must differ
Gamma ray	Correlation and lithology	Sensitive to borehole size
Photoelectric effect	Correlation and lithology	Unusable in barite mud; needs radioactive source; pad device
Induction	Resistivity	For resistivity $< 250\,\Omega$m and $R_{mf}/R_w < 2.5$
Laterolog	Resisitivity	Unusable in oil-based muds
Microresistivity	Resistivity	Unusable in oil-based muds
Density	Porosity and lithology	Pad device
Compensated neutron	Porosity and lithology	Environmental correction required; affected by proximity to wall; underestimates porosity in presence of gas
Sonic	Porosity and lithology	Affected by rock compressibility
Caliper	Hole diameter	
Formation tester	Fluid samples and formation pressure	
Dipmeter	Stratal and structural dip	
Borehole televiewer	Fractures, vugs, and sedimentary structures	
Borehole imaging devices such as UBI®, FMS®, and FMI®, of Schlumberger	Fractures, vugs, and sedimentary structures	
Cased-Hole Logging Tools		
Gamma ray	Correlation and shale volume	Affected by natural radiation
Spectral gamma ray	Correlation and shale volume	
Pulsed neutron	Porosity and S_w	Unusable in freshwater mud; influenced by presence of gas
Compensated neutron	Porosity	Influenced by presence of gas

Source: Alberty (1992).

discussions on the theory and methods of log interpretation. For the nonspecialist, a brief review of the types of modern wireline logs, their applications, and their limitations is presented in Morton-Thompson and Woods (1992). Acoustic imaging and NMR logs have only recently come into widespread use and are not included in Alberty's 1992 compilation. Hodgkins and Howard (1999) present an illustrated discussion on NMR logging in Gulf of Mexico sandstone reservoirs and the handbook by Asquith and Krygowski (2004) presents a variety of methods for calculations from borehole logs including image and NMR logs. Rider's (1996) book is a well-illustrated review of virtually all types of wireline logging procedures, the types of records that are generated by the logging devices, and how geological interpretations are made from the resulting records. Both Asquith and Krygowski (2004) and Rider (1996) include discussions on acoustic and nuclear magnetic resonance imaging, which are not included in the list of logs in Table 2.2. A particularly useful section in Rider's book describes geological interpretations of wireline logs in sequence stratigraphic context.

Traditional methods of interpretation, particularly on older analog records, involve reading values from analog wiggle traces and then making calculations to determine rock and reservoir properties. Graphical methods involve cross-referencing the values read from wiggle traces on nomograms to obtain estimates of rock or reservoir properties. Borehole log data from carbonate reservoirs provides an indispensable aid for stratigraphic correlation, for calculating values for porosity, for estimating lithological composition, especially in mixtures of carbonates and evaporites, for determining fluid saturation, formation resistivity, borehole diameter, structural and stratigraphic dip, and, especially in the case of imaging devices, to detect fractures. Calculations using wireline log data are routinely made to determine saturation (S_w), formation water resistivity (R_w), porosity (φ), density (ρ), and lithology. In the case of terrigenous sandstones, additional inferences can be made about depositional environments based on the shape of the resistivity and SP or gamma ray traces. Gamma ray, or SP, and resistivity readings are sometimes interpreted to represent grain-size trends in siliciclastic sandstones; therefore, by extension, to represent depositional facies characteristics. Resistivity and gamma ray curves that depict "bell" or "funnel" shapes (Figure 2.15) are interpreted as fining-upward and coarsening-upward sandstone textures, respectively. The former could be indicative of a channel-fill sequence, the latter a deltaic sequence. Such interpretations based on log curve shapes enable geologists to create "electrofacies maps" that depict fluvial channels, deltas, and turbidites, among a variety of related facies types. The validity of these interpretations depends on the assumption that shapes of the gamma ray and resistivity curves are proxies for grain size trends, and that the logging engineer made no errors while running and recording the log. In fact, gamma ray and resistivity devices do not measure grain size; they measure natural radioactivity and electrical resistivity. The proxies for fine grain size, higher gamma ray and resistivity values, depend on the presence of clay minerals in the sandstones.

The shapes of gamma ray and resistivity log traces from carbonate reservoirs do not indicate anything about depositional environment, particle characteristics, or pore types. Carbonate reservoir properties are influenced by depositional, diagenetic, or fracture processes, or combinations of those processes as reflected in pore type and pore geometry. It is generally impossible to distinguish depositional or

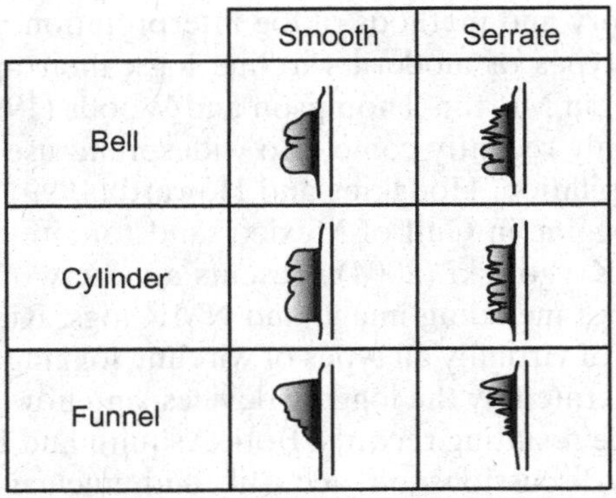

	Smooth	Serrate
Bell		
Cylinder		
Funnel		

Figure 2.15 Log shape classification. When paired with resistivity log traces, these typical gamma ray or SP log curves can be imagined to describe bell, cylinder, and funnel shapes. Alhough these logs do not measure grain size directly, the bell, cylinder, and funnel shapes are routinely used to identify fining-upward, uniform, and coarsening-upward textural trends in siliciclastic sandstones. Maps showing trends in the distribution of log shapes at field scale are known as electrofacies maps. Log shapes are not reliable indicators of texture or facies character in carbonate rocks; consequently, electrofacies mapping is generally limited to sand–shale successions.

diagenetic facies from wireline logs alone; by extension, it is generally not possible to make universally applicable electrofacies maps of carbonate reservoirs. Some methods, such as the Schlumberger SPI™ (secondary porosity index), have been reasonably successful in estimating the different proportions of "primary" and "secondary" porosity with wireline logs. Anselmetti and Eberli (1999) used a similar method that they named the "velocity-deviation log." Their method incorporates sonic and neutron-porosity or density logs to detect variations in log responses that correspond to differences in pore characteristics. These methods have to be calibrated against real rocks and pore characteristics before the operator can be reasonably certain about the results. These methods notwithstanding, depositional and diagenetic pore types in carbonates are not generally detectable by wireline log traces because most carbonate porosity is not simply depositional and interparticle in nature. Acoustic signatures are not unique to one pore type, radioactive minerals common in sand–shale sequences are not usually present in carbonates, and there are no reliable wireline log methods to measure depositional particle sizes in carbonates because carbonate particles are altered or obliterated by diagenesis. It is more useful to focus on methods such as the NMR log that can measure pore geometry in carbonates. Other challenges for the log interpreter include determining reliable petrophysical calculations in carbonates that have a variety of pore types and sizes. Carbonate reservoirs can have bimodal (micropore–mesopore) systems that require careful work to distinguish between effective water saturation, S_{we}, and total water saturation, S_{wt} (Asquith and Jacka, 1992). Calculating a reliable S_w depends on knowing which m (Archie cementation exponent) value to use. In reservoirs with vuggy or moldic porosity, m may be 3 or 4, but in fractured reservoirs it may be close to 1. The presence of certain minerals can influence petrophysical calculations in carbonates, too, as emphasized by Major and Holtz (1997). They found that the presence of gypsum in a Permian carbonate reservoir precluded the use of the neutron and density logs for calculating porosity because the neutron log measured bound water in gypsum as porosity. The low density of gypsum compared to the surrounding dolomite and anhydrite dramatically increased the uncertainty of what values to use in density calculations. Correcting S_{wt} in bimodal porosity and choosing m values for different pore types are discussed in Chapter 3. The occurrence and significance of gypsum in carbonate reservoir rocks is discussed further in Chapter 6.

Many of the oil and gas fields around the globe were discovered decades ago. As new technology and new knowledge become available, studies are made on old fields to evaluate their economic potential as candidates for improved or additional recovery. It is not always possible to find cores or cuttings to use in these studies; therefore one of the most important but challenging tasks for the log interpreter is to determine the lithology of a carbonate reservoir using only borehole logs. This task is not easy nor does it always produce reliable results, especially when several minerals are present in a single reservoir rock. A vivid example from this author's personal experience involved participating as the lead geologist on a team of engineers and geoscientists who were competing with another team to resolve a dispute over well spacing in an infill drilling (field development) program. The dispute centered on whether or not reservoir-quality rocks, mainly dolostones with high percentages of intercrystalline porosity, were present over large areas or were instead widely scattered among nonreservoir-quality baffles and barriers. Lacking cores or cuttings, the teams had to create lithological logs (synthetic rock descriptions) based primarily on wireline log data. The teams reached significantly different conclusions about lithological interpretations from identical logs from one reservoir zone in a single field. Determining the proportions of carbonate and evaporite minerals when they occur together is a well-known problem (Hashmy and Alberty, 1992). However, if several minerals are present in the reservoir rock, as was the case in this author's experience, it is impossible to be completely certain about the presence and relative proportions of limestone, dolostone, evaporite minerals, quartz, and clays from log data alone. In such cases, cuttings or core samples are necessary to make direct determinations of mineral composition to compare with log readings. If samples are unavailable, data from different logs can be crossplotted to derive values that are indicative of a particular rock type. One method in widespread use on carbonate reservoirs is Schlumberger's "M-N" plot, where the value of M derives from a density-acoustic log crossplot and the value of N comes from a density-neutron crossplot. The use of an M-N plot to determine lithology is described and illustrated in Asquith and Krygowski (2004). Rider (1996) points out that on the M-N plot, shale and other minerals become separated into fields and porosity is eliminated. The problem with the M-N plot, however, is that the geological value of the logs is lost and there is, according to Rider, "a tendency to rather obscure cross-plotting in the vain hope of finding a unique 'shale point' or 'mineral point.' These points rarely exist in nature."

Crossplotting can be done rapidly with modern personal computers (PCs) and it is possible to process digital data from several kinds of logs simultaneously making lithological interpretations easier and arguably more reliable. Included among many examples of such interpretive programs in use today are those developed by such companies as Petcom, Landmark, and GeoQuest. Most petrophysicists emphasize that it is important to know the limitations on some of the software applications. Some of the programs can make calculations to depict up to five mineral species in the reservoir being studied; however, the analyst must choose which minerals are most likely to occur in the formation and therefore focus the calculations on those minerals only. Problems result from having to choose which minerals to exclude. For example, consider the case where two petrophysicists are in competition during litigation and have to make choices about which minerals to exclude from a list that includes quartz, clay minerals, anhydrite, dolomite, and calcite. If the petrophysicists exclude different minerals from their calculations, the two outcomes will be different. At minimum, different results will cause uncertainty about the "true" lithology. Consider a situation where reservoir quality is related to dolomite content of the producing formation. The petrophysicist who minimized the importance of dolomite by her choices of which minerals to include in the log calculations also minimized her chances of success in reservoir characterization, flow unit mapping, and economic evaluations.

Rider (1996) discusses multilog quantification of lithology and divides the methods into two main categories: petrophysical multilog analysis and statistical multilog analysis. The former method is "essentially one of solving a number of linked, simultaneous equations, for unknown volumes of chosen minerals or matrices defined by pure, end-member (hypothetical) log responses." If the end members are pure limestone, dolostone, and evaporites, the method works reasonably well because the responses are usually linear. In the presence of shale, the results are unpredictable, however (Rider, 1996). This problem can be tackled by generating computer output of volume percent of each ideal, end-member component. This generates a "CPI" log (computer processed interpretation). A complaint with the method is that the ideal end members are artificially defined absolutes that have little or no relation to fundamental rock properties (Rider, 1996). For example, CPI logs define sandstones on the basis of their quartz content but do not represent fundamental properties such as texture and sedimentary structures.

Statistical multilog analysis involves taking all log responses from a single depth and combining them into a multidimensional set in *n*-dimensional space. The sets are then subjected to multivariate statistical analyses to identify sets that can be grouped into populations of numbers that have some internal similarity and that can be differentiated from other populations of numbers. Think of cluster analysis dendrograms. The next step is to try to relate the different number populations to rock types or synthetic lithofacies. The statistical populations are sometimes called "statistical electrofacies," but as Rider (1996) points out, there is a great distance between what geologists call facies and what is produced by multivariate statistical analysis of log values. In sand–shale sequences this may not be such a problem because siliciclastic grains and shale are relatively easy to distinguish on gamma ray and resistivity logs, among others. But in carbonates where distinctions between constituent components, fabrics and textures, and pore types are not readily distinguishable by borehole log measurements, it is a genuine problem—a problem that can only be resolved with certainty by direct observation of the rocks.

A borehole log that offers great potential for geologists is the NMR (nuclear magnetic resonance) log. The NMR log records the time required for the liquid in a liquid-filled pore to change from an excited state to a relaxed state. This time period is called the relaxation time and is typically known as the T2 relaxation time. An advantage offered by this log is that measurements of relaxation time in liquid-filled pores are a measure of the volume of liquid in the pores. The total liquid volume represents total porosity. Additionally, the liquid volume represents the pore volume that, if samples of the reservoir rock are recovered and examined under the microscope, can be compared with measurements of pore geometry. Pore geometry, the size and shape of the pores in two dimensions, can then be classified according to the triangular, genetic classification of carbonate porosity in order to provide a geological origin for different reservoir pore types (Genty et al., 2007). Ultimately, reservoir zones with dominant pore types can be identified and related to the geological events that produced them. In short, a geological concept can be developed for identifying and mapping specific reservoir zones based on their porosity characteristics.

2.5.2   Tertiary Rock Properties and the Seismograph

The seismograph measures reflected or refracted seismic impulses as they bounce off or pass through layered rocks. Reflection seismology was once used only to identify subsurface structural anomalies. Today, the greatly improved technology and data processing methods make it possible to identify not only structural and stratigraphic features but also, under ideal circumstances, reservoir rock and fluid properties. Seismic stratigraphy, the forerunner of today's sequence stratigraphy, is one of today's most powerful methods for interpreting stratigraphic architecture. Modern data processing techniques for analyzing seismic wave characteristics such as frequency, amplitude, polarity, spatial distribution, and shear wave characteristics enable geophysicists to make vastly more sophisticated interpretations than were possible only a decade ago. The advent of 3D seismology has greatly advanced our ability to interpret subsurface structure, stratigraphy, and even reservoir characteristics. The 3D method is sufficiently powerful that in some settings, particularly in sand–shale sequences, individual depositional bodies such as fluvial channels, deltas, and turbidites can be identified and mapped spatially and targeted for drilling based on whether they contain hydrocarbons, particularly gas. Today's computer technology and software can generate vivid displays of facies architecture—depositional facies or stratigraphic features—but however vivid the displays may be, the technology is not without limitations. The vivid technological displays of reservoir characteristics require contrast in seismic velocities, or differences in acoustic impedance, between the reservoir and its enclosing strata. The seismograph only records acoustic waves that have been reflected from acoustic interfaces in the subsurface; it does not record fundamental rock properties such as texture, grain type, sedimentary structures, or taxonomic diversity. It can not distinguish between depositional, diagenetic, or fracture porosity. Those distinctions have to be inferred by seismic interpreters who, with modern data acquisition and processing techniques, can use reflection characteristics to help identify seismic signatures of reservoirs, particularly gas reservoirs, because gas-filled pores react to seismic pulses much more profoundly than oil-filled pores (Brown, 1999).

Acquiring good seismic reflections from a target reservoir interval depends on the impedance contrast between the target interval and the rocks that enclose it and on the thickness of the target interval as compared to the impulse wavelength.

If a 10-m thick gas reservoir in microporous limestone occurs in a deep (several thousand meters) subsurface limestone interval several times as thick as and similar in lithology to the porous zone, the reservoir will be invisible to the seismograph. The thickness of the porous zone in such a case is below the *limit of separability* or one-quarter the seismic wavelength, as illustrated by Brown (1999). A rule of thumb is that beds of roughly 15–20-m thickness will, on average, be within the required one-quarter wavelength at moderate subsurface depths. The earth acts as a filter and progressively eliminates high-frequency acoustic waves with increasing depth in the subsurface. Small-scale variations in reservoir properties require small wavelengths (higher frequencies) to be "seen." Seismic measurements of reservoir characteristics depend on quality of the reflection data. In other words, for optimal reflection the depth of investigation must be shallow, the signal must be strong and not corrupted by noise, and the acoustic impedance must be sufficient to cause good reflections. The seismograph can be a powerful tool to help identify reservoirs and, in some cases, flow units within reservoirs. Anselmetti and Eberli (1997) studied seismic compressional and shear wave velocities (V_p and V_s) in 295 minicores and found, much as Wang (1997) did in his laboratory study, that different seismic velocities in rocks of equal porosity were the result of different pore types. It is possible, under ideal conditions, to estimate carbonate rock and reservoir properties based on their seismic reflection characteristics. But as Lorenz et al. (1997) emphasize, there are caveats in making subsurface interpretations where the scale of the seismic measurement is larger than the scale of the individual reservoir interval.

SUGGESTIONS FOR FURTHER READING

G. B. Asquith and D. Krygowski's (2004) *Basic Well Log Analysis*, 2nd edition, is a standard reference, especially for using nomograms to obtain values for porosity, saturation, and lithology from wireline log data. This edition includes new material on magnetic resonance imaging and borehole imaging logs. A thorough discussion of logging methods, how the logging tools work, and how to interpret log data is given in M. Rider's (1996) *Geological Interpretation of Well Logs*, 2nd edition. Additional references on seismology include W. M. Telford, L. P. Geldart, and R. E. Sheriff (1991), *Applied Geophysics*, 2nd edition; J. Milsom (2003), *Field Geophysics*, 3rd edition; and R. E. Sheriff (2002), *Encyclopedic Dictionary of Exploration Geophysics*, 4th edition. A widely cited book on 3D seismology is by A. R. Brown (2004), *Interpretation of Three-Dimensional Seismic Data*, 6th edition. A good compilation of research papers specifically on carbonate seismology can be found in I. Palaz and K. J. Marfurt (1997), *Carbonate Seismology*. For those wishing to see color images of carbonate particles, pore types, and effects of diagenesis on carbonate rocks, *A Color Guide to the Petrography of Carbonate Rocks: Grains, Textures, Porosity, Diagenesis*, by P. A. Scholle and D. Ulmer-Scholle (2004), is an excellent starting point.

REVIEW QUESTIONS

2.1. What are fundamental and derived rock properties?

2.2. What is the difference between texture and fabric?

2.3. The carbonate rock classification system in widespread use by the petroleum industry today is the one developed by (a) Grabau, (b) Folk, (c) Embry and Klovan, (d) Dunham, (e) Wright or (f) Riding.

2.4. Porosity and permeability are dependent rock properties. On what do they depend?

2.5. What is a "latent" property?

2.6. What is the difference between total and effective porosity?

2.7. Write the expression for calculating total porosity.

2.8. Permeability has the dimensions of an area. True or false?

2.9. What is the Darcy–Ritter expression for permeability and what does each term represent?

2.10. Can the original Darcy–Ritter expression be used as is to calculate flow rates in hydrocarbon reservoirs? Explain.

2.11. How did Archie illustrate relationships between rock and petrophysical characteristics in his porosity classification of carbonate rocks?

2.12. What is the basic principle behind the Choquette–Pray classification of porosity in carbonate rocks?

2.13. Is there any difference between facies-selective and fabric-selective characteristics? Explain.

2.14. How does Lucia's porosity classification identify relationships between fundamental rock properties and pore characteristics?

2.15. What does a genetic classification of carbonate porosity provide that an objective or purely descriptive version does not?

2.16. Do wireline logs provide direct measurement of rock or pore properties? Explain.

2.17. Which wireline logs should be most useful to detect both fractures and large vugs?

2.18. Are electrofacies maps as reliable and easy to make for carbonate successions as they are in sand–shale successions? Explain.

2.19. Carbonate pore types can be directly identified with the seismograph. True or false? Explain.

2.20. Why does the seismograph detect smaller anomalies at shallow depths than at greater depths?

CHAPTER THREE

PETROPHYSICAL PROPERTIES OF CARBONATE RESERVOIRS

3.1 SATURATION, WETTABILITY, AND CAPILLARITY

This chapter covers reservoir properties and emphasizes those characteristics that strongly correspond to fundamental rock properties enabling geologists and engineers to identify rock properties that serve as proxies or "tags" to aid in mapping and exploiting reservoir flow units, baffles, and barriers. Earlier, we defined *reservoirs* as porous and permeable bodies of rock that contain commercial quantities of hydrocarbons. The storage capacity of a reservoir is determined by its *porosity*, the degree to which fluids can be transmitted through the reservoir is determined by its *permeability*, and the pore volume filled with fluid is defined as *saturation*. *Wettability* is the tendency for one fluid rather than another to be preferentially attracted to a solid. Wettability depends on capillary properties of the fluids and the solid—the phenomenon of *capillarity*.

3.1.1 Saturation

Saturation (S) is defined as the ratio of the pore volume occupied by a fluid, normally water (V_w), oil (V_o), or gas (V_g), to the total pore volume of the reservoir rock (V_p) such that

$$S_w = \frac{V_w}{V_p}$$

$$S_o = \frac{V_o}{V_p}$$

Geology of Carbonate Reservoirs: The Identification, Description, and Characterization of Hydrocarbon Reservoirs in Carbonate Rocks
By Wayne M. Ahr   Copyright © 2008 John Wiley & Sons, Inc.

$$S_g = \frac{V_g}{V_p}$$

Saturations are expressed as percentages totaling 100%, or $S_w + S_o + S_g = 100$. In plain terms, saturation is the amount of fluid in the pores expressed as a percentage of the total pore volume. In conventional literature, water in reservoir rocks is described as *connate water*, or interstitial water remaining from the time of deposition. This interstitial water occupies pores and coats grains. Because of chemical interactions between water and rock during burial, and because formation waters are expelled during compaction or tectonism, reservoir water saturation is really immigrant diagenetic water rather than true connate water. Widely varying amounts and kinds of salts and trace elements in reservoir waters testify to the mobility of basinal liquids, as do the many episodes of late burial diagenetic change that are documented in most reservoir rocks.

Water saturation, S_w, depends on pore and pore throat size, aperture size distribution, and elevation above the free-water level. For oil or gas to enter the reservoir, it must displace the interstitial water. If the pore volume is sufficiently large, oil will displace water and reside in the pore centers (Figure 3.1), but it cannot displace water from small pores or from coatings on grain surfaces. That unmovable water is the *wetting* fluid. Reservoirs may include oil, water, and gas and which phase becomes the wetting fluid is determined by *wettability*, a phenomenon associated with the *capillary properties* of reservoir rocks. Most reservoirs are considered to be water-wet, but oil-wet reservoirs do exist, notably in some carbonate rocks. As water remains in small pores and on grain surfaces, it follows that large pores such as vugs and intergranular pores in coarse-grained rocks have lower values of S_w, and fine-grained rocks have higher values. Oil saturation is just the opposite: lower in fine-grained rocks and higher in coarse-grained ones. Qualitatively, an S_o of about 80% indicates a productive *zone* in the reservoir (Figure 3.2), S_o in the range of 50% represents the *transition zone*, and S_o of 10–20% represents the *water-bearing zone* (Monicard, 1980). In sum, S_o determines which zones are productive and which are not. S_o is represented by $(1 - S_w)$ in oil-water systems and S_w is calculated from

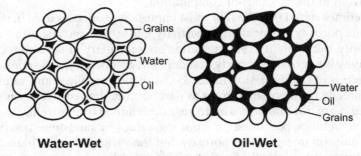

Water-Wet   **Oil-Wet**

Figure 3.1 Idealized representation of water-wet and oil-wet reservoirs with depositional, interparticle porosity. All rocks had to be water-wet originally, but some became oil-wet after hydrocarbon migration, and surface chemical reactions between the hydrocarbons and the pore walls caused the rock to become oil-wet. This is especially true of carbonate reservoirs with oils containing polar organic compounds that react with carbonates.

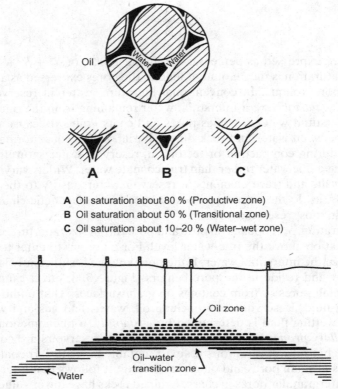

A Oil saturation about 80 % (Productive zone)
B Oil saturation about 50 % (Transitional zone)
C Oil saturation about 10–20 % (Water–wet zone)

Figure 3.2   Oil–water saturation levels in the pores of (a) productive, (b) transitional, and (c) water-wet zones of an ideal reservoir such as the one illustrated in the bottom half of the figure. (Adapted from an illustration in Monicard (1980).)

electric logs or measured in cores. As S_w is used to compute reserve estimates, accurate values are critically important. Before determining S_w to calculate reserve estimates, it is necessary to understand the kinds of data used to compute S_w and which pitfalls to avoid in the methods of computation.

It is sometimes stated that the four most important properties used in petrophysical logging are porosity, permeability, resistivity, and saturation (Asquith and Gibson, 1982). Porosity, permeability, and saturation are secondary or derived rock properties. Resistivity is a tertiary property that depends on porosity, saturation, and electrical properties of the formation fluids. Porosity, permeability, and saturation are measured directly from borehole cores as part of "complete core analysis" routines, but when cores are not available ϕ and S_w are determined from wireline logs. Permeability can sometimes be estimated from log values in sandstone reservoirs with simple and uniform intergranular porosity, but the variety of pore characteristics in carbonate reservoirs generally precludes using wireline log readings to estimate permeability. Resistivity logs provide values for R_t, the "true formation resistivity," where fluids include water and some hydrocarbons; R_o, the resistivity of the porous rock formation at 100% water saturation; R_i, the resistivity of the zone invaded by borehole fluids; and R_{xo}, the resistivity of the flushed zone surrounding the borehole.

Formation water resistivity, R_w, is measured by the spontaneous potential (SP) log. Resistivity and porosity values from logs are routinely used to calculate S_w using the Archie equation. Archie (1942) found that electric current flowing through a porous rock 100% saturated with water follows a twisted path through pores and pore throats such that

$$R_o = F \times R_w, \quad \text{where } F = a/\phi^m$$

or

$$F = \frac{R_o}{R_w} = \frac{a}{\phi^m}$$

where R_o is the resistivity of a porous rock at 100% water saturation, R_w is the resistivity of the water in the pores, F is the Archie *formation factor,* ϕ is reservoir porosity, a is a *tortuosity factor* that varies from about 0.62 to 1.2, and m is Archie's *cementation factor* that can vary from 1.0 to as much as 4.0. The importance of using an accurate m value is illustrated by the fact that a hydrocarbon saturation of 70% calculated with an Archie cementation factor of $m = 2$ (a common choice) will become zero if an m value of 4 is used (Figure 3.3). Values of 3 and 4 are realistic

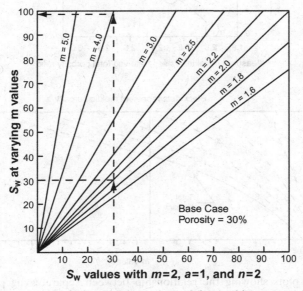

Figure 3.3 Water saturation (S_w) at varying values of the Archie cementation exponent (m) as the vertical axis and S_w at a fixed value of $m = 2$, $a = 1$, and $n = 2$ in a sample with 30% porosity. If $m = 2$, as in the base case, the values of S_w are 30% on both the x and y axes. However, if the true value of m were 4, as in the case of separate vug and moldic pores, the corresponding S_w value would be nearly 100%. In this case the use of $m = 2$ instead of 4 would cause an error in calculating S_w by nearly 70% and would reduce the expected hydrocarbon saturation from 70% to nearly 0. (Adapted from an illustration in Focke and Munn (1987).)

for carbonate reservoirs with a high percentage of separate vug porosity, according to Focke and Munn (1987) and Lucia (1983). Archie m and a values are related to pore characteristics and may vary in different kinds of reservoir rocks. Because the tortuosity factor, a, is not related exponentially to F, and because it has a relatively narrow range of variation around 1.0 for most carbonate reservoirs, it does not have as much impact on S_w calculations as does m. This fact is addressed in Lucia's (1983) porosity classification where the amount of separate vug porosity is compared with m values (Figure 3.4). Focke and Munn (1987) recognized that "*the porosity type is by far the most important parameter in regard to reservoir properties. Archie's own carbonate rock classification, although very useful in wellsite geology, does not include a genetic porosity classification and therefore is not adequate in the study of reservoir properties.*"

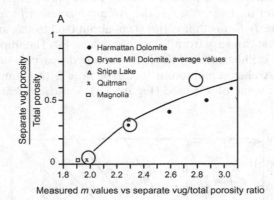

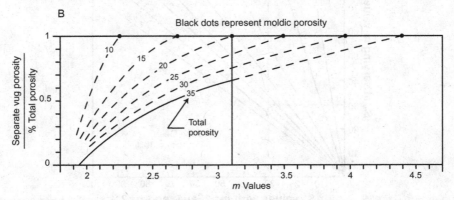

Figure 3.4 Two plots showing the relationship between separate vug porosity and the cementation exponent m. (a) The ratio of separate vug porosity to total porosity is plotted on the y axis and values of m are plotted on the x axis. Data are measured values from samples taken from oilfields denoted by symbols and field names. Note that as the ratio of separate vug porosity to total porosity increases, the value of m increases. (b) An extension of the upper plot in (a) to illustrate the relationship between the proportion of moldic porosity to total porosity and Archie m values. This plot demonstrates that abundant, true moldic porosity acts petrophysically as separate vugs accompanied by increasing values of m. (Adapted from an illustration in Focke and Munn (1987), based on data from Lucia (1983).)

They concluded that the wide variations in data on formation resistivity factor in carbonate reservoirs can be resolved by conducting detailed studies on rock and pore characteristics. According to their findings, rocks with intergranular porosity (depositional) and sucrosic dolomites (diagenetic) generally show m values of about 2. Wackestones, packstones, and rocks with only matrix porosity also show m values of about 2. However, rocks with both matrix and vuggy or moldic porosity show m values greater than 2 depending on the percentage of separate vugs or molds. For example, moldic porosity in oolitic grainstones, such as the Arab D in the Middle East or the Smackover of the Gulf Coast, show m values ranging between about 1.8 at 5% porosity to 5.4 at 30% porosity. Fractured and fissured rocks may have m values less than 2 and theoretically could approach unity.

In carbonate reservoirs, the importance of looking at rocks cannot be overemphasized. Resistivity values from wireline logs substituted into the Archie equation can give totally erroneous results in the absence of some qualifying information about rock and pore characteristics. If S_w is wrong, it is easy to see what an impact the error can have on reserve estimates by changing S_w in the following expression for calculating original oil in place, or OOIP:

$$\text{OOIP} = \frac{7758\,Ah\phi(1-S_w)}{B_{oi}}$$

Here, OOIP is original oil in place, 7758 is a conversion factor (the number of stocktank barrels in a one acre-foot volume), A is the area of the reservoir rock in acres, h is the thickness of the reservoir rock in feet, ϕ is porosity as a decimal, $(1 - S_w)$ is oil saturation as a decimal, and B_{oi} is the formation volume factor for oil at initial reservoir conditions. For estimates of OOIP without knowing the formation volume factor, B_{oi} can be set equal to one.

Archie concluded from his laboratory studies of electrical resistivity in saturated and partially saturated rocks that a single power function relates the electrical resistivity ratio R_t/R_o to saturation. His work showed that S_w in the uninvaded zone that contains water with some hydrocarbons can be determined from the following expression:

$$(S_w)^n = F \times \frac{R_w}{R_t}$$

and that

$$(S_w)^{-n} = I \,(\text{resistivity index}) = \frac{R_t}{R_o}$$

so that

$$S_w = \left(F \times \frac{R_w}{R_t}\right)^{1/n} = \left(\frac{a}{\phi^m} \times \frac{R_w}{R_t}\right)^{1/n}$$

The choice of an appropriate value for n is important because large errors in the value of S_w can result when inappropriate n exponents are used. Laboratory measurements of n range from 1.2 to 3.0 (Sneider, 1988), and it is standard practice to use a value of 2.0 if n is not known. More recently, it has been recognized that, in carbonates with multiple pore types such as fractures, vugs, interparticle porosity, and intraparticle microporosity, n can change when each of the different pore types is subjected to lowering of its original saturation. In other words, different values of n could be used for different ranges of S_w. But in every case, n increases in oil-wet reservoirs. Values of n greater than 8.0 have been reported for oil-wet reservoirs (Tiab and Donaldson, 2004).

Careful choices of m, a, and n values for use in the Archie equation are critical in the determination of S_w and, ultimately, in estimating hydrocarbon reserves. In addition, carbonate reservoir rocks commonly exhibit bimodal porosity systems that may include combinations of intergranular mesopores and megapores with intragranular micropores. Microporosity commonly results when certain forms of diagenesis alter the internal structure of carbonate grains to produce micropores within each affected grain. Micropores are water-filled; therefore averaged values of Archie water saturation for combined intergranular and intragranular porosity may range from 55% to 87% (Asquith and Jacka, 1992). This represents *total water saturation*, or S_{wt}. To determine the actual productive potential in such bimodal porosity systems requires a determination of the *effective water saturation*, or S_{we}. Effective water saturation pertains to the intergranular pores that determine productivity potential. Although estimates of effective and total porosity may be obtained from log data, the actual percentages of microporosity and macroporosity must be determined by examining rock samples. According to Asquith and Jacka (1992), the uncorrected average S_{wt} in an oolitic grainstone reservoir with bimodal porosity was 69.4%. After corrections were made, the average S_{we} was computed to be 34.9%. Clearly, accurate values of S_w are vitally important in reservoir characterization. Values for parameters in the Archie equation must be chosen carefully based on information about rock properties and average saturations (S_{wt}) calculated for bimodal pore systems can lead to miscalculations of reservoir productivity unless corrected, effective saturations (S_{we}) are determined.

3.1.2 Wettability

In qualitative terms, wettability is the tendency for one fluid rather than another to be preferentially attracted to a solid surface. For example, if in an air-filled container a drop of water falls on a glass plate, it will readily spread across the surface and wet the plate. On the other hand, if a drop of mercury falls on the glass plate in identical surroundings, it remains in a spherical, bead-like form because air is the wetting fluid.

In hydrocarbon reservoirs, one must deal with the interface between gas and liquid, between two immiscible liquids, and between liquids and solids. *Adhesion tension* is a function of interfacial tension and determines which fluid will wet a solid. For a system consisting of oil and water in contact with a solid (Figure 3.5), there is a contact angle (θ) which, by convention, is measured through the denser fluid and varies between 0° and 180°. The interfacial tension between the solid and the lighter liquid is represented by σ_{so}, between the solid and the denser liquid by σ_{sw}, and between the two liquids by σ_{wo}. Adhesion tension is defined as

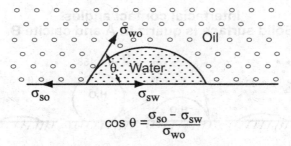

$$\cos \theta = \frac{\sigma_{so} - \sigma_{sw}}{\sigma_{wo}}$$

Figure 3.5 The interfacial tension values (identified by σ) and the contact angle (θ) between the two liquids in a system of oil and water in contact with a solid surface. These relationships provide the basis for calculating values of adhesion tension (A_T) at different liquid–solid interfaces. (Adapted from an illustration in Amyx, Bass, and Whiting (1960).)

$$A_T = \sigma_{so} - \sigma_{sw} = \sigma_{wo} \cos\theta_{wo}$$

A positive adhesion tension indicates that the denser liquid is the wetting phase and an adhesion tension of 0 indicates that both liquids have equal affinity for the solid. If θ is small, the denser liquid will spread readily and coat the solid surface. If θ is large, an external force will be required to make the denser liquid spread across the solid surface. The interfacial contact angles and attendant spreading characteristics of several different liquid systems in contact with quartz and calcite are shown in Figure 3.6. In the case of water and isooctane, water is the wetting fluid for both quartz and calcite. When naphthenic acid is used on both quartz and calcite, it preferentially wets calcite with a contact angle of 106°, but water is the wetting fluid on quartz where naphthenic acid forms a contact angle of 35°. This illustrates that for a water–oil–solid system, it is possible to have an oil-wet or water-wet medium, depending on the composition of the liquids and solids and the surface chemical reactions that may occur. In gas–water, or gas–oil reservoirs, gas is always the nonwetting fluid. Reservoir rocks that were deposited in aqueous environments, as most were, start out as water-wet. However, after oil is trapped in the reservoir, polar organic compounds in the oil can adhere to pore surfaces and, with surface chemical reactions over time, change the reservoir to oil-wet or partially oil-wet (Pittman, 1992). Tiab and Donaldson (2004) point out that the chemically basic nature of carbonates causes them to react readily with organic acids in crude oils, which in turn causes the carbonates to become neutral to oil-wet.

3.1.3 Capillarity

Capillary attraction is defined in the *American Heritage Dictionary* (1992) as "*the force that results from greater adhesion of a liquid to a solid surface than internal cohesion of the liquid itself and that causes the liquid to be raised against a vertical surface, as water is in a clean glass tube. It is the force that allows a porous material to soak up a liquid.*" This interactive force depends on surface tension and on the ability of certain liquids to wet the surface of solids, or wettability. Wettability is manifested by a curvature of the liquid surface that forms a contact angle different

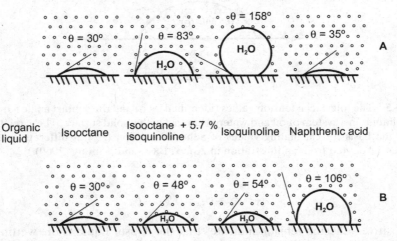

Interfacial contact angles.
Solid surface is quartz in A and calcite B.

Organic
liquid   Isooctane   Isooctane + 5.7 %   Isoquinoline   Naphthenic acid
                        isoquinoline

Figure 3.6   This figure illustrates the different wetting characteristics of four organic liquids on (a) silica and (b) calcite solid surfaces. Different liquids have different affinities for solid surfaces; that is, some liquids preferentially wet (spread over and adhere to) certain solids. Note that three of the four organic liquids have strong preferences to wet calcite but naphthenic acid does not. Some polar organic compounds in oils can transform water-wet carbonate reservoirs to oil-wet ones. (Adapted from illustrations in Amyx, Bass, and Whiting (1960).)

from 90° between solid and liquid. For reservoir specialists, the main reason that capillarity is important is because capillary attraction is the force behind *capillary pressure*. This force depends on the properties of the liquids and solids in the porous medium, on the size and distribution of pore throats, and on the reservoir saturation history. Reservoirs consist of rocks with many different mineralogical compositions and a variety of pore sizes and shapes, and they may contain more than one liquid. The variety of minerals present in carbonate reservoirs is limited, however, so if the fluid properties are known, then capillary pressure characteristics provide an excellent guide to the size and distribution of pores and pore throats. These pressure characteristics are usually expressed graphically as *capillary pressure curves*.

3.2   CAPILLARY PRESSURE AND RESERVOIR PERFORMANCE

Capillary pressure can be thought of as the force necessary to drive a nonwetting fluid through pores already saturated with a wetting fluid. The magnitude of the capillary pressure depends on interfacial tension and the radius of curvature of the fluid interface. Following the example of Amyx et al. (1960), consider a capillary tube of radius r in a beaker of oil and water (Figure 3.7). The pressure in the oil at points A and B is given by P_{oA} and P_{oB}, respectively. The pressure in the water at A and B is given by P_{wA} and P_{wB}. Assume that the beaker is large enough for the

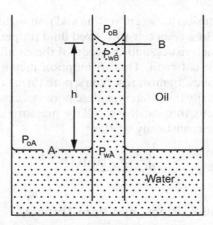

Figure 3.7 Illustration of pressure relationships in a capillary tube partly filled with water surrounded by oil. The capillary pressure is zero at the oil–water interface where the pressures of oil and water, respectively, are equal. The pressure difference across the oil–water interface at point B defines capillary pressure. (Adapted from an illustration in Amyx, Bass, and Whiting (1960).)

interface at A to be a plane and that capillary pressure there is zero so that $P_{oA} = P_{wA}$ at the free-water surface in the beaker. For the pressure at point B, the density of oil and water must be considered such that

$$P_{oB} = P_{oA} - \rho_o g h$$

$$P_{wB} = P_{wA} - \rho_w g h$$

The pressure difference across the interface is the capillary pressure, which must be in equilibrium with gravitational forces if the fluids are in equilibrium and not flowing, or

$$P_c = P_{oB} - P_{wB} = (\rho_w - \rho_o)gh$$

The expression for capillary pressure in terms of surface forces can be obtained by equating the expressions for the upward and downward forces, $2\pi r A_T = \pi r^2 (\rho_w - \rho_o)gh$, respectively, which simplifies to

$$h = gr\left(\frac{2A_T}{\rho_w - \rho_o}\right)$$

If the expression for adhesion tension (A_T) is substituted to obtain an alternative expression for h, which is in turn substituted in the equilibrium capillary pressure expression above, the expression for capillary pressure becomes

$$P_c = (2\sigma_{wo} \cos\theta_{wo})r^{-1}$$

Capillary pressure is a function of adhesion tension ($\sigma_{wo} \cos\theta_{wo}$) and the inverse of the radius (r^{-1}) of the capillary tube. In carbonate reservoirs, adhesion tension depends on interfacial tension between wetting and nonwetting phases and between fluid and solid surfaces. In a reservoir with fixed fluid properties, capillary pressure values reflect pore characteristics, and the radii of the capillary tubes are assumed equivalent to the porethroat radii. That assumption means that pore throats are modeled as cylindrical tubes. In most reservoirs, pore throats are not cylindrical, but may be sheet-like, especially the intercrystalline pore systems characteristic of dolomites. For sheet-like pore throats, the capillary pressure equation, according to Wardlaw (1976), is more realistically written

$$P_c = \frac{(\sigma_{wo} \cos\theta_{wo})}{r}$$

3.2.1 Capillary Pressure, Pores, and Pore Throats

Capillary pressure is inversely related to pore throat radius, but remember that capillary pressure calculations are based on a model in which pore throats are cylindrical tubes. Pore throats have complex geometries so that computed pore throat radius represents the *effective pore throat radius*. Rearranging the expression for P_c provides the equation to compute effective pore throat radius:

$$r_{eff} = \frac{2\sigma_{wo} \cos\theta}{P_c}$$

In this expression, σ is the interfacial tension of the air–mercury system (480 dynes/cm), θ is the air–mercury–solid contact angle (140°), and P_c is capillary pressure in dynes/cm^2 (1 psi = 69,035 dynes/cm^2).

In an air–mercury system, the nonwetting mercury displaces air at low pressures in large pore throats. If the large pore throats are uniform in size (well sorted) and are well connected within the rock (pore–pore throat systems with high accessibility and high coordination numbers), saturation by the nonwetting phase will proceed at low pressure along a flat trajectory until all accessible pores and pore throats have been filled to a limit for that particular rock. A plot of mercury injection capillary pressure and fluid saturation takes on a characteristic shape in response to the manner in which the pores and pore throats are saturated (Figure 3.8). The initial part of the curve reflects the pressure exerted by a nonwetting fluid against a wetting fluid just until the latter is displaced. This initial pressure is known as *entry pressure*, and the pressure at which the nonwetting fluid begins to move wetting fluid from its position in a pore is the *displacement pressure*. It is common to see the terms entry pressure and displacement pressure used synonymously. The vertical axis of the capillary pressure curve reflects pore throat radius in micrometers (µm), with largest values at the origin. The vertical axis also represents the height of a given water saturation above the free-water level (where capillary pressure is zero). Capillary pressure curve trajectories, as they trace fluid saturation at increasing pressure, indicate pore throat *size, sorting,* and *accessibility*. Pressures at points of inflection on the curve represent *threshold pressures*, and a range of threshold pressures on a single curve indicates the presence of several pore throat size clusters in the sample. The terminal saturation value of the wetting phase is referred to as the *minimum unsaturated pore volume*.

CAPILLARY PRESSURE
SATURATION RELATIONS

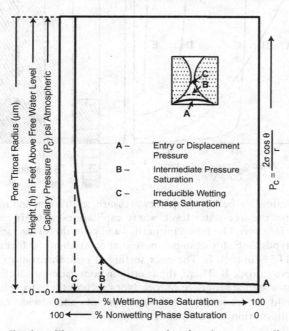

Figure 3.8   An idealized capillary pressure curve showing the entry or displacement pressure at A, nonwetting phase saturation at intermediate pressure B, and irreducible wetting phase saturation at maximum pressure C. Note that the vertical axis of the capillary pressure curve provides data on pressure, pore throat size, and height of oil column above the free-water level (FWL), and saturations at varying pressures. (Adapted from an unpublished illustration with permission from R. M. Sneider.)

Displacement pressure is higher for small pore throats than for large ones, and their respective curve trajectories differ. Examples of these differences in capillary pressure characteristics and saturation levels in response to pore and pore throat geometry are displayed in Figure 3.9. Note the influence of capillary pressure on water saturation in the different rock-pore systems. Capillary pressure measurements are useful indicators of reservoir quality as illustrated in a set of mercury injection capillary pressure (MICP) curves (Figure 3.10) of Smackover Formation (Jurassic) rocks from Arkansas prepared by Bliefnick et al. (1990). Curves 1, 2, 4, and 6 are from reservoir samples, curves 7 and 9 represent "marginal reservoirs," and curves 3 and 8 are from nonreservoirs. Note that reservoir samples exhibit low displacement pressures and low wetting phase saturation at low injection pressure, that is, about 50% wetting phase saturation at 20 psia. Flat trajectories between 50% and 100% wetting phase saturation and about 5 psia indicate that about half of the pore volume is connected by pore throats with effective radii of 1 µm or more. Flat segments on the curves for marginal reservoirs at about 500 psia indicate a second range of yet smaller pore throat sizes. Note the steeper trajectories, higher displacement pressures, and higher wetting phase saturations for any given pressure in the marginal samples, indicating lower accessibility, smaller pore throats, and less pore

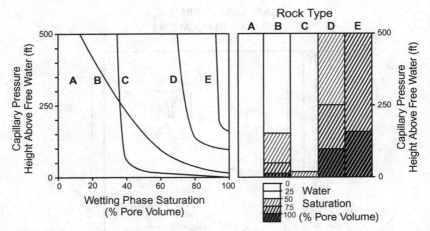

Figure 3.9   The relationship between capillary pressure, saturation, and reservoir character. For any height above the free-water level, where capillary pressure is zero, S_w varies with capillarity, which is influenced by liquid properties and pore throat size and sorting. Among the idealized rock types A–E, for example, note that S_w at about 50 feet above the FWL is 18% in rock A and 95% in rock E. The poor sorting of pore throat sizes is reflected in the broadly sloping curve of rock B. There, the transition zone from productive to water-wet is very broad. If an interbedded sequence of rock types such as these were encountered in the subsurface, one could expect multiple oil–water contacts and a wide range of S_w values. (Adapted from an illustration in Vavra et al. (1992).)

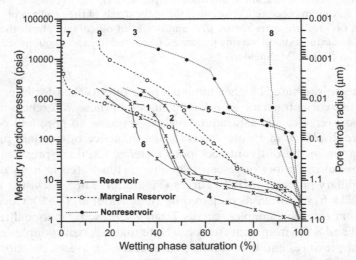

Figure 3.10   Capillary pressure curves based on MICP measurements from Jurassic Smack-over rocks at Walker Creek Field, Arkansas. The capillary pressure curves clearly discriminate between good, intermediate, and poor reservoir quality. Good reservoirs have high oil satura-tion at low capillary pressure, "marginal" reservoirs have lower oil saturation at higher pres-sures, and nonreservoir rocks have high irreducible water saturation and low oil saturation even at high capillary pressure. (Adapted from an illustration in Bliefnick et al. (1990).)

throat size sorting. Nonreservoir curves exhibit characteristics similar to the marginal reservoir samples except that their pressure-saturation trajectories and displacement pressures indicate even smaller pore throat radii, poorer size sorting, and lower pore–pore throat connectivity. The rock represented by curve 8 is a good candidate for a *seal* because it has such a high entry pressure and low nonwetting phase saturation at maximum pressure.

Reservoirs and nonreservoirs can be distinguished based on their respective capillary pressure characteristics. According to Sneider (1988) reservoirs generally have

- Capillary pressure (P_c) less than about 300 psia at 1% bulk volume occupied by mercury
- Mercury saturation greater than 3% bulk volume at $P_c = 1000$ psia (pore throats about 0.01 μm in radius)
- Mercury saturation greater than 3% bulk volume with more than 50% of pore throats larger than 0.05 μm at $P_c = 2000$ psia
- Displacement pressure generally less than 100 psia

Nonreservoirs have

- P_c greater than about 500 psia at 1% bulk volume occupied by mercury
- Mercury saturation of about 2% bulk volume at $P_c = 1000$ psia (pore throat radius ~0.1 μm)
- Mercury saturation of less than about 3% and more than 50% of pore throat radii smaller than 0.05 μm at 2000 psia
- Displacement pressures greater than 100 psia when the rocks are not artificially fractured

Sneider's criteria are one of several methods that can be classified as "rock typing" or flow unit identification and quality ranking. Selected references on other methods of rock typing are included in the Suggested Reading section at the end of this chapter.

MICP curves contain useful information about reservoir rock characteristics. In addition to providing data on pore throat geometry, they can be used to estimate recovery efficiency and permeability. Studies on Smackover reservoirs in Alabama have shown that median pore throat size is strongly correlated with permeability in Smackover reservoir rocks (Kopaska-Merkel, 1991). "Pore facies" constructed from petrographic and stratigraphic data, as described by Kopaska-Merkel and Mann (1993), can then be sorted on their capillary pressure curve characteristics in order to obtain estimates of reservoir recovery efficiency and heterogeneity for the purposes of selecting or grading candidate fields, sectors of fields, or reservoir zones for enhanced and improved oil recovery.

3.2.2 Converting Air–Mercury Capillary Pressures to Oil–Water Equivalents

Air–mercury capillary pressure data must be converted to brine–oil equivalents before applying them to reservoirs. The conversion most commonly used is

TABLE 3.1 Typical Contact Angle and Interfacial Tension Values[a]

System	Contact Angle	Interfacial Tension (dynes/cm)
Air–brine	0°	72
Air–mercury	140°	480
Crude oil–water	0°	35

[a]Typical fluid density ranges, according to Vavra et al. (1992), are gas = 0.00073–0.5 g/cm^3, oil = 0.51–1.0 g/cm^3, and brine = 1.0–1.2 g/cm^3.

Source: Vavra et al. (1992).

$$P_{cR} = P_{cL}\left(\frac{\sigma_{w/o}\cos\theta_{w/o}}{\sigma_{a/m}\cos\theta_{a/m}}\right)$$

where P_{cR} is capillary pressure in the reservoir brine–oil system, P_{cL} is air–mercury capillary pressure from the laboratory, $\sigma_{w/o}$ is the brine–oil interfacial tension, $\sigma_{a/m}$ is air–mercury interfacial tension, $\theta_{w/o}$ is the contact angle of the brine–oil–solid system, and $\theta_{a/m}$ is the contact angle of the air–mercury–solid system. Under ideal conditions the values for contact angles and interfacial tensions should be determined in the laboratory. These measurements are difficult and costly, and according to Vavra et al. (1992), the values in Table 3.1 are commonly used instead.

3.2.3 Height of Oil Column Above Free-Water Level

Once the air–mercury capillary pressure data are converted to brine–oil values, the height of oil above the free-water level ($P_c = 0$) can be calculated. This method can be used to predict saturation for any type of fluid in different kinds of reservoir pore systems at any chosen height in the reservoir. The expression for calculating the height of the oil column is

$$h = \frac{P_c}{0.433(\rho_b - \rho_o)}$$

where h is the height of the oil column in feet above the free-water level, P_c is the capillary pressure value for brine–oil in psi, ρ_b is the specific density of brine in g/cm^3 at ambient conditions, ρ_o is the specific density of oil in g/cm^3 at ambient conditions, and 0.433 is the pressure gradient (specific weight) in pure water expressed as psi/ft.

3.2.4 Evaluating Seal Capacity

If capillary pressure data are available on rocks from sealing formations, the maximum height of hydrocarbon column that can be confined by the seal before leakage is given by

$$h_{max} = \frac{P_{dS} - P_{dR}}{0.433(\rho_w - \rho_{hc})}$$

where h_{max} is the maximum height of hydrocarbon column that can be sustained by the sealing facies, P_{dS} is the brine–hydrocarbon displacement pressure in psi of the sealing rock, P_{dR} is the brine–hydrocarbon displacement pressure in psi of the reservoir rock, and ρ_w and ρ_{hc} are the respective densities of brine and hydrocarbons.

3.3   FLUID WITHDRAWAL EFFICIENCY

Withdrawal efficiency is a measure of the ease with which a rock will give up nonwetting fluids, usually mercury, after high pressure injection followed by pressure release. Mercury withdrawal efficiency in laboratory experiments is at least phenomenologically related to hydrocarbon recovery efficiency in carbonate reservoirs. In fact, for water-wet reservoirs, the same factors of pore geometry that contribute to increased mercury withdrawal efficiency also contribute to increased hydrocarbon recovery efficiency in subsurface reservoirs (Vavra et al., 1992).

Visual examination of pores and pore throat sizes, shapes, and coordination numbers provides information with which to evaluate withdrawal efficiency and predict hydrocarbon recovery efficiency. Pore and pore throat geometry are examined on resin casts made from carbonate reservoir rocks impregnated with resin and then dissolved in acid. The remaining pore casts in the insoluble resin are studied under a microscope or with the scanning electron microscope (SEM) in the manner described by Wardlaw (1976). Estimates of withdrawal efficiency require injection at increasing pressure increments followed by pressure release and withdrawal of the nonwetting fluid. Ordinarily, this procedure involves mercury injection and withdrawal from cleaned rock samples. The procedure for obtaining MICP curves has developed largely from the pioneering work of Purcell (1949) and is described in detail in Amyx et al. (1960). Studies by Pickell et al. (1966), Wardlaw and Taylor (1976), and Wardlaw and Cassan (1978) applied mercury injection and withdrawal techniques to investigate withdrawal efficiency. Injection curves are known to petroleum engineers as *drainage curves* because the wetting fluid is being drained as the nonwetting fluid is injected. Withdrawal curves are known as *imbibition curves* because the wetting fluid is imbibing pore space as the nonwetting fluid is withdrawn (Figure 3.11). On injection, the wetting fluid is displaced (drained) from the rock. On withdrawal, the wetting fluid displaces some of the nonwetting fluid and some of the nonwetting fluid remains trapped in the pore and pore throat network. Efficient withdrawal equates to low imbibition; that is, if the rock returns a high percentage of the nonwetting fluid as pressure is decreased, it has high withdrawal efficiency or high *recovery efficiency*. Recovery efficiency is defined by Wardlaw and Taylor (1976) as "the ratio of the volume of mercury withdrawn from a sample at minimum pressure to the volume injected at maximum pressure before the pressure was reduced." This can be expressed as

$$W_E = \left(\frac{S - S_R}{S} \right) \times 100$$

to convert to percent. Here, W_E is withdrawal efficiency, S is the volume of mercury at maximum pressure before pressure was reduced, and S_R is the volume of mercury withdrawn from the sample at minimum pressure.

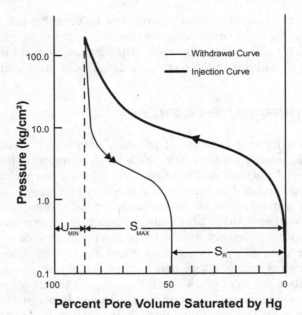

Figure 3.11 Capillary pressure curves showing the drainage (injection) and imbibition (withdrawal) curves. S_{max} and S_{min} are volumes of mercury at maximum injection pressure and minimum withdrawal pressure, respectively. U_{min} is the minimum unsaturated pore volume that can be compared to irreducible water saturation in oil–brine systems. Withdrawal efficiency [$100 (S_{max} - S_{min})/S_{max}$] can be used as a qualitative indicator of how rock properties influence reservoir recovery efficiency. For more accurate estimates of how petrophysical rock properties predict reservoir recovery efficiency, the reservoir fluid properties and drive mechanisms must also be considered. (Adapted from an illustration in Wardlaw and Taylor (1976).)

For constant fluid properties, recovery efficiency depends primarily on (1) pore/ pore throat size ratio, (2) pore-to-pore accessibility, (3) heterogeneity of pore distribution, and (4) pore throat roughness. The pore/pore throat size ratio is a very important factor in estimating recovery efficiency because large pores connected by small pore throats are difficult to drain. As nonwetting fluids are withdrawn from a system of large pores and small pore throats, the nonwetting fluid column breaks apart in the narrow throats, leaving a large amount of fluid isolated in the large pores. This phenomenon is called *snap-off* (Yu and Wardlaw, 1986a,b). Pore-to-pore accessibility is determined by coordination number, or the number of pore throats that connect with a pore. Heterogeneity of pore distribution is an index of how uniformly pore and pore throat sizes are distributed throughout the reservoir rock. A group of large pores and pore throats (a low coordination number geometry) surrounded by uniformly distributed small pores and pore throats (higher coordination number) will have lower recovery efficiency than the surrounding small pores (Figure 3.12). Heterogeneity depends to a large extent on the origin of the pore system. In a pore system created by depositional processes, pores and pore throats are distributed according to the depositional texture and fabric of the rock. For example, a rudstone lens surrounded by bioclastic grainstones around a reef would

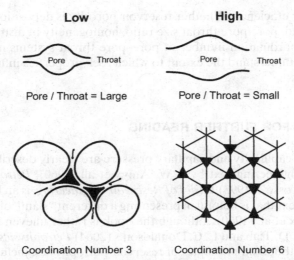

Figure 3.12 Rock properties that influence fluid recovery efficiency in reservoirs: pore/pore throat size ratio and coordination number. Large pore throats and large pores provide high recovery efficiency. Large pores will not drain efficiently through small pore throats. Coordination number refers to the number of pore throats that connect each pore. The more pore throats there are to connect each pore, the greater the recovery efficiency, all else being equal. (Adapted from an illustration in Wardlaw (1979).)

present large intergranular pores with low coordination numbers surrounded by smaller intergranular pores with higher coordination numbers. Although the rudstone has large pores, it will have low recovery efficiency because snap-off will prevent the large pores from draining through the surrounding small pore throats. Finally, pore throat roughness is calculated as an empirical parameter derived from a three-stage graphical procedure described by Kopaska-Merkel (1991), who found that the roughness value, designated as the *a* factor, did not correlate with median throat size, mercury recovery efficiency, or permeability. The "*a* factor" was developed by Shively (1991) in his efforts to find a descriptor of pore throat shape; however, as the *a* factor has not been shown to correlate strongly with ordinary reservoir rock and pore properties, it will not be considered further here. Pore throat characteristics are important. Examining resin pore casts with scanning electron microscopy is one way that provides direct observational data to compare with drainage and imbibition capillary pressure curve characteristics.

In summary, pores are connected by pore throats that vary in size and shape depending on their geological origin. The number of pore throats that connect a single pore is the coordination number for that pore-to-pore throat system. Intergranular pores consist of curvilinear surfaces connected by pore throats that may be modeled as cylinders, while intercrystalline pores tend to be tetrahedral and exhibit sheet-like pore throats (Wardlaw, 1976). Large, disconnected pores such as separate vugs, shelter pores, constructed voids in reef rocks, and widely spaced fractures may contribute greatly to total porosity but they depend on fluid transmission through the small connected pores in the host rock matrix to remain filled. Such a rock type exhibits moderate to high porosity but comparatively low permeability and low recovery efficiency. Whether reservoir porosity is depositional, diagenetic, or fracture formed, pore/pore throat size ratio, homogeneity in distribution of pore geometry, and coordination number of pore–pore throat systems must be investigated in order to understand the extent to which rock properties influence reservoir fluid behavior.

SUGGESTIONS FOR FURTHER READING

Fundamentals of capillarity and capillary pressure are clearly described in the standard reservoir engineering text by J. W. Amyx et al. (1960), *Petroleum Reservoir Engineering*. R. Cossé's (1993) *Basics of Reservoir Engineering*, is an English translation of a French text that is helpful in presenting a different "slant" on some material covered by Amyx et al. (1960), although the book is rather uneven in its treatment of subject matter. D. Tiab and E. C. Donaldson's (2004) *Petrophysics*, 2nd edition, is a voluminous textbook on principles of reservoir petrophysics, including some geological topics. It is arguably more useful as an encyclopedic reference than a text.

"Rock typing" is a way of identifying pore and pore throat characteristics that correspond to highest storage capacity combined with highest capacity to flow: in short, a way to identify rock samples and group them into categories with best, intermediate, and poorest flow unit quality. The paper by C. Genty et al. (2007) in which NMR measurements are used to identify genetic pore types in carbonate reservoirs is one of a variety of these methods. F. J. Lucia's (1995) paper includes his method for identifying three petrophysical rock types based on porosity, permeability, particle size, and MICP characteristics. G. W. Gunter et al. (1997) describe a method that revolves largely around the Winland R35 criterion, as does the method presented in Martin et al. (1997).

REVIEW QUESTIONS

3.1. Write an expression that defines water saturation, or S_w, and define S_w in words.

3.2. What do the exponents *a*, *m*, and *n* represent in the Archie equation and what are the ranges of representative values for each?

3.3. Calculations of S_w in carbonate reservoirs with 100% moldic porosity require carefully chosen values of *m* that differ greatly from those in ordinary detrital carbonates or in terrigenous sandstones. What is a representative *m* value for reservoirs with 100% moldic porosity?

3.4. What is meant by the "transition zone" in a reservoir?

3.5. All rocks were either deposited in water or passed through the water table during burial so that all original pores were filled with water. How is it possible that some reservoirs can be oil-wet?

3.6. In a naphthenic acid and water system, naphthenic acid forms a contact angle of 35° on quartz surfaces and forms a contact angle of 106° on calcite surfaces.

Is naphthenic acid the wetting fluid in either case? If yes, which one and why?

3.7. What is the expression for adhesion tension—A_T?

3.8. Explain the physical meaning of A_T in plain language.

3.9. Write the expression for capillary pressure, P_c, and explain the meaning of each term. Is there a relationship between P_c and wettability? What is it?

3.10. What are three different kinds of information that can be obtained from MICP curves?

3.11. What does a "flat" mercury pressure trajectory through a given range of pressures reveal about pore throat sorting in that portion of the sample?

3.12. You have air–mercury MICP data from the laboratory and you want to use it to calculate the height of an oil column in your reservoir. The laboratory P_c (displacement pressure) of the reservoir rock is given as 1200 psi Hg/air, oil in the reservoir is 38° API oil with a density of 0.876 g/cm³, and water in the reservoir is brine with a density of 1.110 g/cm³. What is the height of the oil column?

3.13. Is it possible to test the capacity, or "sealing strength," of the rock that confines fluids in a reservoir? How?

3.14. What is meant by "snap-off"?

3.15. In what ways do the following pore system attributes influence reservoir quality: accessibility, coordination number, pore/pore throat ratio, and pore throat roughness?

CHAPTER FOUR

STRATIGRAPHIC PRINCIPLES

This chapter is about ways in which rock and reservoir characteristics can be combined to formulate geological models of carbonate reservoirs. In turn, each model represents a geological concept on which to base exploration and development strategies. Before the models can be mapped and exploited, and before flow units, baffles, and barriers can be fully understood, the models must be put in stratigraphic context—the larger-scale stratigraphic section and structural setting in which the reservoir model occurs. Part of this chapter is devoted to basic principles of stratigraphy and sedimentology to help students understand the differences between depositional platform types, between facies and formations, and between time, rock, and time–rock units. Other parts focus on basic principles of stratigraphic correlation and sequence stratigraphy to illustrate how individual rock and petrophysical measurements can be placed in stratigraphic context. Finally, this chapter introduces the concept of ideal depositional successions to set the stage for Chapter 5, in which ideal depositional successions are the central theme.

Finding reservoir rocks in sedimentary basins requires some knowledge about basin architecture—the geological evolution of different basin types and the structure and arrangement of strata within them. If we know something about the structure and arrangement of stratigraphic units in basins, then we can make reasonably accurate predictions of the sizes and shapes of carbonate reservoir bodies within those units. Reservoir rocks only make up a small percentage of the total basin fill; consequently, successful exploration and development require knowledge about how the reservoir bodies are distributed with respect to source and seal rocks. Basins, source rocks, reservoirs, seals, and overburden are elements in the *petroleum system* as defined by Magoon and Dow (1994). Finally, depositional rock properties are usually altered during burial and new rock and pore characteristics are super-

Geology of Carbonate Reservoirs: The Identification, Description, and Characterization of Hydrocarbon Reservoirs in Carbonate Rocks
By Wayne M. Ahr   Copyright © 2008 John Wiley & Sons, Inc.

imposed on the depositional template. The timing during burial and the mechanisms by which diagenesis has modified the depositional template will be discussed in Chapter 6. Knowing where to look and how to recognize clues that can lead to finding porous and permeable rocks is the key to successful exploration and development. It is beyond the scope of this book to provide detailed discussions of basin origin and evolution. Fortunately, an extensive literature already exists on those subjects. The focus of this chapter is on the fundamentals of stratigraphy, particularly those that relate to carbonate reservoir rocks.

Stratigraphy is the study of layered rocks. It is commonly referred to as the mother science of geology. The study of reservoir geology depends entirely on the application of stratigraphic principles to interpret the origin of a reservoir, predict its spatial dimensions, and map individual flow units. We learned in Chapter 1 that carbonates are intrabasinal in origin and consist mainly of biogenic or chemical constituents. The anatomy—the size and shape—of all depositional and many hybrid carbonate reservoirs conforms directly or indirectly to the geometry of depositional bodies. Those bodies accumulated or grew (in the case of reefs) on a variety of seafloor bathymetric configurations.

All depositional surfaces, regardless of their physiography, or bathymetry, are called *platforms*, and each platform type has its own distinguishing stratigraphic characteristics. Reservoir and nonreservoir strata accumulate in predictable patterns and locations on platforms depending on how platform physiography interacted with environmental dynamics—generally meaning wind or water movement. Platform sediment accumulations are organized in depositional facies that occur in predictable, lateral arrays from shoreline to basin floor. These arrays may be cyclically repeated—vertically stacked—over time. Depositional facies are also called lithogenetic units because each represents sediment deposited under one set of environmental conditions. Facies boundaries may not correspond with time boundaries, however. Facies boundaries may coincide with boundaries of reservoir porosity and permeability if the origin of the porosity is depositional or facies-selective diagenetic. Some kinds of fractures may also correspond in part or in total to facies boundaries because brittle behavior is commonly related to stratal thickness, grain size, and stratiform mineral occurrences. To establish a sense of scale and stratigraphic perspective, we begin with the types of carbonate depositional platforms and their characteristic facies arrays.

4.1 CARBONATE DEPOSITIONAL PLATFORMS

Few classifications for carbonate platforms existed before the 1970s. Some of the early schemes include those of Wilson (1970, 1974), Ahr (1973), and Ginsburg and James (1974). But it was not until an integrated scheme to classify a variety of platform types was presented by Read (1982, 1985) that the stage was set for the development of present-day terminology. In current usage, carbonate *platform* is the informal term for all depositional surfaces upon which shallow-water carbonate facies are deposited. Depositional surfaces include submerged continental margins, submerged perimeters around islands, or the shallow margins of intracratonic seas. Platform therefore includes any depositional surface that cannot immediately be assigned to a more specific category (Tucker and Wright, 1990; Burchette and

Wright, 1992). Platform geomorphology varies, especially in the amount and regularity of slope on the depositional surface. It is the interaction of the oceanic hydrological regime with platform geomorphology that determines the anatomy of the depositional facies on the platforms, the location of the facies on the platforms, and the rock types exhibited in the facies. Unbroken slopes from shore to open sea are swept by open ocean waves and currents; those with slope breaks occurring within fair-weather wave base will have "high-energy" facies at the slope break and restricted circulation behind it. It is the interaction of oceanic hydrodynamics with antecedent platform geomorphology that determines facies characteristics and evolutionary trends in platform anatomy. Some platforms slope continuously from beach to basin without a pronounced break in slope; they are called *homoclinal ramps* (Figure 4.1). Note that the original definition of ramp does not specify any angle of slope for ramps, although some later authors impose a slope of 1° or less for ramps. Actually, the slope on ramps may vary depending on antecedent topography. Ramps with well-defined slope changes on their distal margins are called *distally steepened ramps* (Figure 4.2). There are no facies changes associated with slope changes on distally steepened ramps because the distal steepening occurs at depths below the depths at which oceanic waves and currents affect bottom sedimentation. Other platform types may have steeper or flatter slopes, but they differ markedly from ramps in that they have a major break in slope accompanied by distinct facies changes some distance from shore. At the change in slope, water depth increases markedly over a relatively short lateral distance, forcing the hydrologic

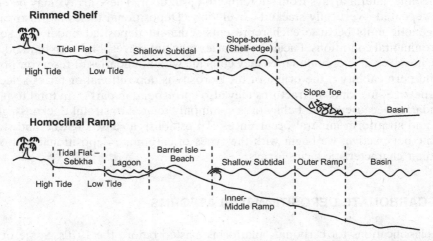

Figure 4.1   Examples of rimmed shelf and homoclinal ramp platform types. These models represent the most commonly encountered, end-member platform geometries. Rimmed shelves typically have reef rims but sand–wave complexes of carbonate grainstones can also act as hydrologic baffles (rims). An example of such a sand–wave rim is shown in Figure 5.5. All shelves have pronounced slope breaks accompanied by facies changes that mark the shift from shelf interior to slope or slope toe. Homoclinal ramps may be attached to the mainland shores or more commonly, as in this illustration, a barrier island–beach complex is separated from the mainland shore by a lagoon. The seabed on homoclinal ramps passes laterally from the strandline to basinal depths without a pronounced break in slope. Facies changes are gradational rather than abrupt.

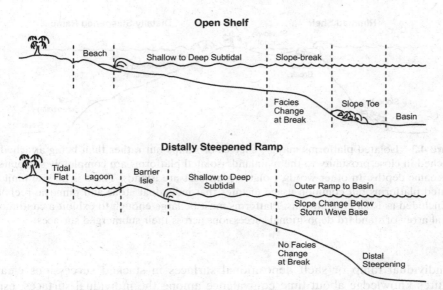

Figure 4.2 Open shelves and distally steepened ramps have slope changes along their margins. Open shelves have no rims but they have laterally persistent slope breaks accompanied by equally persistent facies changes. It is the persistent facies changes at the slope break that distinguish open shelves from distally steepened ramps. In low-energy settings, the facies characteristics of open shelves may be similar to those on rimmed shelves, except that topographic rims are absent. Such open shelves still exhibit facies changes at the slope break but they may also exhibit poorly developed strandplain facies—beaches or barriers—similar to those described in the Lower Cretaceous of Texas by Stricklin (1973). Distally steepened ramps have slope changes that occur at depths below fair-weather wave base and there are no facies changes at the slope change.

regime to change significantly. These differences in oceanographic conditions on either side of the slope break cause bottom sedimentation to vary greatly on either side of the slope break. Platforms with slope breaks and topographically prominent rims are called *rimmed shelves* (Figure 4.1). Shelves may be rimmed or open (Figure 4.2), depending on the presence or absence of reefs, banks, remnant topography, or grainstone accumulations at the slope break. Ramps and shelves may be "attached" to the mainland from beach to basin, or they may be large islands surrounded by oceanic depths, in which case they are classified as *isolated platforms* (Figure 4.3). An example of an isolated, rimmed shelf is the Great Bahama Banks and the Balearic Platform of Spain is an isolated ramp. Platforms are two-dimensional depositional surfaces outlined by bathymetric contours. Ramps and shelves can also be defined on the basis of the facies progressions that occur from shore to basin across the respective platforms. Along transects from shore to basin, each sedimentary bed is a two-dimensional record of deposition at an instant in geological time. For sedimentation to continue without change in water depth, there must be some "accommodation" provided by sea-level rise, by platform subsidence, or both. As sedimentation continues, successive depositional surfaces are "stacked" in a stratigraphic column representing the third dimension. The fourth dimension is the chronostratigraphic record represented by the stacked depositional bodies. Recognition

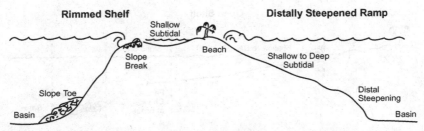

Figure 4.3 Isolated platforms may be shelves or ramps but rather than being attached or detached in close proximity to the mainland, isolated platforms are completely surrounded by oceanic depths. In other words, isolated platforms are islands. There is no size limit for isolated platforms but common usage dictates that individual atolls and pinnacle reefs are not included as isolated platforms. Platforms must be large enough to exhibit a continuous, lateral array of standard depositional successions across their submerged surfaces.

of individual ramp or shelf depositional surfaces in stacked successions clearly requires knowledge about time equivalence among the individual surfaces, especially where facies have migrated laterally and vertically. Facies may migrate up or down depositional platforms as they follow rising or falling relative sea level. If facies are correlated only on the basis of lithology, then they will cut across time surfaces, a condition known as *diachroneity*. One of the objectives in sequence stratigraphy is to identify time surfaces and the depositional units that are bounded by them.

4.1.1   Rimmed and Open Shelves

Rimmed shelves are shallow marine platforms that have pronounced slope breaks some distance from shore. The slope break is marked by rim-forming reefs and banks that may consist of rigid reefs and mounds, as on the South Florida shelf that extends seaward from the Florida Keys. Shelf rims may consist of submarine dunes and grainstone "sand waves," as on the Great Bahama Banks. Such rims are constructional; that is, they were formed by depositional processes that shaped the shelf edge. In other cases, shelf rims may be inherited from periods of destructional exposure and weathering (Purdy, 1974). Regardless of their origin, shelf rims interact with the hydrologic regime. Incoming waves, tides, and currents from the open sea are refracted, reflected, or translated by the rims so that the shelf interior is shielded from vigorous water movement except during the most severe storms. This situation can only exist in water shallow enough for the rim to interact constantly with inbound, fair-weather waves and currents. In Florida and the Bahamas, the rimmed margins are partly or totally exposed at low tide. The depth to which reefs or grainstone banks can interact with waves and currents is determined by wave climate and rim bathymetry. Oolite grainstone banks and hermatypic coral reefs do not form rims in water deeper than a few meters. For example, the average maximum water depth on the Florida and Bahama shelves is about 10 meters.

Rims on rimmed shelves serve as baffles, or "sieves," to filter and reduce the power of incoming oceanic waves, tides, and currents. Geologists commonly use the term "high energy" to describe settings in which wave power is great or in which current velocity is high. In the case of rimmed shelves, the rim environment has the highest "energy" regime (greatest amount of hydraulic power) on the platform. Sediments on the seafloor adjacent to the rim are subject to more winnowing, sorting, and transportation than those in other zones on the shelf. *Shelf edges*—the rim and its immediate surroundings—exhibit depositional facies generally characterized by grainstones and framestone/bindstone reefs. These accumulations lack mud matrix and can have high depositional porosity. Inboard from the rim, the character of the depositional facies changes progressively from grainstones and reefs to mud-rich facies of the shallow subtidal and tidal-flat domains. This predictable facies pattern enables the explorationist to predict the location and dimensions of facies with highest and lowest depositional porosity. In fact, it is possible to subdivide the entire platform into "cells" or zones where each cell is represented by depositional facies that exhibit a limited variety of textures, grain types, sedimentary structures, and biota. This idea was developed by Wilson (1975) as the "standard microfacies" concept and modified by Ahr (1985) to form three-dimensional depositional models on ramps and shelves. If enough is known about the age of the shelf, it is also possible to predict the kinds of fossils that will be present in the various paleoenvironmental cells.

Open shelves have a pronounced slope break accompanied by facies changes, but they lack shallow-water rims that interact with fair-weather waves, tides, and currents. The absence of rims may be the result of cool water temperature that prevented reef growth and inorganic carbonate precipitation. Most reef-building organisms, particularly hermatypic corals, thrive in water temperatures above 18 °C and large-scale inorganic precipitation is limited to warm waters. In other words, most reefs and grainstone accumulations form in tropical environments instead of temperate or cool ones. Another reason for the absence of rims at shelf edges is that the breaks in slope may occur at a depth too great for prolific growth of reef organisms or for inorganic carbonate precipitation. The environment where biogenic and chemogenic production of carbonate sediments is at its maximum productivity is sometimes referred to as the "carbonate factory." Sediment production in tropical waters is dependent largely on photozoan organisms; consequently, sediment production typically decreases with water depth at a rate depending on the depth of light penetration, among other environmental variables. For tropical environments, the optimum depth for carbonate sediment production is about 10 meters, according to Wilson (1975). Production may occur at depths of 90 m and more in temperate environments (Fornos and Ahr, 1997, 2006; James and Clarke, 1997), or where heterozoan organisms (those not dependent on photosynthesis) are the primary contributors. Finally, the zone of maximum sediment production on open shelves may not coincide with the zone of sediment retention. Sediment produced in shallow water may be transported to deep water by powerful waves and currents. In some high-energy, temperate environments both production and retention of carbonate sediments occur at water depths exceeding 100 meters (James et al., 1992). In such settings, the edge of the open shelf may occur at tens of meters to over 100-m depth and still be accompanied by laterally persistent differences in sedimentary characteristics (facies) above and below the slope break. On open shelves with extremely vigorous wave climates, such as present-day southern Australia, there may be little or no sediment accumulation in shallow water because powerful waves and currents erode and transport it offshore (James et al., 1992).

4.1.2 Homoclinal and Distally Steepened Ramps

Homoclinal ramps are submerged platforms that extend from shore to basin without a sharp increase in slope. Ramp slope angles may vary from one platform to another, but the unchanging characteristic is the absence of a slope break. Monotony in seaward slope and absence of a topographic "filter" to intercept incoming ocean waves, tides, and currents cause shorelines on ramps to be high-energy zones where grainstones and even carbonate gravels accumulate. Seaward transects of homoclinal ramps pass across increasingly muddy facies until basinal depths are reached. In other words, the usual facies progression across a ramp is from grainstones on the inner ramp (near shore) to mudstones on the outer ramp (distal portion of the ramp). If the shoreline is attached directly to the mainland, the strandplain facies on a homoclinal ramp will be beaches and coastal dunes. If longshore drift transports carbonate sands and gravels parallel to the shoreline forming a barrier island just offshore, lagoons will be ponded behind the barriers and muddy or evaporitic tidal flats may line the protected lagoonal shores.

The absence of a slope break in shallow water is a deterrent to the formation of buildups that might otherwise develop a topographic rim. Reefs develop on ramps, but they occur typically as patch reefs rather than laterally continuous reef trends. Reefs built by photozoans develop on hard seabeds in shallow, warm water where warm temperatures favor the formation of aragonitic skeletons and where nutrients and oxygen are abundant. Hard bottoms that are elevated above the surrounding seafloor are sometimes associated with antecedent highs such as those over salt domes, fault blocks, or erosional remnants. In short, some knowledge is required about paleobathymetric highs and lows on ramps in order to understand and predict locations of reef growth. Some buildups are constructed by heterozoans, by cements, or by chemogenic processes associated with seafloor vents and seeps. These features may develop in deep-water settings that are not related to topographic highs and by organisms that do not depend on sunlight, warm water, or abundant oxygen in near-surface water agitated by breaking waves. Because they are unrelated to otherwise easily detectable bathymetric or depositional features, deeper water buildups are somewhat more difficult to find in the subsurface.

Distally steepened ramps, like homoclinal ramps, exhibit high-energy facies near shore that pass to lower-energy, muddy facies with increasing depth. In other words, facies differences from shore to basin represent changes from mobile to stable substrates—from grainy to muddy seabeds. Distally steepened ramps differ from homoclinal ramps by exhibiting slope breaks at some distance from shore. They differ from open shelves in that slope breaks on distally steepened ramps do not have facies changes that coincide with the steepening (deep-water slope breaks), as is true of open shelves. For this situation to exist, slope changes on the ramp must occur below the depth normally swept by the fair-weather hydrologic regime. If carbonate buildups nucleate on slope changes along distally steepened ramps, they will not be laterally persistent and will not be accompanied by continuous facies differences that coincide with distal steepening. Such a situation is unusual and would be difficult to distinguish from an open shelf break. Steepening is a topographic feature but it is not accompanied by facies changes. Homoclinal ramps and rimmed shelves are end members of a range in platform shapes. Distally steepened ramps and open shelves are variations in form and they may have similar geological characteristics. In fact, some researchers do not make the distinction between open shelves and distally steepened ramps. For consistency and simplicity in definition of platform types, this book classifies those with slope breaks accompanied by facies changes as shelves and those without slope breaks or with slope breaks that are not accompanied by facies changes as ramps.

4.2 ROCK, TIME, AND TIME–ROCK UNITS

Geologists distinguish between rock units, time units, and the composite time–rock units. Stratigraphy based on rock layers only is called lithostratigraphy. Stratigraphy based on rock units that have corresponding time significance is chronostratigraphy. Sequence stratigraphy incorporates time–rock units in a system of stratal architecture. Time units are simply measures of geological time, usually given in millions of years abbreviated as Ma. Time units in the geological record are ranked and named according to the length of time they cover. Units that span the longest time are called eons. In descending order of time span, the remaining units are eras, periods, and epochs. The Phanerozoic Eon includes the Paleozoic, Mesozoic, and Cenozoic Eras. Each of those eras includes periods and each period includes epochs. We live in the Holocene Epoch of the Quaternary Period of the Cenozoic Era in the Phanerozoic Eon.

4.2.1 Rock Units

Rock layers represent the three-dimensional record of earth history. Ideally, each millimeter-scale lamina represents the sedimentary record at an instant in geological time, but bathymetric charts of modern seas show that depositional surfaces are rarely, if ever, flat and a single sedimentary lamina may represent a longer or shorter time depending on rate of sedimentation and efficiency of preservation. In other words, a lamina does not necessarily represent a time plane. Nearly every marine environment has bathymetry that may be depositional, tectonic, or erosional in origin. Depositional topography may result from the growth of carbonate buildups such as reefs, mounds, or skeletal sand accumulations. Tectonic features result from faulting and folding, and erosional topography includes scoured or karst depressions as well as remnant hills and ridges. Submarine topography may be passively involved with sedimentation, such as a topographic depression being filled with detrital carbonates, or it may be dynamically interactive, as in the case of topography being created or accentuated by active reef growth. Depositional surfaces are not flat and they may change constantly as they interact with the environment. Lithogenetic units are rock units that have a common origin and share many descriptive characteristics. They may or may not be bounded by time surfaces. The practical solution to dealing with rock units was developed in North America by Schenk and Muller (1941), whose pioneering concepts were incorporated in the Code of Stratigraphic Nomenclature and the International Guide to Stratigraphic Classification. Rock units are mapped without being associated with time and they are assigned to a hierarchy of scale that includes, from large to small, group, formation, member, and bed. Formations are the smallest distinctively mappable rock units and they have no time significance. They may include diachronous facies or the formations themselves may not conform to time boundaries. A group is more than one formation, formations are subdivided into members, and members are divisible into lentils and tongues, or beds. Formations and members are generally identifiable at reservoir scale, sometimes so easily identifiable that they can be correlated for great distances. Traditional subsurface correlations based on wireline log signatures or sample "tops" are lithostratigraphic rather than chronostratigraphic. Of course, correlations based only on rock units can lead to errors in identifying flow units. One of the main reasons for the current popularity of sequence stratigraphy is that it focuses on chronostratigraphic rather than lithostratigraphic correlations. Lithostratigraphy groups rocks of common character but not necessarily of the same age, a potential flaw that can cause one to overlook internal flow barriers or baffles that conform to time surfaces. Geologists commonly try to establish correspondence between wireline log characteristics and lithology in order to draw subsurface correlation sections and to generate a variety of subsurface maps. This traditional method has a built-in potential for error, however, because precise determination of individual flow units within a reservoir body usually requires some time reference, especially if the reservoir incorporates time-transgressive facies. At a larger scale, over hundreds of kilometers, for example, correlation of individual rock units without time reference is hardly possible because of the major differences in depositional characteristics that exist across large areas over time.

4.2.2 Time Units

Relative geological time is determined by fossils. For example, the trilobite *Olenellus* is useful worldwide as an indicator of Early Cambrian age. Such fossils, commonly called marker fossils or index fossils, ideally have short ranges and widespread distribution in the rock record, enabling the geologist to use them to identify specific time intervals within geological periods. In some situations, a single fossil species or genus does not provide enough information to make a precise age determination, but in association with other fossil species or genera in the rocks, the age of the assemblage can be determined. Accurate stratigraphic correlation depends on geochronology—establishing the age of the strata. Sometimes fossils may not be present, but accurate correlations can still be made with isotopes, fission tracks, or other geochemical methods. For example, a layer of volcanic ash represents an event—an instant in geological time—that precisely separates in time the beds below and above the ash layer. Unconformities represent gaps in the rock record produced by erosion or nondeposition and they can be used as indicators to mark the age of first deposition above the unconformity. Beds below the unconformity may be widely different in geological age because some parts of the unconformity surface may have been eroded more deeply than others. Consider a regional unconformity that covers hundreds of square kilometers, slopes seaward, and then becomes gradually submerged by a relative rise in sea level. The advancing sea will submerge the unconformity bit by bit, at first depositing a body of shallow-water sediments. Then, with further sea-level rise, the shallow-water sediments at the first depositional site will be covered with a younger layer of deeper-water deposits. As that deeper-water deposit was being laid down, a coeval shallow-water facies was deposited landward from the first one. The upslope migration of these progressively younger deposits illustrates the concept of *time-transgressive facies* (Figure 4.4). That is, facies may be

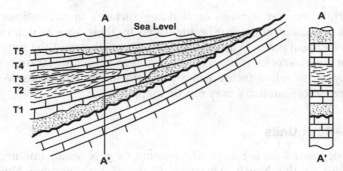

Figure 4.4 An illustration of time-transgressive facies and Walther's law. Note the way depositional facies migrate landward during transgressive sea-level phases and move seaward during regressions. At an arbitrary starting time, T1, the limestone facies migrates landward and deposition continues through T2 time but without change in facies identity. This is called time-transgressive deposition. The vertical column A–A' shows what would be found if a borehole were cut at the point marked by the line A–A'. From the top of the column downward, the sandstone overlies the limestone. The same facies are present as time equivalents in a seaward direction along the surface, an expression of Walther's law (Walther, 1894).

identical in rock properties, may have formed in identical depositional regimes on a platform, but they are not equivalent in age. The importance of these facts becomes evident when one attempts to reconstruct the shape of an ancient platform from a three-dimensional succession of layered rocks.

"Absolute" geological time is based mainly on ratios of stable daughter isotopes and their unstable parent radioactive isotopes. For example, an isotope of uranium, ^{238}U, decays to ^{206}Pb through a series of steps. The rate at which ^{238}U decays is known precisely; therefore if the Pb/U ratio of a rock sample is known, its age can be calculated. Other geochemical techniques are available to determine the age of rocks. The point is that these methods can provide age measurements in years of actual time before the present whether fossils are present or not. The precision with which age dates can be determined varies depending on the method used, the amount of substance present for analysis, the purity of the substance, and the age of the rock. Absolute ages are determined with margins of error as great as 5% or more for dates of 10^8 years before present. Relative time based on index fossils is even less precise and dates based on any particular fossil may vary by many millions of years. Geochemical methods are not often used to determine the ages of reservoir rocks because mineral species that contain datable isotopes are not always present and it is expensive and time consuming to make the measurements. In general practice, correlation for reservoir mapping and sequence stratigraphic applications can be accomplished with marker or index fossils. Of course, the marker fossils must be present. If they are absent, the precision of stratigraphic correlation along a specific time surface decreases dramatically with distance between points of reference (wells or outcrops), the likelihood of miscorrelation increases, and accurate sequence stratigraphic applications are difficult if not impossible. A basic rule in sequence stratigraphy is that sequences are bounded by unconformities and their correlative conformities. The amount of time represented by the sequence-bounding unconformities must be established by identifying the ages of the beds above and below the surfaces. Correlative conformities or bedding surfaces in successions where the entire record is present must also be identified by their ages as determined from fossils or geochemical dates. Without knowing the geometry of the time surfaces in successions of layered rocks, it is not possible to reconstruct the stratigraphic history of basin filling, or to relate facies arrays to platform configurations. Without geochronology, reservoir anatomy may not be mappable.

4.2.3 Time–Rock Units

Rock layers deposited during a specific amount of geological time are time–rock units according to the North American Code of Stratigraphic Nomenclature. Geological time is divided into eons (Cryptozoic and Phanerozoic), eras (Paleozoic, Mesozoic, and Cenozoic), periods (Cambrian through Quaternary), epochs, and ages. Rocks deposited during eons, eras, periods, epochs, and ages are designated as *eonothems, erathems, systems, series*, and *stages*, respectively (Table 4.1). For example, all rocks deposited during the Quaternary Period comprise the time–rock unit known as the Quaternary System, including time–rock units of shorter span belonging to the Pleistocene Series and the stages that correspond to the ages of the Pleistocene Epoch. Time–rock boundaries are defined by relative or absolute age dates, not by rock characteristics. It is important to remember that time surfaces are not always parallel to rock unit boundaries.

4.3 CORRELATION

Stratigraphic correlation is the process of linking age-equivalent or lithologically similar strata in two or more different places. These links are formed in several ways, including (1) by establishing physical continuity, (2) by establishing identical age, (3) by identifying beds that occur in the same order within a sequence, or (4) by matching wireline log or geophysical characteristics and assuming that those signatures are reliable substitutes for direct observations of lithology and chronology. Correlations based on log and geophysical similarities are the least reliable and generally represent the kind of work done in the subsurface when outcrops, borehole samples, or cores are not available. Physical continuity of beds can be determined by "walking-out" or physically tracing the beds from one location to another. The same geological age of beds in different locations can be established by finding identical marker fossils or assemblages of fossils at each location or by determining

TABLE 4.1 Rock, Time, and Time–Rock Units

Lithostratigraphic (Rock) Units	Geochronological (Time) Units	Chronostratigraphic (Time–Rock) Units
Group	Eon	Eonothem
Formation	Era	Erathem
Member	Period	System
Bed	Epoch	Series
	Age	Stage

the absolute age of the rocks from radioactive isotopes. Correlations based on relative geological time may be made more precise when it is possible to identify widespread features that represent "instantaneous" events such as ash beds and lava flows. Beds that appear in the same order within a depositional sequence can be correlated without the aid of fossils or absolute age dates. The Triassic Period, for example, takes its name from three distinctive rock units that always appear in the same order over much of Europe north of the Alps, particularly in Germany. A bright red sandstone called the Bunter is always overlain by shelly limestone called the Muschelkalk, which is always overlain by variegated shales and salt beds called the Keuper. The principle that enables correlation of Triassic beds over hundreds of square kilometers in Europe can be applied to bed sequences of any age at any location provided that the order of appearance of the beds remains constant in the sequence.

Subsurface stratigraphic correlation can employ the same methods described previously but in practice, most field correlations are based on wireline log "kicks" (Figure 4.5). Similarly, log traces are identified as datums for making structural and stratigraphic cross sections, structure maps, net pay, net sand, and interval isopach maps. Some types of logs that are commonly used in these applications are listed in Table 2.2. It is particularly important to remember that logging devices provide information about tertiary or latent rock properties. They are not representations

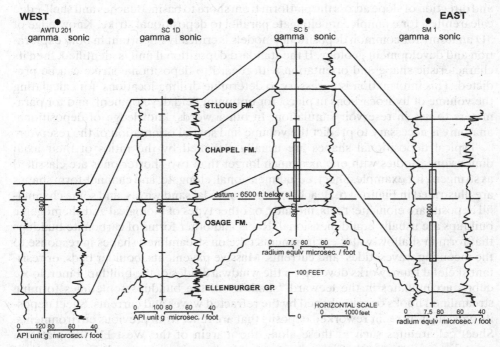

Figure 4.5 A subsurface structural cross section illustrating correlation by borehole log signatures. In this example the top and bottom of each formation can be identified by characteristic "log kicks" on the gamma ray and sonic log pairs. The structural datum is indicated by the dashed horizontal line. (From Ahr and Walters (1985).)

of rock properties per se, and log correlations have no time value. Chronostratigraphic information must be determined independently and superimposed on
lithostratigraphic correlations in order to distinguish between age-equivalent and
time-transgressive rock units. Likewise, geophysical information such as seismograph records is not a direct representation of rock properties. Seismic traces, like
wireline log traces, have no time significance. This fact is vitally important to the
worker who may try to establish direct correspondence between seismic stratigraphy and sequence stratigraphy. Without independent time markers superimposed
on the seismograms, no specific age can be assigned a priori to seismic reflectors. If
ages are not known, sequences cannot be correctly identified or mapped. Independent geochronology is usually determined from age dates established in boreholes
along, or projected into, seismic lines.

4.4   ANATOMY OF DEPOSITIONAL UNITS

Depositional sedimentary bodies have characteristic 3D shapes. In most of these
shapes, the dimensions of length, width, and thickness are different. Shapes may be
elongate, flattened, or otherwise streamlined in response to physical and biological
sedimentary processes. In addition, the long axis of sedimentary bodies may be oriented parallel or at some angle to depositional strike. Depositional strike is the
compass direction perpendicular to depositional dip. Depositional dip is the amount
and direction of slope across the platform from shore to basin. Beaches and shelf-edge
calcarenites, for example, are elongate parallel to depositional strike. Knowledge of
3D anatomy for common depositional models is critically important in both exploration and development geology. If the standard depositional unit is identified, then its
characteristic shape and orientation with regard to depositional strike can be predicted. This information is necessary to determine drilling locations, for calculating
the volume of hydrocarbons in place, for optimum field development, and for parameters to use in reservoir simulation. In other words, knowledge of depositional
anatomy is necessary to predict the volume and spatial orientation of the reservoir.

Typical depositional shapes are usually classified by the ratios of their axial
dimensions. Bodies with one axis much longer than two shorter ones are classified
as stringers, for example. Sheet, equidimensional, elongate, and channel-form shapes
are illustrated in Figure 4.6. Beaches, dunes, subtidal sand waves, and some channel-
fill deposits are elongate; reefs, mounds, or other types of biological and chemogenic
buildups are usually equidimensional. Reefs and other forms of carbonate buildups
that form in shallow-water environments take on streamlined shapes in response to
the incoming waves, tides, and currents. Massive pavements, boulder beds, or resistant skeletal frameworks develop on the windward sides of the buildup. Finer, loose
debris accumulates in the leeward "shadow" of the buildups sometimes forming
streamlined hooks or spits shaped by the refracted waves and currents. Sheet deposits typically represent reworked deposits that are relict from previous environments.
Sheet calcarenites such as those along the margin of the West Florida platform
are relict grainstones that have been reworked on the drowned Pleistocene shelf
(Figure 4.7).

When depositional bodies represent one or two episodes of sedimentation in a
specific cell on a platform, it is not difficult to predict the size of the bodies, their

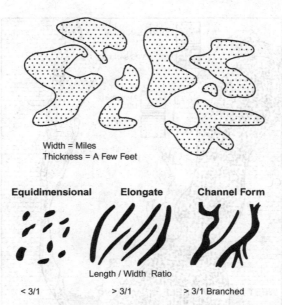

Width = Miles
Thickness = A Few Feet

Equidimensional   **Elongate**   **Channel Form**

Length / Width Ratio

< 3/1   > 3/1   > 3/1 Branched

Figure 4.6   Sheets or blankets: typical shapes of depositional bodies deposited in carbonate environments adapted from an illustration in Potter (1963). Sheet forms are usually a few feet thick at most and may cover many square miles in area; they usually represent reworked material deposited during previous lowstands on outer parts of platforms. Elongate shapes are typical of strandplain deposits, reefs and mounds are usually more-or-less equidimensional in plan, and channel forms may be found in tidal deltas, turbidites, and density current deposits.

spatial orientation, and the porosity and permeability distribution within them. A case history is helpful to illustrate this concept. Pleistocene oolites occur in a wide belt at present water depths of about 100 m off the western coast of Yucatán. Oolites and shallow-water benthic organisms identify these Pleistocene facies as trans-gressed remnants of shoreline deposits similar to the modern barriers and beaches on the northeastern coast of Yucatán at Isla Cancun and Isla Mujeres, among others. Knowing that beaches parallel depositional strike, it is possible to predict the ori-entation of the body, and data on the dimensions of the beach at one or two locations enables predictive maps to be made on the size and shape of the potential reservoir. These are external, large-scale characteristics that relate beaches to their laterally equivalent facies across the entire depositional platform. Two or more cores of the Pleistocene oolites at different locations on the Campeche Bank should provide enough information to construct a map to predict the size, shape, and location of the oolite facies trend as long as the bathymetry remained relatively constant. The oolite body will extend along depositional strike following the paleobathymetry of the Pleistocene shoreline. Any changes in bathymetry such as local highs would cause deviations in the shoreline trend, emphasizing that it is always necessary to take bathymetry into account when making facies maps.

  If several episodes of sedimentation are repeated and a compound depositional body forms within one cell on the platform, it may be difficult to separate strata that represent each episode. If some of the strata are nonreservoir rocks (baffles or

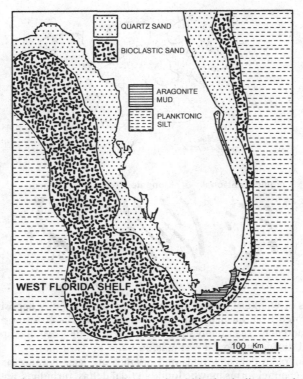

Figure 4.7 An example of a sheet-form deposit of bioclastic lime sand that extends over hundreds of square miles where grain-rich deposits of Pleistocene age cover much of the modern West Florida shelf. They have remained unburied and spread across wide areas because post-Pleistocene sedimentation rates have been low and reworking by waves and currents has kept the seabed mobile. (Adapted from an illustration in Purser (1980).)

barriers), they can compartmentalize the reservoir or act as permeability barriers. It is especially difficult to dissect deposits where there is little contrast between sedimentary properties in layers of different ages or episodes of sedimentation. Reefs and homogeneous oolite sand bodies are good examples of depositional units that develop episodically. They are not easy to dissect into time slices because depositional rates vary in space as the bodies form and because there may be few systematic differences in rock properties from one layer to another. Unconformities may further complicate the problem by removing different amounts of rock at different parts of a depositional body, leaving a truncated surface that exposes rocks of different age and possibly different petrophysical characteristics. An example of a compound depositional body with both reservoir and nonreservoir zones exists in Jurassic oolite buildups at Overton Field, Texas. Here, reservoir porosity corresponds with diagenetically altered oolites. Nonreservoir zones are represented by beds of unaltered oolites. There are few, if any, distinctive differences in depositional rock properties to provide data for field development or for finding another prospect in the trend. The solution to the problem lies in knowing the geological cause of the diagenetic alteration. Alteration commonly has subtle connections to depositional anatomy. In this case, both depositional anatomy and diagenetically produced microporosity were influenced by antecedent topography. The present structural configuration of the area has been modified by syn-to-post-diagenetic salt doming. Oolite sand bodies on crests of antecedent structural highs around the perimeter of a buried basement feature are altered and productive; those not on that paleo-high are unaltered and tight. Although older oolite strata in one of the wells are presently high on a salt structure, they are nonporous and were not diagenetically altered to become microporous. This apparent paradox is resolved by understanding that the microporosity in the productive, younger oolites was formed shortly after deposition above the relatively static, basement high and before salt doming lifted the older oolites to their present-day position. These subtle differences in sand body anatomy associated with antecedent topography can be mapped to facilitate optimum field development. The cause–effect relationships between topography, localized deposition of oolites, diagenesis, and reservoir porosity provide the foundation of a *geological concept* to use in exploration for additional prospects.

4.4.1   Facies, Successions, and Sequences

The word facies derives from the Latin *facia*, meaning face, countenance, or exterior appearance. In geology, facies is mainly used to describe the fundamental rock properties that characterize depositional units smaller than member rank. It is purely a rock-stratigraphic term with no time value. Metamorphic rocks may be classified into different metamorphic facies on the basis of the mineral assemblages they contain. Biofacies are determined on the basis of fossil content. In sedimentology, facies are determined on depositional rock properties such as texture, constituent composition, and sedimentary structures. Sometimes the differences between sedimentary facies are not obvious to the unaided eye. Thin section petrography may be used in such cases to identify differences in microscopic rock properties such as constituent percentages, microscopic pore characteristics, or biological microstructures. Facies defined on petrographic data are called *microfacies*. Extensive discussions of microfacies analysis can be found in Flugel (1982), and some pioneering applications of thin section petrography to establish facies boundaries in modern carbonate sediments include Ginsburg (1956), Purdy (1963a), and Logan et al. (1969). In his study of modern Bahamian carbonates, Purdy (1963a) determined microfacies boundaries by applying a statistical procedure called factor analysis to constituent point-count data. The resulting "reaction groups" provided objective, but rather abstract, microfacies boundaries. In general practice, facies boundaries are determined on estimated percentages of constituents, general textural trends, and sedimentary structures.

Facies, or more precisely lithofacies, form in depositional environments where hydrological and biological processes create genetically distinct associations of texture, constituent composition, and sedimentary structures, or *lithogenetic units*. In carbonate reservoirs, those rock properties are generally not distinguishable with wireline logs; consequently, the reservoir geologist must examine cuttings or cores to accurately identify carbonate facies. If the process–response mechanisms that produced the depositional facies are known, then it is possible to work backward from data on rock properties to interpret both depositional environment and anatomy of the depositional body. Sometimes facies are named for the environment in which they formed. For example, lagoonal facies, tidal-flat facies, or reef facies are common terms in ordinary practice. More formal procedures require that facies names reflect only rock properties, not environmental terms. A name such as "intraclastic, skeletal packstone facies," for example, does not have explicit environmental significance. This distinction is made to avoid subjective or interpretive terms in facies names, but in common practice geologists do not always make these distinctions. It is important to recognize that facies are considered in this book to be two-dimensional (2D) features that may not represent all of the variations in sedimentary character that can occur in one environment through some span of time until the original facies is replaced by a new one as the environment shifts in response to changes in relative sea level. Some workers will not agree with this usage, but for teaching purposes, it is valuable to consider all the variations in fundamental rock properties in one environment as a depositional succession, not as one facies. If all variations in rock properties in a beach–dune complex comprised a facies, for example, then each variation (lower shoreface, middle shoreface, upper shoreface, and dunes) would become a subfacies. I find it simpler to work with depositional successions that include the usual variations in one vertical "package."

Depositional succession (also called depositional sequence) is therefore defined as the vertical array of distinctive lithological and biotic characteristics at a single site such as a borehole or an outcrop. Because the vertical dimension is included, depositional successions only occur in the rock record; modern sedimentary facies are two-dimensional "blankets" that represent a single, short time. The word "succession" is defined as the process of following in order or in sequence. Succession, in the geological sense, describes the vertical array of rock properties at a location over an unspecified, but generally short and continuous time span. A short time span in this case can be on the order of 6000 years or less, as illustrated by the Holocene successions that have accumulated in the various environments on carbonate platforms since sea level reached its present height. It is critical that successions represent essentially unbroken timelines, because gaps in stratigraphic continuity represent unknown amounts of time and time–rocks in the sedimentary record. Clearly, the Holocene successions will have different thicknesses, textural trends, sedimentary structures, and constituent compositions, depending on the environment in which they were deposited. Finally, the term succession is used partly to distinguish successions from standard microfacies and to avoid confusion with *sequence*, as defined by Sloss (1963), namely, *"rock stratigraphic units of higher rank than group, megagroup, or supergroup, traceable over major areas of a continent and bounded by unconformities of interregional scope."*

In the following pages, idealized depositional successions are illustrated as sketches that resemble borehole cores arbitrarily about 10m long (the length of a standard core barrel), showing the typical sedimentary characteristics for each environment on ramps and shelves. Thicknesses of the successions depend on accommodation, time, and rate of sedimentation, but the ideal successions are rock units without time connotation. Depending on relative sea-level history, accommodation, and depositional rates, the successions can migrate landward or seaward with advancing or retreating relative sea level. For example, Holocene barrier islands on the northeastern coast of Yucatán may be about 10m thick, representing deposition during only a few thousand years since present sea level was reached. However, where sediment production and accumulation are less than on barrier islands, such as the shallow subtidal environment on the Yucatan ramp, the postglacial sediment blanket varies in thickness from just a few centimeters to about 1.4 m (Logan et al., 1969). The shallow subtidal Holocene blanket on the Florida shelf varies in thickness from zero (bare Pleistocene rock) near the shore at Key Largo to about 5 m in the White Bank sand wave buildup (Enos and Perkins, 1977). These variations in sediment thickness reflect accommodation, sediment source, and mechanisms of sediment dispersal and retention. Sedimentary successions have stratigraphic contacts that mark a beginning and an end, and the successions may be cyclically repeated as time progresses. There may be a variety of rock properties in the vertical dimension, but the variations in those rock properties are generally regular and predictable.

Ideal depositional successions, like two-dimensional facies, are defined on fundamental rock properties and biota. They differ from facies in being explicitly linked to depositional environments, in the emphasis placed on their vertical dimensions, and by exhibiting rock properties that can vary widely, but predictably, from base to top of the succession. The beach–dune example mentioned earlier consists of several subenvironments and their characteristic deposits—dunes, upper shoreface, middle shoreface, and lower shoreface. Those deposits are stacked vertically to form an ideal beach–dune or barrier island depositional succession. Each of the subenvironments is associated with a set of characteristic rock properties. If the beach–dune complex builds seaward (regressive barrier) then the horizontal arrangement of the subenvironments is repeated in the vertical (Figure 4.4) according to Walther's rule (Walther, 1894). This vertical array represents all of the subenvironments in the barrier complex and comprises a *standard depositional succession*.

4.4.2 Environmental Subdivisions and Standard Depositional Successions

Ideally, facies are defined objectively based on their fundamental rock properties regardless of depositional environment. However, Wilson (1975) interpreted associations of rock properties, biota, and depositional environments to define nine *standard microfacies* that represent nine environmental subdivisions of an ideal platform (Figure 4.8). Application of the standard microfacies concept to the rock record enables one to interpret environments of deposition directly from rock and fossil characteristics at any location on the standard platform and to predict the locations and characteristics of other facies on the platform. This method of interpreting depositional environments in carbonate rocks is commonly practiced today and it can be a powerful tool for predicting the locations of reservoir rocks. The concept of standard microfacies was a major advance in our understanding of the relationships between environment, sedimentary characteristics, and biota. Wilson's (1975) method provides reliable results as long as it is recognized that his idealized platform is a rimmed shelf. Note that beaches, dunes, or barrier islands that typify ramp shorelines do not normally occur on rimmed shelves and are not included in Wilson's nine standard microfacies.

Standard microfacies are two-dimensional (2D) blankets that represent the sedimentary cover on a platform sector during a short time (maybe a few hundreds to thousands of years—a "blink" in geological time). As time and sedimentation continue, the sedimentary blankets evolve into stacked assemblages (successions), each

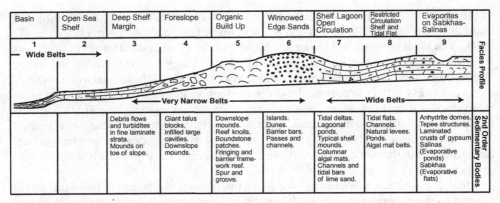

Basin	Open Sea Shelf	Deep Shelf Margin	Foreslope	Organic Build Up	Winnowed Edge Sands	Shelf Lagoon Open Circulation	Restricted Circulation Shelf and Tidal Flat	Evaporites on Sabkhas-Salinas	
1	2	3	4	5	6	7	8	9	*Facies Profile*
← Wide Belts →									
				← Very Narrow Belts →			← Wide Belts →		
		Debris flows and turbidites in fine laminate strata. Mounds on toe of slope.	Giant talus blocks, infilled large cavities. Downslope mounds.	Downslope mounds. Reef knolls. Boundstone patches. Fringing and barrier frame-work reef. Spur and groove.	Islands. Dunes. Barrier bars. Passes and channels.	Tidal deltas. Lagoonal ponds. Typical shelf mounds. Columnar algal mats. Channels and tidal bars of lime sand.	Tidal flats. Channels. Natural levees. Ponds. Algal mat belts.	Anhydrite domes. Tepee structures. Laminated crusts of gypsum Salinas (Evaporative ponds) Sabkhas (Evaporative flats)	*2nd Order Sedimentary Bodies*

Figure 4.8 The Wilson standard carbonate platform with its nine environmental subdivisions and their accompanying standard microfacies. The concept of standard platform geometry with characteristic facies representing different environmental subdivisions paved the way for modern methods of sedimentary facies analysis and sequence stratigraphy. (Adapted from an illustration in Wilson (1975).)

of which has distinctive rock and fossil characteristics. These vertical successions vary as accommodation, sediment input, sediment retention, hydrological regime, and biotic characteristics change over time at the site of deposition. For example, the typical shallow subtidal facies on both ramps and shelves consists of mudstones and wackestones with occasional patches of packstones. If conditions are right, patch reefs may develop in the shallow subtidal environment, thereby introducing distinctly different facies characteristics compared to the laterally equivalent mudstones and wackestones. These variations must be considered when ideal successions are constructed to serve as depositional models. Depositional successions are records of sedimentation during the accumulation of stacked, 2D microfacies within each platform sector. Variations in sedimentary and biotic characteristics notwithstanding, standardized depositional successions can be used in subsurface reservoir geology as long as they record rock and fossil characteristics that occurred in the same platform sector over time and with more or less constant environmental conditions. For example, tidal flats and lagoons typically occur behind barrier islands on ramps. As long as the hydrologic regimes, bathymetry, and subsidence allow sedimentation to continue without dramatic change, the resulting successions will provide unique records of tidal-flat deposition. The interplay between subsidence and sedimentation will dictate how the typical tidal-flat microfacies—lagoonal, intertidal, and supratidal microfacies—are stacked. Remember that in chronostratigraphic context, individual depositional successions are assumed to represent short time intervals, continuous sedimentation, and limited spatial migration of environmental sectors on platforms. During high-amplitude, basin-scale changes in relative sea level, individual microfacies (stacked, individual blankets that make up successions) may migrate considerable distances up or down depositional dip to stay in step with their shifting environments. These migrating facies can be time-transgressive, as we have discussed earlier; consequently, it is very important to identify time markers when doing correlations between successions at field or platform scale.

Ideal depositional successions for siliciclastic sandstone environments were described and standardized by LeBlanc (1972). Before publishing his ideal depositional successions, he used them for many years as illustrations in his classes at Shell Development Company. He illustrated and described successions for eolian, fluvial, deltaic, coastal interdeltaic, and marine environments in such a way that practicing reservoir geologists could use them as aids to identify and map reservoir sandstones in three dimensions with a minimum of subsurface information. To an extent, the same method of grouping common lithological attributes in vertical successions and assigning them to depositional environments can be done with carbonates. The differences between carbonates and siliciclastics that were discussed at the beginning of this book must be kept in mind.

A general review of carbonate depositional facies can be found in the book *Facies Models*, a compendium of papers edited by Walker and James (1992). They discuss facies that represent a variety of depositional environments, and in most examples, vertical successions are also described. Those authors do not emphasize the distinction between two-dimensional facies and three-dimensional successions and the model successions are not always linked to depositional settings on more than one kind of platform. Subsurface geology in exploration and production makes extensive use of geological concepts to help predict occurrences of reservoir rocks with limited information. The concept of standard depositional successions is particularly helpful in this respect because (1) standard depositional successions are depicted in the way they would appear in cores or detailed lithological logs from cuttings, and (2) standard successions are associated with specific environmental sectors on ramps and shelves. A greatly simplified but somewhat incomplete representation of carbonate depositional successions is presented in Ahr (1985).

Just as standard microfacies depict lithological characteristics in the nine environmental cells across Wilson's (1975) idealized platform, the greatly simplified depositional successions in Ahr (1985) portray vertical assemblages of rock properties formed over time at specific cells on any carbonate platform, although that paper did not include discussions of slope environments, the variety of basinal deposits that are commonly found, or the distinction between beaches attached to the mainland and detached barrier islands with lagoons and tidal flats behind them. Standard depositional successions—think of them as depositional models—presented here represent the typical variety of fundamental rock properties that can develop in each environmental sector on platforms over time, but the successions do not have precise chronostratigraphic values. Because of the variations in the kinds of rock properties that occur in each succession, consideration has to be given to the amount of unexpected variability that can exist in the successions. Some license has to be taken to generalize these successions in order to make them more applicable to platform configurations of essentially any geological age and any place on the globe. In order to formulate such generalized models, ramps and shelves must be divided into environmental subdivisions, or sectors, beginning at the shoreline and ending in the deepest part of the basin. In this book, all platform types—ramps, shelves, and isolated platforms—can be divided into seven generalized environmental zones or sectors, each with standardized depositional successions, or models, that represent typical stratigraphic sequences that might be found in borehole cores. Sketches of these idealized successions and the various platform types appear at the end of Chapter 5.

The seven standardized successions presented in this book are designed to allow for some variability. For example, the beach–dune succession may be attached or detached from the mainland. Detached beach–dune successions occur as barrier islands that may have updip equivalent lagoon and tidal-flat successions. Lagoonal successions typically consist of burrowed, wavy bedded mudstones and wackestones with low-diversity biota. In some cases, storm washover fans may be incorporated in the lagoonal succession. Tidal flats may or may not have extensive evaporite deposits like those in the Trucial Coast sebkhas. Tidal flats may or may not have extensively developed channels in the intertidal zone. Shallow subtidal successions may or may not include patch reefs and associated grainstones; slope-break successions on rimmed shelves may consist of reef trends along the break or of grainstone buildups with few, if any, reefs. Slope-toe successions may include slumps, debrites, grain flows, or coarse turbidites, and basinal successions may appear as microlaminated zones, rhythmites, or turbidites. Seven generalized environmental zones were chosen for simplicity, brevity, and especially to reduce the number of possibilities one must identify from unknown successions in borehole cores or "lith logs" from cuttings. Some environmental zones produce more complex patterns of sedimentation than others. In those cases, it is necessary to have "supplemental" successions to provide more complete coverage of the range in depositional style. The seven basic environmental sectors are: (1) the shoreline environment consisting of beaches with or without dunes and that may be attached to or detached (as a barrier island) from the mainland; (2) the intertidal or tidal-flat environment updip from lagoons (on ramps) or shallow subtidal waters (on shelves); (3) the lagoonal environment behind barrier islands detached from the mainland and consisting of strike-parallel, elongate islands with beaches plus or minus dunes; (4) the shallow subtidal environment; (5) the slope-break environment that characterizes shelves; (6) the slope and slope-toe environments that comprise the slope below the shelf-slope break and the base, or toe of slope below the break; and (7) the basinal environment that extends from the toe of slope on shelves to the greatest basinal depths on ramps or shelves. This is a greatly simplified system by design. Basinal environments on ramps may encompass environments extending from outer neritic to bathyal regimes. It is not possible to predict every variation in facies characteristics that may occur on any given platform within any given environmental zone. That notwithstanding, years of personal experience and extensive review of the literature have convinced me that this simplified system enables one to identify depositional textures, sedimentary structures, biota, and constituent components common to specific depositional successions, to associate the successions, or models, with environmental zones or positions on ramps or shelves, and to make reasonable predictions about the types of laterally equivalent successions in updip and downdip environmental cells.

Most of the ideal successions are associated with narrowly defined locations on platforms and with implicit constraints on variability of environmental conditions that characterize each zone, or sector. Limited variability in environmental conditions suggests limited variability in lithological characteristics of the associated depositional successions. Some of the environmental zones are not so narrowly defined and may have rather widely varying lithological characteristics depending on platform geomorphology, bathymetry, and hydrological regime. For example, monotonous bathymetry and depth in the subtidal zone (the neritic environment) on low-to-moderate-energy platforms will be accompanied by rather monotonous, mud-dominated depositional successions (wackestones and mudstones). Platforms with the same bathymetry and depth but with a very high-energy environment may be swept clean of detrital sediment, leaving a bare rock surface as on the high-energy platform off southwestern Australia (James et al., 1992).

Variations in bathymetry can cause unexpected facies to occur in the usually monotonous neritic, wackestone–mudstone terrain. If bathymetric highs or lows are present on the otherwise monotonous flats, then facies variations will be present in and around the highs and lows that are not the same as the ordinary mudstones or wackestones. Grain-dominated facies or even reefs may be present on the highs and be surrounded by the more monotonous neritic flats and thick sections of mudstones may fill in low areas. In fact, bathymetric highs may have been so shallow that they were exposed as islands or they may have been shallow shoals surrounded or blanketed by grain-rich facies that have fundamental rock properties nearly identical to shelf-edge or barrier island grainstone belts. Without knowing that a grainstone or reef was deposited on an isolated high within a broader neritic zone, it would be difficult to tell the difference from a rock sample alone between it and an identical succession deposited on a shelf edge or in the subtidal segment of a strandplain succession. In such cases, seismic data or extensive subsurface geological data are invaluable aids to help differentiate between local anomalies and regional trends in facies character.

Studies of modern and ancient platforms generally confirm that standard depositional successions can be associated with specific depositional environments on ramps, rimmed shelves, and open shelves. For example, shoreline grainstone successions are typical of ramps but they are absent on rimmed shelves because rimmed shelf shorelines lack breaking waves and strong currents required to winnow and concentrate grainstones. Stronger hydrological conditions occur at the shelf-slope break on rimmed shelves, well away from the shoreline. Grainstones may occur along the shorelines of some open shelves depending on the power of the surf and on the supply of grains to be deposited. Tidal flats and lagoons are common on rimmed shelves and where they occur behind detached barrier islands on ramps. Tidal flats are mud-dominated systems that require sheltered or restricted circulation that occurs behind barrier islands and on nearshore zones far inboard from the agitated environment of the slope break.

The shallow subtidal environment is common to all depositional platforms, as is the basinal environment. Oceanography textbooks define the neritic environment as the zone that extends from below mean low tide to a depth of 200 m, but that depth is below the principal zone of carbonate production virtually everywhere. In fact, 200 m is about the maximum depth of the Persian Gulf, where it represents the basinal environment. For our purposes the shallow subtidal or neritic environment is the zone inhabited by an abundant and relatively diverse benthic fossil population that contains mainly photozoans (perhaps including reef organisms), and that is bounded on the landward side by tidal flats or beaches and on the seaward side abruptly by slope breaks (shelves) or gradationally by outer-ramp environments (ramps).

The shelf-slope break is present only on open and rimmed shelves—by definition. Distally steepened ramps may have changes in slope, but those changes are not accompanied by facies changes. That is, the sea may deepen rather abruptly but the fundamental properties of the sediments landward and seaward of the slope change are identical. In general, this requires that the distal steepening has to occur at depths great enough so that there are no differences in the hydrological regime or the sediment supply system on either side of the slope change. In other words, there is no difference in wave or current activity or in sediment sources on either side of the slope change. There must be a difference in the hydrological regime coincident with the slope break on shelves and there may be differences in sediment sources or mode of supply on the inboard and outboard sides of the shelf-slope break. These differences in the hydrologic environment and sediment input produce different facies inboard and outboard of the slope break. Likewise, the slope environment is characteristic of shelves only, not of ramps. Slopes extend from the slope break to the slope toe at the inboard margin of the basin. No slope angle is included in the definition; it may vary from steep to gentle, depending on substrate stability, the hydrological environment, and tectonic stability. Sloping surfaces are not places where gravity lets sediment accumulate; therefore one can think of slopes either as the source of sediment supplied to the basin floor, as the "slide" down which the sediment was transported, or both. The site of ultimate deposition is the base, or toe, of the slope. Deposits on the toe of the slope, still included with slope deposits in this text, consist of the mass wasted material from the slope break and the slope such as slumps, rock slides, debrites, grainflows, proximal turbidites, and rock falls. Essentially everything is out of its original place, it is probably out of its original up–down orientation (if it had one), and it is out of its original, shallower-water environmental setting, probably interbedded with or encased by pelagic mud.

There is no specific depth range for basinal environments because they occur in such a wide range of depths. Oceanographers define the bathyal environment as the zone that extends from 200 to 1000 m in the modern oceans. They define the environment deeper than 1000 m as the abyssal zone. Rarely, if ever, do carbonate sediments accumulate—especially in reservoir facies—in bathyal or abyssal depths at least partly because carbonates dissolve below the CCD, which is in the abyssal depth range. As for the range of depths that can be considered basinal, the deepest parts (basinal depths) of the modern Persian Gulf are only 200 m, but basinal depths off the Great Bahama Bank are over 1 km. We can define the basinal environment therefore as the environment that exists at the greatest depth where carbonate sediments accumulate normally and on a regular basis within any particular basin. Environments in which carbonates do not accumulate but where siliciclastic oozes, clays, and other noncarbonate sediments accumulate are not "carbonate depositional environments." Basinal environments of all types are not affected by surface waves, tides, or shallow-water currents. They are characterized by the absence of light, by dysoxia to anoxia, and by low taxonomic diversity. True, deep-water deposits consist mainly of a fine rain of pelagic detritus including pelagic skeletal detritus, clays, extraterrestrial "dust," and organic matter. Organic matter may be well preserved in deep-water settings because they are usually characterized by reducing and aphotic conditions where water motion is limited in most cases, except for density currents including contour and turbidity currents. Basinal successions include laminites (millimeter-scale microlaminated beds), distal turbidites, density flow deposits, and rhythmites (cyclically repeated coarse-to-fine beds of millimeter to centimeter scale).

4.5   SEQUENCE STRATIGRAPHY

Sequence stratigraphy puts depositional rock bodies in a chronostratigraphic frame-work. It is a way to identify sedimentary bodies by their position in the stacks of bedded rocks deposited in fluctuating environments between one major stratigraphic boundary, such as a major erosional unconformity (Type 1 Sequence Boundary), and the next.

4.5.1   Definitions and Scales of Observation

We have described standard microfacies as the sediments that blanket the surfaces of environmental sectors across platforms. Ideal successions consist of vertically stacked arrays of microfacies that characterize each environmental zone on ramps and shelves. Both standard microfacies and depositional successions represent deposits in limited geographical areas, or sectors, on platforms. These model succes-sions are sketched to represent idealized versions of cored intervals or outcrop sections. Interpreting the depositional history of large areas requires correlation of many such depositional successions over large distances. It also requires knowledge of the relationship between rock units and time units. In short, there is a hierarchy of scale in time and space in stratigraphy. There are methods to describe this hier-archy, to define boundaries between the units within it, and to define the genetic significance and geological history of the entire package. Currently, the method of choice is sequence stratigraphy.

Modern sequence stratigraphy evolved largely from petroleum industry research on the geometry of depositional bodies interpreted from seismic records, or *seismic stratigraphy*. The historical roots and fundamental concepts of seismic and sequence stratigraphy can be found in Vail et al. (1977a,b) and Brown and Fisher (1977). Sequence stratigraphy can be defined as "*the subdivision of basin fills into genetic packages bounded by unconformities and their correlative conformities*" (Emery and Meyers, 1996). In ordinary language, that means that sequence stratigraphy is a method of correlating depositional bodies that share a common origin and that are bounded by surfaces that have some time value. In today's literature there are several variations on that definition, but this one was chosen because it is concise and relatively easy to understand. Sequence stratigraphy differs fundamentally from lithostratigraphy because it is based on time–rock units instead of rock units. An illustration of this fundamental difference is shown in Figure 4.4, where facies with identical lithological characteristics are deposited as time-transgressive lithofacies. The consequences of these differences for the reservoir geologist or engineer are illustrated in Figure 4.9, which shows the difference between lithostratigraphic and sequence-stratigraphic correlation of basinward-prograding grainstone successions. Each grainstone body and its lateral equivalents represent deposition during a series of prograding, high-frequency depositional sequences, or HFS. The reservoir is compartmentalized along boundaries within the chronostratigraphic HFS boundar-ies. When sequence-stratigraphic interpretations are employed, the compartmental-ization is recognized. On the other hand, if chronostratigraphic relationships are ignored and correlations are based only on lithostratigraphic similarity, as would commonly be done with wireline log traces, the different grainstone units would be correlated as if they were three laterally continuous reservoir zones. Such a layer-

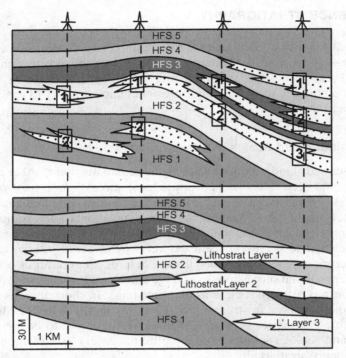

Figure 4.9   A sketch showing the consequences of erroneously correlating by rock type only (lithostratigraphic correlation), neglecting or not recognizing chronostratigraphic correlations. Four wells are shown penetrating high-frequency sequences (HFS) of grainstones (dot pattern) and mudstones (darker tones without the dot pattern). The chronostratigraphic (sequence-stratigraphic) correlations shown on the top part of the figure correctly identify the HFS chronostratigraphic units. Lithostratigraphic correlations identify only the grainstones and because they have similar lithology in each of the four wells, they can be erroneously correlated as continuous strata in a "layer-cake" form. Grainstones identified with the number 1 would be correlated as a single layer-cake unit, as would those marked by numbers 2, 3, and 4. The resulting layer-cake lithostratigraphic correlation is shown on the lower half of the figure. The layer-cake correlations will lead to confusion when it is discovered that the grainstones are not in flow communication. (Adapted from an illustration in Tinker (1996).)

cake correlation would lead to the erroneous expectation that each of the grainstones would be in flow communication across HFS boundaries. The consequences of such errors in correlation have obvious and significant impact on determination of well spacing, volumetric calculations, field development, and enhanced recovery operations.

Posamentier and James (1993) point out that sequence stratigraphy is a tool for the study of geological data rather than an end in itself. They note that a variety of stratigraphic methods exist. Some of the methods include *allostratigraphy* (subdivides the stratigraphic succession on disconformities, or breaks in the rock record without regard to their origin or time value), *genetic stratigraphy* (subdivides stratal successions on maximum flooding surfaces and condensed intervals), *sequence stratigraphy* (divides successions at unconformities and their correlative conformities—a system in which time value of the sequence is fundamental), and *parasequence stratigraphy* (recognizes shallowing-upward successions characterized by flooding surfaces at their upper boundaries). These methods have in common that (1) they recognize cyclicity in the rock record, (2) they depend on establishing a time-stratigraphic framework, and (3) they focus on the natural boundaries that delineate stratigraphic successions. These boundaries are unconformities, disconformities, conformities, and flooding surfaces.

Cyclical sedimentation results when relative sea-level changes systematically from high to low to high, and so on. Relative sea-level change can be caused by changes in global ice volume (glacioeustatic change) and by tectonic processes of regional or local scale. Some points for argument among geologists are the causes of relative sea-level change, the extent to which sea-level change is global (eustatic) or regional (tectonic), and what mechanisms govern the periodicity of sea-level change (e.g., changes in basin volume, changes in global ice volume, and forces related to variations in the Earth's orbit). The periodicity of sea-level change determines the *order* or time rank of the cycle. First-order cycles span 200–300 Ma and are caused by major plate tectonic movements that may open basins or break up continents. Depositional onlap and offlap at cratonic scale are produced by these cycles. Second-order cycles have durations of 10–50 Ma and are related to changes in ocean basin volumes by tectonism, changes in global ice volume, or both. Second-order depositional sequences may be hundreds to thousands of meters thick. Third-order cycles are thought to be driven by changes in ice volume; they represent relative sea-level changes on the order of 50 m or less. High-frequency, climatically driven cycles of 20–400 ky are thought to be caused by periodic fluctuations in the Earth's orbital characteristics referred to as Milankovich cycles. The amplitude of relative sea-level change in these cycles may range from 100 m to 10 m depending on whether the sea-level change happened during "icehouse" (extensive continental glaciation) or "greenhouse" (limited continental glaciation) times, respectively.

The natural boundaries that delineate sequence-stratigraphic successions are breaks in the rock record produced by erosion or nondeposition. Unconformities are surfaces of erosion or nondeposition that represent gaps in time and that usually have discordant relationships with bedding above and below. Disconformities also represent breaks in continuity of deposition, but disconformities are surfaces that are parallel with beds above and below. Conformities, following the definition in Van Wagoner et al. (1988), are surfaces that separate younger beds above from older ones below without evidence of erosion or nondeposition and along which no significant hiatus is indicated. They may also define marine flooding surfaces as surfaces that separate older from younger strata and across which there is evidence of abrupt deepening. Maximum flooding surfaces and condensed intervals separate the transgressive phase from the highstand phase of a stratigraphic sequence.

We have already discussed the time ranking of cycles in sea-level change. The spatial scales of observation in seismic and sequence stratigraphy range in area from entire platforms to meter-scale flow units within fields. First-order sequences may be thousands of meters thick and occupy entire basin margins. Each systems tract, the lithogenetic association that formed on a platform during lowstand, transgressive, or highstand relative sea level, may incorporate many facies or ideal depositional successions such as those described earlier in this chapter. Flow units, on the other hand, may be less than 1 m thick, may extend only over a few acres, and may occupy all or only part of one ideal depositional succession. Flow units may occupy all or part of parasequences that, in turn, could occur as field-scale (or smaller), shallowing-upward cycles. On platforms with complex bathymetry, shallowing-upward cycles may not cover large areas. Grainstone buildups on small antecedent highs are a good example. On the other hand, monotonously flat platforms may have shallowing-upward successions that extend over many square kilometers. An example of such a large, relatively flat platform existed during Permian times from northern Texas through Kansas. During this time, the Permian Chase Group that contains the reservoir for the Guymon–Hugoton gas field was deposited (Siemers and Ahr, 1990; Olson et al., 1997; Mazzullo, 1994) Parasequences can sometimes be grouped into *parasequence sets*, where vertical trends in parasequence thicknesses and lithology display *stacking patterns*, in the manner discussed by Van Wagoner et al. (1990). These diagnostic trends in thickness and lithology are used to interpret the history of relative sea-level change and the relationship between parasequences and sequences at field scale or larger.

4.5.2 Sequence Stratigraphy in Carbonate Reservoirs

Carbonate sequence stratigraphy differs from siliciclastic sequence stratigraphy mainly because carbonates are intrabasinal in origin and siliciclastics are extrabasinal in origin. Early work on seismic and sequence stratigraphy focused on terrigenous clastic depositional systems in which the sediment source is extrabasinal and continental sedimentary processes such as fluvio-deltaic sedimentation and coastal progradation play dominant roles in the formation of stratigraphic architecture. First attempts to force the models for siliciclastic sequences on carbonates resulted in confusion and misinterpretations. Carbonates are intrabasinal in origin and are largely independent of continental influence. Instead, carbonate sequences consist mainly of in situ marine biogenic and chemical sediments created by what is informally called the "carbonate factory." Carbonate sedimentation reacts differently to changes in relative sea level than do siliciclastics, sites of maximum carbonate production are different on ramps and rimmed shelves, and carbonate production occurs at different depths and in different volumes on temperate and tropical platforms. Ramps and shelves exhibit different sequence architecture. A major step forward in carbonate sequence stratigraphy was made by Sarg (1988), who recognized the essential differences in sedimentation and sequence architecture of siliciclastics and carbonates. Later, Handford and Loucks (1993) illustrated the major differences between carbonate sequences on ramps, shelves, and detached (isolated) platforms. Their sequence-stratigraphic model for a rimmed shelf is illustrated in Figure 4.10.

4.5.3 Sequence Stratigraphy in Exploration and Development

Carbonate reservoirs are porous and permeable bodies of rock that contain commercial quantities of hydrocarbons. Pore systems may be depositional, diagenetic, fractured, or combinations of all three. Recognition and mapping of reservoir boundaries and internal flow units and flow barriers requires examination of cores

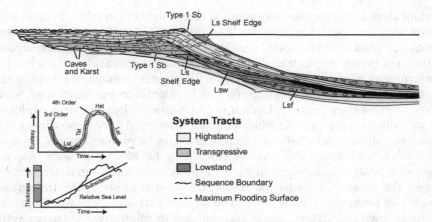

Figure 4.10   An illustration of the sequence-stratigraphic evolution of a tropical rimmed shelf in a humid climate as illustrated by Handford and Loucks (1993). As relative sea level and subsidence progress, depocenters migrate seaward or landward, depending on the direction of relative sea-level change. Sequence-stratigraphic terminology for depositional successions formed at low sea level stands is LST, or lowstand systems tracts, those at highstands of sea level are HST, and those deposited during relative sea-level rise are TST, or transgressive systems tracts. LS indicates lowstand. LSW and LSF represent lowstand wedge and lowstand fan, respectively. (Adapted from an illustration in Handford and Loucks (1993).)

or cuttings, borehole logs, and geophysical data. It also requires an ability to recognize facies patterns, to synthesize depositional models, and to formulate geological concepts to incorporate the information into a working plan for exploration and development. Small reservoirs may become only one-well fields. Others, such as the giant Ghawar Field in Saudi Arabia, may extend over thousands of square kilometers. In the end, it is the individual flow units or the intervals that exhibit optimum combinations of porosity and permeability with the highest recovery efficiency that have the greatest influence on the economic value of a field. Flow units are rarely larger than parasequence scale and may, in fact, be contained within a single parasequence. An entire reservoir may not incorporate much more than a few stacked parasequences or parasequence sets. In view of this range in the size of productive segments within carbonate sequences, it is important to keep in mind how sequence-stratigraphic analyses can be used to find and develop reservoirs. One of the most important bits of information needed to determine the size and shape of a depositional body is the shape (bathymetry) of the antecedent platform. If it was monotonous, flat, and extensive, then depositional successions will probably be rather thin, extensive, and relatively monotonous. If bathymetry varied greatly over small areas, depositional successions, parasequences, and flow units will vary in size and complexity accordingly.

For the explorationist, the fundamental problem is locating potential reservoir facies within sedimentary basins. Reservoirs with the greatest amount of depositional porosity normally are present in rocks with high grain/mud ratios and in some reefs and mounds. The most extensive grainstone deposits are usually present near the shoreline on ramps and at the slope break on rimmed shelves; therefore the explorationist can focus on those zones in ramp and shelf sequences. Some important steps in exploration for potential reservoir rocks involve (1) determining platform geometry, whether ramp, shelf, or other; (2) identifying the systems tracts in which potential reservoir facies should exist; (3) tracing the spatial arrangement (locations) of those systems tracts through cycles of change in relative sea level; and (4) determining where in the systems tracts the greatest volume of potential reservoir rock exists. Platform architecture can be identified by seismology, surface geology, or subsurface geology. Until it is determined to be otherwise, exploration wells are drilled on prospects in which porosity is assumed to be depositional; therefore beaches, dunes, and shelf-edge sand bodies may be primary targets. Lowstand, transgressive, and highstand systems tracts can be identified using sequence-stratigraphic analyses, and detailed study of the systems tracts can determine if, for example, the greatest depositional reservoir potential exists in the transgressive, highstand, or lowstand systems tracts. The sequence stratigrapher will determine if deposits in lowstand systems tracts accumulated in relatively low-energy settings causing them to have high mud content; or if the potential reservoir facies in the high-energy sectors of the transgressive systems tract are less voluminous than those in the highstand systems tract owing to the short time of sedimentation in any given space during the transgression. Armed with these analyses and interpretations, the geological concept for exploration can be completed.

Chronostratigraphic correlations of individual HFS in field development studies, for example, will identify the location and spatial arrangement of discrete depositional bodies as previously illustrated in Figure 4.9. Subsurface seismic and geological structure maps can be compared with the spatial distribution of the individual reservoir bodies to determine the volume of reservoir facies that is enclosed in a structural trap, for example. If the trap is not structural, other kinds of studies are done, but for purposes of this example, only structural traps are considered. When trap structure and reservoir architecture are examined simultaneously, optimized field development can proceed. As more information is acquired, refinements can be made to identify and rank individual flow units within the field.

SUGGESTIONS FOR FURTHER READING

Additional information, illustrations, and discussions on generalized sequence-stratigraphic models for the different kinds of carbonate platforms can be found in *Sequence Stratigraphy* by Emery and Meyers (1996) and *Carbonate Sequence Stratigraphy* by Loucks and Sarg (1993). More general discussions of depositional environments, facies, and stratigraphy can be found in *Sedimentary Environments: Processes, Facies, and Stratigraphy,* 3rd edition, by Reading (1996) and *Principles of Sedimentology and Stratigraphy*, 3rd edition, by Boggs (2001). Carbonate platforms evolve with time. Some of these changes are discussed in Schlager's, 1981 paper on the paradox of drowned reefs and carbonate platforms in Schlager and Ginsburg's 1981 discussion of the Bahama carbonate platforms, and in Manfrino and Ginsburg's (2001) description of the Plio-Pleistocene depositional history of the upper Bahama Bank. Precisely documented links between sea-level change, carbonate sequence stratigraphy, and patterns of diagenesis are described by Pomar (1993) and in Pomar and Ward's (1999) review of reservoir-scale heterogeneity in Miocene depositional and diagenetic facies on the Island of Mallorca.

REVIEW QUESTIONS

4.1. What is a "carbonate platform"?

4.2. Name at least three varieties of platforms and explain how they are, or are not, different.

4.3. What typically makes up the "rim" on rimmed shelves?

4.4. What is meant by "standard microfacies"?

4.5. Does the "carbonate factory" operate to the same depths in tropical and temperate seas? Explain.

4.6. What is the difference between lithostratigraphy and sequence stratigraphy?

4.7. Do "formations" have time significance?

4.8. What is considered to be the "basic mappable unit" in lithostratigraphy?

4.9. What is the difference between "absolute" and "relative" geological time?

4.10. Does a seismic reflection (a line on a seismic profile) represent a time surface? Explain.

4.11. How would you describe the depositional anatomy of a barrier island?

4.12. According to the usage in this book, what is a "standard depositional succession"?

4.13. Name the seven standard depositional successions that characterize all carbonate platforms.

4.14. Do all seven standard depositional successions occur on both ramps and shelves? Explain.

4.15. Give a concise definition of "sequence stratigraphy."

4.16. What are examples of "natural boundaries that delineate stratigraphic successions"?

4.17. What are some causes of cyclicity in the stratigraphic record and what are some examples of cycle duration (length of time occupied by cycles)?

4.18. There are significant differences between terrigenous sandstone and carbonate sequence-stratigraphic relationships. Why?

CHAPTER FIVE

DEPOSITIONAL CARBONATE RESERVOIRS

This chapter focuses on depositional porosity, depositional environments, and depositional rock properties. Different depositional environments that exist across ramps and shelves produce distinctive depositional successions. The fundamental rock properties that characterize each of these depositional successions can be synthesized to create seven *ideal depositional successions*. For ease in visualization, each ideal depositional succession is described and illustrated to resemble borehole cores. Most successions represent shallowing-upward deposition and most of them can be divided into several parts (the microfacies defined earlier) to enhance their individual character and make it easier to identify them in the field or in borehole cores. Review questions on volumetric calculations, on exploration strategies, and on the best methods for field development in different facies are included at the end of the chapter and selected case histories of depositional reservoirs are presented in Chapter 8.

Depositional reservoirs are those in which reservoir porosity is the result of depositional processes only. Older literature might classify this kind of porosity as "primary." Recognizing depositional reservoirs depends on the ability to distinguish depositional porosity from diagenetic and fracture porosity, along with the various hybrids of end-member pore types that are usually present. Purely depositional porosity is rare but depositional texture, fabric, or mineralogy usually have so much influence on diagenesis that altered reservoirs can be exploited as if they were governed by depositional porosity only. For example, it is common to find reservoirs with fabric-selective, texture-selective, or facies-selective porosity and permeability even though the rocks have been altered by diagenesis. These are *hybrid reservoirs* because the pore characteristics reflect both depositional and diagenetic influences. In order to confirm that hybrid reservoirs have both depositional and diagenetic

Geology of Carbonate Reservoirs: The Identification, Description, and Characterization of Hydrocarbon Reservoirs in Carbonate Rocks
By Wayne M. Ahr   Copyright © 2008 John Wiley & Sons, Inc.

rock properties, one has to examine rock samples—borehole cores or cuttings—first-hand. There is no indirect method of measurement or logging yet known that can distinguish between depositional and diagenetic porosity. Once the key depositional attributes are identified from rock descriptions, they should be incorporated into subsurface structural and stratigraphic cross sections and maps that include facies descriptions so that the spatial distribution of reservoir attributes can be predicted. Ordinary wireline logs do not measure rock properties that discriminate between pore types in carbonate reservoirs. Neutron, density, and acoustic logs can be used to calculate total porosity but those logs cannot discriminate between carbonate pore types or pore origins. Conventional wireline logs do not distinguish between depositional facies in carbonates either, because logs cannot detect differences in carbonate grain types, depositional textures, fossil content and diversity, or most sedimentary structures. It therefore follows that one *cannot* make electrofacies maps of depositional or diagenetic facies in carbonate reservoirs. As mentioned earlier, when enough well control is available in an established field, log signatures of depositional carbonate rock and pore characteristics can be identified with some success, especially if reservoir porosity is facies selective and relatively free of complications or partitioning by diagenesis. In such cases, so-called electrofacies can be identified and mapped with reasonable success. The "rock typing" methods of Lucia (1995), Gunter et al. (1997), and Martin et al. (1997) for distinguishing between flow units, baffles, and barriers in carbonate reservoirs have also been comparatively successful but the method does not distinguish between depositional, hybrid, and diagenetic pore types—the key information needed to predict the distribution of reservoir zones in stratigraphic space. Large-scale sedimentary structures such as eolian crossbeds and talus beds that dip off the flanks of carbonate buildups can be detected with dipmeter and with imaging logs. Imaging logs can also detect large vuggy pores in carbonate reservoirs. The borehole log that has great undeveloped potential to discriminate between carbonate pore types is the NMR log. Much research is being done on this subject as this book is written and there is strong evidence that the NMR log may become a powerful application for identifying pore types in carbonate reservoirs both by origin and by pore geometry (Genty et al., 2007).

Depositional facies maps made from sample or core examination can be used as proxies for maps of effective porosity if reservoir porosity is depositional in origin. Seismic data alone is not useful to discriminate between depositional and diagenetic reservoirs, but seismologists can identify structural features that may have influenced depositional or diagenetic patterns and in a few cases where impedance contrasts are sufficient, they can extract seismic attributes that can distinguish between reservoir and nonreservoir rocks. Borehole logs and other petrophysical measurements such as capillary pressure and NMR measurements are very important for identifying flow units, baffles, and barriers within reservoirs when used in conjunction with sample or core descriptions. Such integrated data is especially useful for establishing a quality ranking system for reservoir flow units. Using the different types of data, one can determine a relative quality or rank of the flow units based on their combined porosity and permeability values, their capillary pressure characteristics, including median pore throat diameters, NMR values for pore geometry, and their dominant genetic pore types identified by direct observation. The genetic classification of porosity links pore types to geological events during depositional and burial history; therefore it can also be useful in tying flow units, baffles, and barriers to the larger stratigraphic architecture, enabling one to generate predictive maps and sections of ranked flow units at field scale.

5.1   DEPOSITIONAL POROSITY

Depositional processes are those by which sedimentary constituents accumulate to form rocks. Depositional sedimentary constituents include detrital grains, crystalline precipitates (as precipitated micrite), and biogenic material such as skeletal components and microbialites. Some cements are so intimately associated with microbialites that for practical reasons they can be included as depositional components. Carbonate rocks are detrital, biogenic, or chemical in origin. Detrital carbonates exhibit four basic pore types (see Figure 2.13): (1) intergranular, (2) intragranular, (3) shelter or keystone, and (4) fenestral pores. Intergranular pores, those between grains in a detrital rock, may be both fabric and facies selective. Porosity is highest in rocks with the least mud; therefore the Dunham rock classification is a proxy for this pore type. Petrophysical characteristics of intergranular pores can be estimated using Lucia's (1983) classification for interparticle porosity and his x-y plot of porosity, permeability, and particle size. If the relationship between petrophysical properties and Dunham rock classification can be established, and if it is consistent within facies, then the Dunham classification alone can be used to describe both petrophysical and depositional characteristics of each flow unit in the reservoir.

Intragranular pores, those within grains, may exist naturally in skeletal grains or in diagenetically altered grains of any origin. Usually they occur in porous, skeletal allochems, which make up varying amounts of detrital sediments. Bryozoans, for example, have internal pores, as do corals, sponges, stromatoporoids, and many other reef-building organisms. Some mollusks such as rudistid clams have large internal cavities that may contribute greatly to total porosity, but little to permeability. In a bioclastic grainstone reservoir consisting of sand-sized bryozoan fragments, the bryozoans may contribute significantly to total porosity (Ahr and Walters, 1985). In that case, intragranular porosity corresponds with the distribution of skeletal grainstones and packstones and, in turn, with facies maps.

Shelter and keystone pores in detrital rocks are not common enough to be significant contributors to total reservoir porosity. Shelter pores are formed when large grains such as bivalve segments act as shelters or "umbrellas" and prevent detrital grains from filling pore space beneath the shelter of the large grain. Keystone pores are formed when the pounding of breaking waves expels air from beach sands. As sand grains are repacked under this hydraulic pounding, the odd grain may fall in a position similar to the keystone in an arch and prevent the sand packing arrangement from reaching maximum density.

Fenestral or "bird's-eye" pores result from desiccation or from expulsion of gas during decay of organic matter in muddy sediment. The pores may be millimeter to centimeter in size, they are elongate and planar in a direction parallel to bedding, and they are especially common in tidal-flat environments where sediments are alternately wet and dry. In such cases, fenestral pores become facies selective and the facies map is interchangeable with the reservoir porosity map. Fenestral pores may not be in flow communication with each other; consequently, they may have the petrophysical properties of small, separate vugs. If they are touching and are not filled with cement, they may present high permeability values at comparatively low porosity. A useful procedure is to plot permeability as a function of porosity ($\log k$ vs ϕ) when the data are available. This simple plot provides a great deal of insight into the relationship between ϕ and k, especially if it is used together with accurate rock descriptions.

Depositional porosity in reef organisms consists of intraparticle or intraskeletal pores. Interparticle porosity exists mainly between skeletal constituents and between detrital grains derived from the breakdown of the reef structure by hydrologic action or bioerosion. Reef detritus may be gravel or sand sized and comprises the loose fill between skeletal framework elements in reefs, around the perimeter of the reef mass, or between reef masses. Depositional porosity is commonly higher in the detrital fill than within the skeletal framework. Reef rock classifications, as we have discussed, focus on the different textures and fabrics of reefs and carbonate mounds. These schemes can be helpful to identify pore types and to map the spatial distribution of specific pore categories, but they may not be useful in predicting reservoir performance. Reef rock petrophysical properties are particularly complex because the rock fabrics represent skeletal anatomy, oriented growth fabrics, and detrital textures. Each rock type in a specific reef will have a predictable range of pore characteristics, however. Because of the variability in reef rock texture and fabric, reef reservoirs must be evaluated case-by-case. There is no single classification that is universally applicable to predict reef reservoir characteristics.

5.2 DEPOSITIONAL ENVIRONMENTS AND PROCESSES

Most carbonate reservoir rocks are marine in origin; consequently, identification of individual depositional environments is a matter of dividing the marine environment into smaller sectors that have enough distinctive attributes to stand alone as discrete subenvironments. Coastal dunes are included because they are usually part of the beach and nearshore marine environment. Lacustrine carbonates are common in the geological record, but they are not usually hydrocarbon reservoirs although nonmarine carbonates may contain large quantities of hydrocarbons as "oil shales." According to North (1985), nearly half of the world's exploitable oil-shale resources are in lacustrine carbonates of the Eocene Green River Formation. Green River oil shales cover about $42,000\,km^2$ in Colorado, Utah, and Wyoming. Rather than shales, the Green River rocks are actually thin-bedded, bituminous limestones but some zones in the Green River Formation are known for their fossil fish remains rather than their bitumen content. Nonskeletal carbonate grains such as pisoids and oncoids are present in Green River rocks, and some shoreline deposits around the Great Salt Lake in Utah consist of oolite grainstones. For the most part, however, carbonate reservoir rocks formed in the marine environment. Nonmarine reservoirs are also known in mainland China, but there is not much published information on them in English.

Oceanographers have already used arbitrary water depth to divide the marine environment into ecological zones such as the neritic, bathyal, and abyssal zones. But these depth zones cannot be identified in the rock record. Identification of ancient depositional environments requires the ability to recognize them by their distinctive, fundamental rock properties. Oceanographic research has shown that carbonate platforms are either ramps or shelves and that each platform type has a characteristic array of lithofacies defined on fundamental rock properties that characterize each depositional environment from the shoreline to the depths of the basin. The distinguishing environmental processes for each environmental zone are assumed to vary only within narrow limits, although the boundaries between zones are gradational at best. Some examples of the environmental zones and accompanying lithofacies are described in oceanographic studies of modern carbonate platforms, where workers have identified both the prevailing environmental conditions and their accompanying sedimentary facies; they include the South Florida shelf (Ginsburg, 1956), the Bahama Banks around Andros Island (Purdy, 1963a,b), the Yucatán Platform (Logan et al., 1969), Shark Bay, Western Australia (Logan et al., 1970), and the Balearic Platform (Fornos and Ahr, 1997, 2006).

Ancient depositional environments are reconstructed by studying the texture, constituent composition, sedimentary structures, and fossil assemblages that occur in each subenvironment on ancient carbonate platforms and comparing the interpreted successions with modern examples. Rock properties produced by sedimentary processes in each depositional setting over time comprise unique depositional successions that can then be used to identify ancient depositional environments. Depositional successions are 3D bodies of rock illustrated as idealized borehole cores or outcrops that represent vertical stacks of 2D depositional microfacies. Think of these microfacies as time slices of rocks with specific textures, grain types, sedimentary structures, and fossils. The slices are stacked one on top of another to make up the 3D succession that accumulated over time in each specific subenvironment until relative sea level shifted and the subenvironment migrated updip or downdip, became emergent, or was drowned. The 3D successions and their characteristic rock properties determine the amount and type of depositional reservoir porosity, the size and shape of the reservoir body, and, ultimately, the economic value of the reservoir.

Carbonate ramps and shelves can be divided into seven subenvironments or sectors: (1) the attached or detached beach (or barrier island) environment with or without dunes; (2) the tidal-flat environment and its associated lagoons or adjacent subtidal waters; (3) the lagoonal environment behind detached barrier islands (which includes beaches and dunes); (4) the shallow subtidal or neritic environment; (5) the slope-break environment; (6) the slope, or toe of slope, environment; and (7) the basinal environment. The location of each environment with respect to the others—their juxtaposition on the platform—is predictable for ramps and shelves; therefore when the location of any one environment is known and the platform type is known, the locations of the corresponding laterally equivalent successions are predictable. The environmental characteristics of each of these zones are discussed in the following paragraphs. Those discussions are followed—zone-by-zone—with descriptions of the fundamental rock properties that correspond to each environment. Finally, each ideal depositional succession on ramps and shelves is illustrated to resemble what one would see in borehole cores or outcrops.

5.2.1   The Beach–Dune Environment

Beaches mark the boundary between the mainland and the sea or they may be separated from the mainland by lagoons. Those that extend from the mainland directly into the sea are called *attached* because they are extensions of the mainland. Those separated from the mainland by lagoons are *detached* beach–dune complexes, or simply barrier islands. Dunes are usually present immediately updip from beaches in carbonate environments but they may be absent if the sediment supply is limited, if the eolianites migrate inland, or if the winds are not strong enough on a regular basis to move sand off the beaches. Some reservoirs in ancient beaches and dunes have probably gone unrecognized because of a lack of knowledge about their depositional characteristics. A variety of papers describing carbonate eolianites ranging in age from Paleozoic to Holocene can be found in Abegg et al. (2001a). Kerans and Loucks (2002) describe the attached beach in the Lower Cretaceous Cow Creek Formation of Central Texas and present useful interpretive illustrations of beach anatomy and sedimentary structures.

Processes that create beaches and dunes have been studied in modern environments where cause–effect relationships between processes and products—fundamental rock properties—are reasonably easy to identify. The studies have shown that the beach environment has three distinctive subdivisions (from bottom to top): the lower, middle, and upper shoreface divisions. The environmental characteristics of these divisions depend on the hydrodynamic regime and the prevailing climate at that geographic location. Open ocean waves are the main engine that drives shoreline sedimentary processes. Waves are formed by momentum transfer from wind to water as wind shears across the water's surface. Winds, of course, are part of the global atmospheric circulation system, but they are also influenced by regional climatic events such as tropical storms associated with low-pressure circulation and the "northers" that accompany atmospheric high-pressure fronts. The amount of momentum transfer from wind to water and the resulting wave climate depends on the distance (fetch) that the wind shears over the water's surface, the length of time the wind blows (duration), and the velocity of the wind. A storm wind that blows over a long fetch for a long time at high speed will generate very large waves that can rise to great heights in the breaker zone. In deep water, ocean waves (known as swells in the open sea) take a form in cross section that is similar to a sine curve. These swells, called Airy waves after the mathematician George Airy, have little effect on deep sea sediments, but as they approach shallow water, they undergo a shoaling transformation that changes oscillatory motion in the water column beneath each wave to surging solitary waves, to breaking waves, and finally to longshore currents. Over-steepened solitary waves break in the surf zone and run up the sloping beach face. Water from breaking waves that does not infiltrate the beach sands returns downslope as backwash. Water returning to the sea after passing through the breaker zone is shunted laterally in a stepwise fashion forming a longshore drift or longshore current that runs parallel to the beach face. Sometimes this drift accelerates to the extent that rip currents are produced to move water perpendicular to the beach face.

Beach environments extend to water depths where fair-weather wave motion just moves sediments by oscillatory motion generated by swells as they just begin to feel the bottom and interact with loose sediment. The depth at which this oscillatory motion reaches the bottom with enough velocity to move sand-sized sediment depends on grain size, wave climate, and beach slope, among others. In the NW Gulf of Mexico, where the fair-weather wave period averages about 4 seconds and waves average about 4 feet in height, fine terrigenous sand is moved by oscillatory motion from surface waves to average depths of about 10 m. This lowest division of the beach succession is called the lower shoreface zone and it is dominated by oscillatory motion from Airy waves passing overhead and from burrowing by bottom dwellers. The middle shoreface environment is dominated by unidirectional currents rather than oscillatory wave action. The currents are formed indirectly by waves and are known as longshore currents. They create bar–trough topography in water depths that vary from about 1 to 4 or 5 m, depending on the power of the wave climate and its resulting longshore currents. In the NW Gulf example the middle shoreface environment extends from knee-deep to about 2- or 3-m depths. Finally, the upper shoreface environment is the zone of breaking waves where the inclined beds of the beach face are formed. The physical processes that dominate the upper shoreface zone are breaking waves with run-up and backwash that move up, across, and down the sloping beach face. Storms and accompanying high tides and large waves can create a storm berm landward of the upper shoreface surface and in front of the adjacent dunes, if dunes are present.

Carbonate dunes are usually confined to coastal zones adjacent to beaches because beaches are their source of sediment. However, dunes can migrate tens of kilometers inland when deflation exhumes or rips up carbonate grains and supplies them to the shifting dunes (Abegg et al., 2001). In contrast, siliciclastic desert eolianites can occur in the middle of continents where they are not dependent on a coastal sand source. Even though deflation can produce particles for dune growth, the land surface must be amenable to erosion. It is unlikely that indurated mudstones and wackestones would yield much carbonate sand to winds blowing over their surfaces. Availability of sediment is a primary limiting factor for dune construction because carbonates are produced in always-wet environments. As we have already mentioned, carbonate particles are not easily transported over long distances because they are soft and there is no aqueous carbonate sand factory in deserts. This usually limits carbonate eolianites to coastal settings. Depositional porosity in beaches and dunes is probably greater than that in almost any other kind of carbonate succession, but because early cementation is limited in beaches and especially in dunes, they seem to be particularly susceptible to early compaction and attendant porosity loss (Abegg et al., 2001).

5.2.2 Depositional Rock Properties in Beach–Dune Successions

Fundamental rock properties include texture, grain types (composition), sedimentary structures, and fossil content (diversity and type). Beach–dune textures depend on grain types in most cases. Carbonate beaches in modern tropical environments are composed primarily of nonskeletal grains such as ooids, pellets, clasts, and peloids, but beaches in modern temperate climates generally consist of skeletal grains because the carbonate factory produces different constituents in temperate and tropical environments, as was first pointed out by Chave (1967), Lees and Buller (1972), and Lees, (1975). Oolitic and peloidal beaches are typically fine to medium sand sized, but bioclastic dunes may consist of fine, medium, or coarse sand-sized particles. Sorting is usually very good in beaches and dunes, particularly in the dune and upper shoreface sectors where breaking waves, swash, and backwash continually abrade, sort, and polish constituent grains that are supplied to beaches and dunes alike. Beach and dune textures are grainstones for practical purposes. The middle shoreface environment represents a transition between breaking waves on the landward shore and the deeper water of the shallow neritic domain; consequently, sedimentary textures are less well sorted and may contain a mud fraction. These rocks may include packstones along with grainstones. The lower shoreface environment is mud-rich and dominated by wackestones to mudstones.

Grain types in beach–dune complexes may consist of nonskeletal sands, skeletal sands, or both. Most tropical beaches, except those around reefs, consist of nonskeletal sands rich in aragonitic components such as ooids and peloids. As we already mentioned, temperate to cool-water beaches usually consist of bioclastic sands without aragonitic constituents.

There are distinctive sedimentary structures that mark the berm, upper, middle, and lower shoreface zones in beaches. The typical beach profile consists of a nearly level berm at the storm high-water line, a seaward-sloping zone called the upper shoreface that may include ridges and runnels parallel to the shoreline, a longshore bar and trough zone called the middle shoreface, and a burrowed, transitional zone between beach and open sea called the lower shoreface, or shoreface toe. Berms result when storm waves erode and redeposit shoreface sands above mean high-water level. They are intimately associated with the upper shoreface environment, where the dominant sedimentary structures are low-angle, inclined, parallel to subparallel beds. Rill marks and ripple crossbeds are common but difficult to identify in borehole cores or limited outcrop exposures. Middle shoreface sedimentary structures mainly reflect the strong unidirectional scour of the longshore current system. Concave-upward, festoon, or trough crossbeds are the typical structure, and they represent migrating large-scale ripples that move with the longshore current. Carbonate sand is transported parallel to the shoreline to produce ridges and swales, or runnels. Middle shoreface topography is marked by distinctive submarine ridges and troughs oriented parallel to the shore. Lower shoreface sedimentary structures are characterized by small-scale (centimeter) oscillation ripples and bioturbation. The ripples are formed by the oscillatory motion induced by the passage of deeper-water swells. The bioturbation in this case is in the form of burrows, mainly feeding traces of shallow neritic marine organisms. Most feeding traces are horizontal to subhorizontal, but because the burrows are commonly large (centimeter size and larger) in this environment, they are relatively easy to identify in slabbed borehole cores.

Sedimentary structures in dunes can be gigantic in scale. Carbonate dunes are reported to range in height from 2 or 3 m to over 30 m and may have total formation thicknesses that range from 6 to over 300 m (McKee and Ward, 1983). A distinguishing characteristic of dunes is their large-scale sedimentary structures, particularly the crossbeds in the lower parts of dunes. McKee (1966) measured individual dunes in the White Sands National Monument, New Mexico and found that they have tabular crossbeds stacked to heights exceeding 10 m and extending over many tens of meters horizontally. According to his findings, the windward portions of dunes are rarely preserved. Large tabular and spillover crossbeds in the lower parts of dunes pass vertically into smaller scale features that include trough crossbeds and current ripple marks. Individual trough crossbed sets in middle and upper dune segments may be 1 m or more thick and their orientation may deviate from the average orientation of the larger tabular crossbeds below, depending on paleo-wind directions. This general tendency to exhibit large, tabular and spillover crossbeds

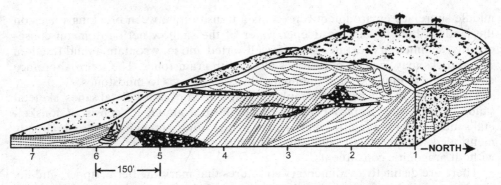

Figure 5.1   Sketch of a Pleistocene dune on New Providence Island, Bahamas showing large-scale spillover and foreset beds in the lower part of the dune passing upward to smaller scale trough crossbeds. Note the change in character from large, convex-upward and inclined bedding at the bottom to smaller scale, concave-upward (troughs) toward the top of the dune. Imagine what one might see in a 15-cm diameter × 10-meter long borehole core through such a large feature and if the entire body of the dune were above the oil–water contact, estimate the OOIP volume in only the one dune. (Adapted from an illustration by Ahlbrandt in McKee and Ward (1983).)

near the base, intermediate to smaller scale, trough crossbeds in midsections, and small scale ripple marks in upper sections is useful in developing model successions for carbonate dunes. Ideal dune sequences are divisible into three components: (1) large-scale tabular and spillover crossbeds at the base, (2) intermediate scale, trough crossbeds in the middle, and (3) small-scale trough crossbeds with asymmetrical ripples at the crest (Figure 5.1).

Although they are not mechanically formed sedimentary structures, diagenetic characteristics can be very useful to distinguish dunes from other crossbedded grainstone successions. Dune deposits are sometimes distinguishable in thin section by the presence of vadose diagenetic features. Chapter 6 deals exclusively with diagenetic reservoirs, but vadose diagenesis is discussed briefly here because this distinctive attribute is a tool to aid in recognizing eolianites. Vadose diagenesis includes the chemical changes that take place in rocks above the water table. Dunes may rise many meters above the fresh or marine water table on the coastline so that the only time water percolates downward through the dune pores is during and after a rain. Usually the pore walls and pore throats in dune rocks are dry. As rain falls and passes through the pores, they become temporarily filled with a mixture of water and air. The surface tension of water causes it to migrate as a film along grain surfaces until the film reaches a point of contact between grains. There, a meniscus forms as the water film bridges from one grain to another. Over time, as the water is saturated with $CaCO_3$, it precipitates cement in the form of the meniscus at grain contacts (Figure 5.2). In some instances, pendulous cements form where droplets of water clung to the lower surface of grains. These meniscus and pendulous cements may be recognizable in ancient dunes and provide another intrinsic property of eolian facies to simplify the task of distinguishing dunes from other carbonate grainstone buildups.

Beach and dune biota vary depending on climate and environment. Tropical environments produce aragonitic constituents and phototrophs. Temperate environ-

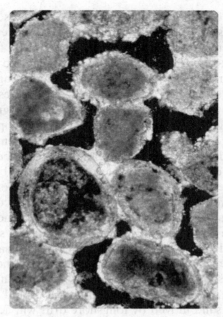

Figure 5.2 Photomicrograph of ooids and peloids with meniscus cement typical of the vadose diagenetic environment in a Holocene eolianite from Cancun, Mexico. Note that the meniscus cement occurs only at grain contacts. The grains are 1.5 mm in diameter.

ments produce calcitic constituents and heterotrophs. In both cases taxonomic diversity is usually high because the beach setting is typical of ramps and some open shelves, where normal oceanic temperature, salinity, and nutrients reach the shoreline unimpeded by rims at the platform margin. Diversity notwithstanding, the beach environment is a difficult place to live except for those animals able to burrow to escape being swept away by breaking waves and swash. Some plants thrive in dunes, but only those with root systems adapted to anchor the plant against being swept away by waves or storms. Most skeletal constituents in beaches are derived from the nearshore subtidal environment. In the temperate setting of the western Mediterranean Sea around the Balearic Islands, *Posidonia* seagrass meadows grow at tens of meters water depth and provide an environment for large populations of mollusks, echinoderms, bryozoans, red algae, benthic foraminifera, and calcareous epiphytes. Fragments of those organisms make up nearly the entire sand fraction in beaches and dunes along the shoreline (Fornos and Ahr, 1997, 2006). In the tropical Caribbean Sea around Isla Cancun, Yucatán, aragonitic ooids are being formed in the nearshore environment and then transported onto the beaches and dunes. Trace fossils typical to coastal dune environments include plant roots, burrows by land-dwelling animals, and footprints of creatures that walked or crawled over the dunes. Plant roots penetrate directly downward through the upper dune facies forming sharp angles with inclined crossbeds. Rhizocretions (also called root casts, rhizomorphs, or rhizoliths) are commonly associated with dune exposure surfaces and, along with caliche, provide additional features for the top of the idealized dune sequence. Body fossils commonly found in dune deposits include air-breathing snails, and skeletons of land-dwelling animals may be present. It is rare to find skeletons and body fossils in borehole cores because their remains, if fossilized at all, are so widely scattered within dunes.

We know that beach–dune successions may be attached to the mainland or they may be detached as barrier islands. In either case, the distinguishing fundamental rock properties are the same. The major difference for the reservoir analyst to recognize is whether or not the beach is detached because fine-grained, lagoonal sediments or evaporites might be present to form the updip trapping facies for barrier island reservoirs. If the beach is attached, either dunes or continental facies will be immediately updip and their trapping potential may not be good. If dunes are present, they may have excellent potential as reservoir rocks but little value as trapping facies. Unfortunately, most carbonate dunes lose their porosity and permeability to diagenesis and rarely become productive reservoirs, according to Abegg et al. (2001). Barrier islands and barrier spits are formed when sand is moved parallel to the shore by longshore currents. Bathymetric irregularities such as points on the mainland or submerged shoals serve as nucleation sites for sand accumulation. As the longshore drift system continues, the sand deposit is extended in up-current and down-current directions, forming barrier spits or barrier islands. If the bathymetry adjacent to a mainland shore is sufficiently monotonous and the shoreline is comparatively straight, the sand moved by longshore drift will accumulate directly on the mainland beach. Beach sediment texture depends largely on the kind and origin of sedimentary constituents. Ooids and peloids are common in the fine to medium sand range, but a wide range of sizes may be typical of bioclastic beaches. As we have already noted, the texture of bioclastic sands may reflect more about the skeletal anatomy of the constituent organism than it does about a history of abrasion and transportation.

Before undergoing burial diagenesis, beach rocks may be subjected to marine, freshwater, or vadose diagenesis, depending on local hydrology and on the location of the beachrock sample on the beach profile. Samples taken above the storm high-water level exhibit vadose diagenesis, those consistently saturated with seawater exhibit marine diagenesis, and buried segments that are in contact with the freshwater table exhibit meteoric phreatic diagenesis. Mixing-zone diagenesis is sometimes evident in zones where seawater and freshwater are combined. Most beachrocks show some degree of marine cementation. Diagenetic characteristics indicative of freshwater, vadose, and marine environments and the distinctive vertical arrangement of sedimentary structures provide the intrinsic evidence to distinguish beach deposits from dunes as well as from shelf margin sand waves and tidal bars. The external relationships that distinguish beaches and barrier islands are their lateral facies relationships. Either lagoons or continental coastal plains lie immediately updip and the shallow subtidal zone of the open sea lies immediately downdip.

Fundamental rock properties are intrinsic characteristics. Beach–dune successions also have extrinsic or external characteristics that help distinguish them from other carbonate successions. For example, beach–dune successions are elongate parallel to depositional strike and may extend for many miles. Barrier beaches may be dissected along strike by tidal inlets and their accompanying tidal deltas. Beach sequences formed during sea-level stillstands are only a few meters thick and may extend seaward from the foredune or berm area 2–3 km to the toe of the lower shoreface, depending on the slope angle. Most barrier islands consist of repeated depositional episodes stacked obliquely in the direction of sea-level advance or retreat. Regressive beach and barrier complexes are normally thicker than transgressive beaches because the rate of sedimentation is greater than the rate of erosion during shoreline progradation. Stacked beach sequences may be tens of meters thick in the aggregate.

Three offset barrier island sequences were described from the Jurassic Smackover Formation at Oaks Field, North Louisiana (Erwin et al., 1979). Each is about 0.8 km wide and extends about 3 km along depositional strike. These regressive barriers range in thickness from about 10 to 15 m, they are laterally offset, and they are enclosed in tight, siliciclastic facies that seal this stratigraphic trap. The barrier islands consist of well-sorted oolitic grainstones that exhibit vadose, meteoric phreatic, and mixing-zone diagenesis—mainly cementation. The reservoir produces from cement-reduced, intergranular porosity that ranges from about 1% to 15% (Erwin et al. 1979). Thirteen of the 18 wells drilled in this trend were productive because the explorationist projected the barrier island facies along depositional strike, minimizing the number of dry holes.

5.2.3 Tidal-Flat and Lagoon Environments

Tidal flats are mud-dominated environments because they are protected from open ocean waves and currents. They are sinks for lime mud transported to, or formed in, sheltered environments behind barrier islands or wave-filtering rims on rimmed shelves. Widely cited modern examples of tidal flats occur behind barrier islands in the Persian Gulf (Figure 5.3) and on the leeward side of Andros Island in the Bahamas (Figure 5.4). Ancient examples are present on the shallow interior of the restricted platform that existed during deposition of the Permian San Andres Formation of Texas and New Mexico and over much of the tidally influenced or *peritidal* setting that covered vast areas on the ramp-like North American Craton during Cambro-Ordovician times. Tidal flats offer limited potential for the formation of depositional porosity because they are mud-dominated systems. Tidal flats in arid climates like the Persian Gulf usually include evaporite deposits; those in wet climates like Andros Island in the Bahamas do not. These low-energy flats lack waves and strong currents that concentrate grainy sediments and they are inhospitable places for reef organisms to grow; consequently, depositional porosity is limited to small areas of grainy sediment accumulations in tidal channels and to zones where fenestral porosity is common. Fenestral porosity is formed mainly by desiccation of lime mud and, in part, by decomposition of algal mats. Tidal-flat environments can be subdivided into three zones: (1) lagoonal or open ocean environments at their downdip margins—the always-wet, subtidal zone; (2) channeled areas between open water and dry land—the intermittently wet, intertidal zone; and (3) the area above normal high tide—the mostly dry, supratidal zone.

The subtidal or always-wet zone may be a lagoon ponded behind a barrier or a shallow reach of the open ocean depending on the physiography of the tidal flat. The character of the always-wet zone sediments depends largely on climate and on the volume of water exchanged between the lagoon and the open sea. Restricted circulation and dry climate result in deposition of evaporites and relatively low rates of lime-mud accumulation. Open circulation and wet climate limit the amount of

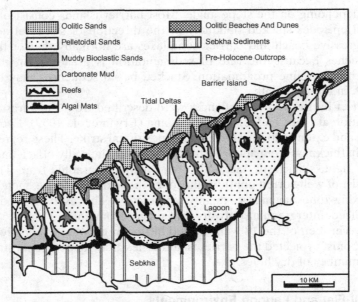

Figure 5.3   The tidal-flat or sebkha environment of the Trucial Coast near Abu Dhabi, United Arab Emirates. Note that small barrier islands partially enclose the lagoon and tidal channel zone. This tidal-flat environment would not exist on an open ramp without the protection of the barrier island. The hot, arid climate of the Trucial Coast is conducive to the formation of evaporites in the upper intertidal and supratidal zones. (Adapted from an illustration in Purser (1980).)

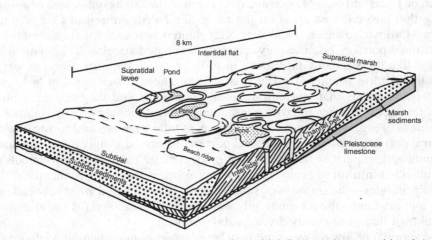

Figure 5.4   An idealized model of the channelized tidal flats on the west side of Andros Island, Bahamas. The channels provide access to incoming tides that flood the channel belt and sometimes flow over the channel margins to fill the ponds, or lakes, that lie between them. The wet climate of the Bahamas favors algal mats, grasses, mangroves, and small palm trees. Evaporites are absent and channels are deeper and more widespread than on the sebkhas of the Persian Gulf. (Adapted from an illustration in Shinn et al. (1969).)

evaporite deposition but favor accumulation of lime mud. Tidal currents move water and lime mud onto the intertidal zone between the always-wet and almost always-dry (supratidal) zones. Ebb tides allow the water to drain off leaving the mud behind. Over time, the tidal flats where mud accumulates in this fashion can build up and outward. In other words, they are progradational.

The ebb and flood of tidal currents typically scour channels between the upper reaches of the intertidal zone and the adjacent subtidal zone. Depending on the tidal prism (the volume of water exchanged during the tidal cycle), the channels may be less than a meter in depth. If the tidal prism is large and current velocity in the channels is great, tidal channels may be considerably deeper than one meter. The greatest channel depth is nearest the subtidal zone and channels become smaller landward. As with all natural channels, meanders form if the channeled zone is large enough in area. Meandering tidal channels superficially resemble delta distributary channels with small point bars, cut-banks, and levees. Levees are the highest topographic feature along the channels, they may be above the level of high tide, may be subjected to desiccation, and may be extensively burrowed and vegetated. Between channels and levees are almost always-wet hollows commonly called ponds. Ponds are receptacles for lime mud brought in during flood tide. Pond margins, depending on climate and tidal range, may be intermittently wet and dry. Especially adjacent to levees, desiccation may produce algal stromatolites, mud cracks, and fenestral porosity.

Finally, above the intertidal zone is the nearly always-dry, supratidal zone where desiccation is the main fair-weather environmental process. Periodic flooding occurs during storms, exceptional tides, or during heavy rains. Because this zone is usually above the reach of the mud-carrying tides, the sedimentation rate is low, vegetation and algal mats may grow where moisture permits, and in dry climates where evaporation causes interstitial waters to become supersaturated, evaporite precipitation occurs. Gypsum and rarely anhydrite are precipitated in the upper sedimentary layers of the intertidal–supratidal transition zone. Mud cracks, fenestral pores, and stromatolites are products of periodic wetting and drying in the upper intertidal and lower supratidal zones. High tides or storms erode the mud-cracked surfaces to produce intraclasts, or "flat pebbles." These clasts, along with fossil shells and pellets, are the principal grain types found on modern and ancient tidal flats. Flat-pebble conglomerates and intraclastic grainstones or packstones with modest amounts of intergranular porosity exist on tidal flats but their volume is usually too small to be considered seriously as exploration targets. In most cases, tidal-flat reservoirs produce from diagenetically enhanced porosity rather than from depositional porosity. Altered tidal-flat deposits are among the most common examples of reservoirs described in the literature. Roehl (1967) discussed an example of an altered Paleozoic tidal flat in the Williston Basin, Montana. He drew parallels between this subsurface reservoir and the modern facies array on the tidal flats of Andros Island in the Bahamas.

5.2.4 Depositional Rock Properties in Tidal Flat–Lagoon Successions

Tidal flat–lagoon depositional successions reflect the three main environmental subdivisions we have already discussed—namely, the subtidal, intertidal, and supratidal regimes. Depositional successions in tidal flats may vary from one location to another if there are differences in climate, physiography, and, to a lesser extent, differences in hydrologic regimes at the different places. For example, subtidal (shallow neritic or lagoons behind barrier islands) facies typically contain mixtures of in situ marine and allochthonous lagoonal allochems along with constituents derived from the intertidal and supratidal zones but that were swept off the flats by strong ebb tides or storm currents. Hurricane-strength storms can produce grainy washover fans that are incorporated into the otherwise muddy, burrowed, lagoonal sequence. Subtidal successions exhibit muddy depositional textures typical of low-energy settings and they exhibit marine phreatic diagenesis because the environment is always wet. Subtidal facies may have differences in taxonomic diversity because open marine settings have higher diversity than lagoonal or restricted marine settings. Rates of sedimentation vary depending on the amount of mud imported or produced in place and on the volume of allochthonous constituents brought down from the adjacent intertidal or supratidal regimes. The volume of imported sediment varies with storm frequency, tidal range, and local hydrologic regime. Subtidal sequences adjacent to tidal flats have many distinctive characteristics that make them relatively easy to recognize regardless of age or location.

Intertidal facies vary greatly but predictably. Incised tidal channels may vary in number and dimensions across tidal flats but they are always marked by their sharp erosional bases and fining-upward channel-fill sequences. Channels only a few tens of centimeters deep may not display obvious fining-upward sequences. Repeated successions of shallow channels filled with flat-pebble conglomerates and intraclastic packstones are characteristic of the Cambro-Ordovician rocks around much of the Transcontinental Arch in North America, for example (Lochman-Balk and Wilson, 1958). The form of these pebble-filled zones is sometimes hardly recognizable as channels. Larger channels like those on Andros Island in the Bahamas are flanked by levees or curvilinear ridges covered with desiccated algal mats, tunneled with bioturbation, and blanketed with pelleted micrite. Fecal pellets are common constituents on tidal flats of almost all geological ages. The pellet producers are typically high-spired browsing gastropods that ingest organic-rich mud and excrete compacted pellets. Ponds between leveed channels may vary in size but they nearly always contain extensive deposits of pelleted micrite and relatively continuous algal mats that have not been cracked by desiccation. In dry climates, ponds may be ephemeral and the once-wet hollows may be filled with evaporites such as gypsum or halite. Evaporitic pond deposits on the interior of broad, shallow shelves may extend for many square kilometers. Ultimately, these thin, widespread evaporite beds can be barriers or baffles to fluid movement in the subsurface.

Supratidal facies are characterized by desiccated algal mats, flat-pebble intraclasts derived from fragmented algal mats and mud cracks, stromatolites, bioturbation, root traces, fenestral porosity, and crusts of evaporite minerals. In arid climates such as the modern Persian Gulf, the supratidal zone, commonly known as the *sebkha*, is partially covered with siliciclastic sands blown off the adjacent Arabian desert. Interstitial evaporites are common in the shallow subsurface of the upper intertidal–lower supratidal zone, or sebkha. If subsurface precipitation of the evaporites, mainly gypsum and anhydrite, continues for a long enough time, the growth of evaporite crystals displaces the surrounding carbonate sediment to form peculiar and characteristic deformation called enterolithic structures. Gypsum crystal rosettes may continue to grow at the surface and in the shallow subsurface until the rosettes coalesce to form a feature commonly called "chicken-wire" fabric. In the subsurface, at depths of 3000 feet or more according to Hardie (1967), the chicken-wire fabric consists of anhydrite ($CaSO_4$); at the surface it consists of the stable mineral gypsum ($CaSO_4 \cdot 2H_2O$). In wet climates such as Andros Island in the Bahamas, large deposits of interstitial evaporites are absent, but thin dolomitic crusts have been found around elevated hummocks in the supratidal zone (Shinn et al., 1969).

5.2.5 The Shallow Subtidal (Neritic) Environment

The shallow subtidal or neritic environment extends from the lowest low tide line to an arbitrary depth of 200 m. On large platforms, the shallow subtidal zone can cover very large areas and the sedimentary characteristics of this vast expanse can vary considerably. Unlike the tidal-flat environment, where variability is predictable within the three main subdivisions of the zone, the shallow subtidal zone may be full of surprises for the uninformed explorationist or engineer. Shallow subtidal environmental characteristics and resulting depositional rock properties depend on the following main factors: (1) bathymetry, (2) hydrological regime, (3) carbonate productivity, and (4) taxonomic diversity.

The shallow subtidal environment can be defined bathymetrically as the always-wet zone that extends from lowest low tide to the slope break on shelves and from lowest low tide to the outer ramp environment on ramps. Because there is no physiographic marker on ramps to denote the lower limits of the neritic environment, an arbitrary water depth or a significant change in facies character is commonly used to mark its end, but facies characteristics, trace fossils, and taxonomic diversity may vary markedly from the inner ramp to the outer ramp environments. Oceanographers traditionally put the upper and lower boundaries of the neritic environment at the low tide mark and about 200 m, respectively. The pelagic or open water zone deeper than 200 m is classified simply as the oceanic environment, although some texts refer to the benthic zone that extends from 200 to 1000 m as the bathyal or archibenthic environment. The benthic zone below 1000 m is classified as the abyssal or abyssal-benthic environment. These are arbitrary boundaries that are not detectable in the rock record; consequently, they are not very helpful for reservoir studies. Most geologists now divide carbonate ramps into inner, middle, and outer ramp environments (Burchette and Wright, 1992). Those authors describe the typical sedimentological and stratigraphic characteristics of each environmental subdivision. The beach–dune, tidal-flat, and lagoonal environments discussed previously are part of the updip margin of the inner ramp environment. There is no depth designation that distinguishes inner from middle, or middle from outer ramp. Those distinctions are made on facies characteristics and biota, which change gradually as depth increases. The changes are gradual so that the transitional zones between inner-to-middle and middle-to-outer ramp have overlapping characteristics. Depending on the ramp slope angle and environmental hydrodynamics, the shallow neritic environment can include the entire inner ramp and all or part of the middle ramp. The middle ramp environment differs from the inner ramp setting in that it is characterized by lower taxonomic diversity, lower grain/mud ratio, and fewer patch reefs (except where antecedent topographic highs present suitable habitat). But middle ramp facies commonly include organisms, sedimentary structures, and depositional textures that are typical of the shallow, open ocean rather than the slope or basin;

therefore both inner and middle ramp settings could be included in the shallow neritic environment. Evidence of wave and current action, presence of photozoans, including reefs, in tropical environments, and high taxonomic diversity are all characteristic of the middle ramp setting. Much of the outer ramp extends below the depths that are commonly affected by surface waves and currents, below ample light penetration, and usually below the zone of optimum carbonate production. The outer ramp environment may be in the bathyal rather than the neritic environment, although some basins such as the modern Persian Gulf are so shallow that they do not reach the required depths of more than 200 m to fit the strict definition of bathyal. The outer ramp represents the deep-water transition to the basinal environment. In that sense, it can be somewhat similar to the upper slope environment on open shelves, but the outer ramp environment is very unlike the slope environment on rimmed shelves because ramps lack pronounced slope breaks and their tendency to submarine mass wasting that gives rise to slumps, rock falls, debris flows, and similar gravity-driven slope deposits. Ramps and some open shelves lack those types of deposits.

Hydrological characteristics that influence carbonate sedimentation on ramps and shelves are (1) the hydrodynamics of the environment, mainly the amount of wave and current energy, (2) nutrient content, (3) salinity, (4) temperature, and (5) water clarity. The shallow subtidal environment exists on both ramps and shelves but the facies on rimmed shelves and ramps are different. Inner parts of ramps and open shelves are unprotected from incoming ocean waves and storms while rimmed shelf interiors are protected by the rim at the slope break. Except for the beach environment, fair-weather wave action in the neritic environment has only limited effect on bottom sediments; that is, the swells pass through the neritic environment and form oscillation ripples on bottom sediment, depending on wave period and water depth. If the water is too deep, passing swells have no effect on bottom sediments. Currents are driven mainly by wind and tides, although as waves pass through gaps in shelf rims, they are diffracted and translated into currents. Density currents, turbidity currents, and geostrophic currents are uncommon in the shallow neritic zone. Tidal effects are most pronounced in areas where tidal currents are focused by natural inlets and around abrupt changes in bathymetry. Storms may affect the entire shallow ramp or shelf, depending on storm severity. Storms can cause massive movement of bottom sediments, shoreline erosion, and strong currents at all depths across platforms, especially ramps and open shelves. As powerful storms pass inland and the storm surge recedes, strong return flows may manifest as subtidal density currents that transport shoreline sediments considerable distances seaward. These deposits are distinctive and usually easily recognizable as anomalies on the otherwise muddy neritic seabed.

Restricted interiors behind shelf rims are subject to wide swings in temperature and salinity during seasonal changes. Shallow waters on rimmed shelf interiors are also protected from oceanic currents, tides, and waves. In contrast, ramps and open shelves are swept by incoming waves, currents, and tides that fertilize and oxygenate the water while maintaining normal marine salinity and temperature. These more favorable environmental qualities make the neritic environment on ramps and open shelves better suited for greater rates of biogenic sediment production than the sheltered interiors of rimmed shelves; therefore, from the reservoir geology viewpoint, shallow interiors of both ramps and open shelves are likely to exhibit patch reefs, bioclastic buildups, or offshore grainstone banks but the subtidal interior of rimmed shelves may not. Finding the potential reservoir facies is then a matter of identifying antecedent topography on platforms because relict highs are generally sites for high-energy deposits and relict lows will be sinks for mud-dominated sediments.

Carbonate productivity and taxonomic diversity depend on bathymetry, climate, and hydrological characteristics. Optimum carbonate production (the carbonate factory) in tropical climates extends to a depth of about 10 m and the maximum depth of sediment production is about 200 m (Wilson, 1975). In temperate oceans carbonate production extends to depths of as much as 100 m because productivity in the temperate carbonate factory is based on *heterozoans*, a term coined by James (1997) to identify cool-water associations such as echinoderms, bryozoans, foraminifera, and red algae, among others. In contrast, photozoans—organisms that require sunlight—occupy sunlit zones in clear-water, tropical carbonate environments. The depths where the carbonate factory is in full production in both tropical and temperate settings are about the same as the depths where most waves, persistent currents, and storms occur. It follows that the inner ramp environment in tropical and temperate climates will have high grain/mud ratios (grain-rich sediments), large numbers of benthic organisms per unit area, good potential for patch reef growth, and generally high taxonomic diversity. In traditional oceanographic terminology the zone from high tide to about 60-m depth is known as the eulittoral environment, where sessile organisms and plants flourish. This depth range corresponds roughly with the zone of optimum carbonate production on temperate, low-energy ramps (Fornos and Ahr, 1997, 2006). Taxonomic diversity is an important clue to recognize when trying to distinguish neritic successions formed in restricted environments (rimmed shelf interiors) from those formed in open marine environments (ramps and open shelves). Restricted environments are stressful to most normal marine life so that few organisms, especially stenothermic and stenohaline (low tolerance to temperature and salinity change) benthic organisms, can flourish under those conditions. This means that rimmed shelf interiors commonly have depositional successions with high mud content, low taxonomic diversity, or even low skeletal grain content. The combination of arid climate and restricted circulation usually leads to hypersalinity, and under extreme conditions, this can result in evaporite precipitation. The shallow interior of restricted shelf environments is also subjected to low water temperature and hyposalinity in wet, temperate climates. This will result in low taxonomic diversity and perhaps in low carbonate productivity, especially the aragonitic constituents that require tropical conditions.

5.2.6   Depositional Rock Properties in Shallow Subtidal Successions

Standard facies descriptors such as depositional texture, sedimentary structures, constituent composition, and taxonomic diversity are determined by the interactions among hydrological characteristics, bathymetry, and productivity. Hydrological characteristics, as we have noted, include wave and current activity, water temperature, oxygenation, salinity, nutrients, and water clarity. Environmental quality is a way of describing hydrological characteristics based on whether the hydrological regime is favorable to carbonate productivity and taxonomic diversity or not. Bathymetry is simply seabed topography. Bathymetry in the ordinary neritic environment, whether it is a ramp, rimmed shelf, or open shelf, can be monotonous. Flat or gently sloping surfaces do not offer much resistance to the passage of waves, tides, and currents in low- to moderate-energy settings. This means that there is little winnowing to concentrate grainy sediments. Grainy facies are characteristic of the strandline on ramps and some open shelves, and the slope break on rimmed shelves. For grainy facies to exist in the neritic zone requires some topographic feature to focus waves, tides, and currents. If suitable habitat exists and environmental quality is good, patch reefs may develop to create local bathymetric highs that interact with incoming waves and currents. Bioclastic sands are commonly derived from the breakdown of reef organisms and from opportunistic dwellers in the reef habitat. These sands accumulate on and around bathymetric highs to form local grainstone and packstone accumulations, which along with the patch-reef buildups stand in stark contrast to the muddy facies that characterize most of the neritic seabed away from the highs. Most shallow neritic depositional successions consist of monotonous wackestones and mudstones with two possible variants: (1) patch reefs and their flanking grainstones and packstones, or (2) grainstone shoals formed as winnowed lag deposits on paleo-highs. If commercial reservoirs with depositional porosity exist in the ordinary shallow neritic environment, they will almost certainly be either patch reefs or grainy facies on paleo-highs. Examples of such reservoirs on isolated highs surrounded by subtidal mudstones and wackestones are the Mississippian skeletal grainstones at Conley Field, Hardeman Basin, Texas (Ahr and Walters, 1985), the James Limestone (Cretaceous) reefs on a "turtle" structure in the East Texas salt basin, and many of the Smackover (Jurassic) oolite buildups on salt structures in the ancestral Gulf of Mexico salt basins from Texas to Alabama. Porosity enhancement by diagenesis, and especially by fracturing, can transform the monotonous neritic mudstones and wackestones into excellent reservoirs, however. At the other end of the environmental spectrum, extreme climatic conditions such as those in the "roaring 40s " latitudes in the southern hemisphere, fierce winds, waves, currents, and storms sweep detrital sediments out of the shallow subtidal zone. A modern example of this situation exists off the coast of southern Australia. There is no deposition in the shallow subtidal zone on this open shelf because the hydrologic regime is so powerful. Coarse sediments are winnowed and redeposited by strong waves and currents at depths of up to 140 m (Nelson et al., 1982; James et al., 1992; Boreen et al., 1993). Clearly, it is the low- to moderate-energy platforms on which significant deposits of shallow subtidal facies will accumulate.

Chalks Most chalks were deposited in the subtidal shelf or inner and middle ramp environments in epicratonic basins, or in shallow seas that flooded continents, especially northwestern Europe, western North America, and the Middle East during Middle to Late Cretaceous times. Although most chalks are shallow marine deposits, some were deposited in basinal settings. To avoid duplication, the discussion on chalk deposits is included in a later section on basinal facies.

5.2.7 The Slope-Break Environment

Modern bathymetric charts show that rimmed carbonate shelves like those in South Florida and the Bahama Banks have slope breaks at 10 m or less, well within the highly oxygenated, nutrient-rich, euphotic zone. Slope breaks on tropical rimmed shelves have high rates of carbonate productivity and high taxonomic diversity because the rim-forming process is a feedback loop in which favorable environmental quality promotes biological productivity that generates a reef rim. A fundamental characteristic of rimmed and open shelves is that the shelf-slope breaks have persistent and distinctive facies changes paralleling the slope break. This distinguishes both rimmed and open shelves from distally steepened ramps, which do not have facies changes that parallel abrupt changes in slope. Facies changes do not occur at slope changes on distally steepened ramps because the distal steepening occurs in water deeper than that at which fair-weather waves and currents interact with the bottom to influence sedimentation. Distal steepening on ramps occurs below the greatest depth at which the carbonate factory operates. Generally, the sediments that accumulate around deep-water slope changes on distally steepened ramps are allochthonous mudstones and wackestones. In strong contrast, slope breaks on tropical, shallow-water rimmed shelves are dominated by high-energy deposits. Shelf rims build nearly to sea level and act as barriers to incoming oceanic waves and currents. The rims are buffeted by breaking waves and strong currents so that only the most resistant reef structures can withstand the pounding. Coarse rudites usually accumulate on the seaward sides of rims, and grain-rich successions accumulate in the immediate lee of the rims. Rim-forming organisms do not appear to thrive in temperate climates nor do temperate seas favor the precipitation of aragonite; consequently, there are probably no temperate rimmed shelves. Another platform configuration with a slope break is the open shelf. When one compares modern high-energy open shelves such as the Lacepede shelf of Australia with low-energy rimmed shelves like the one in South Florida, it is immediately clear that great differences exist in where and how much sediment accumulates around the open shelf and rimmed shelf slope breaks. These differences exist because the dynamics of the respective hydrologic regimes are so different. It is not enough to identify the slope break and attendant facies changes that distinguish shelves; it is also necessary to determine if the environment of deposition was high or low energy, and if it was a tropical or temperate regime.

5.2.8 Depositional Rock Properties in Slope-Break Successions

Slope-break successions are asymmetrical or polar deposits. That is, grainy facies, carbonate gravels, and massive reef growth characterize the seaward side of the slope break, facing the prevailing winds and incoming oceanic waves and currents. Lower-energy facies accumulate on the leeward side of the slope break. If reefs develop at the slope break, growth patterns of the reef-building organisms also reflect the polarity of the environment. Massive skeletal forms develop on the high-energy, windward side of shelf-margin reefs and more delicate skeletal forms develop in the protected zones behind the reef. Spur and groove features, also known as buttresses and chutes, may mark the windward side of some framestone reefs. These gaps in the reef help diffuse and absorb incoming wave and current energy. The area behind the shelf-edge reef trend is usually blanketed by bioclastic grainstones and packstones derived from the breakdown of reef skeletal structures, or from the benthic organisms that lived on and near the reefs. In tropical environments the detrital grains may be oolitically coated. The modern White Bank shoal of South Florida is a bioclastic sand–wave (grainstone) complex that extends about 2 km

Figure 5.5 Air photo of shelf-edge grainstone buildups on the Great Bahama Bank. These buildups, or sand–wave complexes, form the rim on an isolated, rimmed shelf. Spillover lobes indicate direction of transport from open ocean to bank interior (right to left) during tidal exchange. The sand–wave complex borders a precipitous slope break that is swept by powerful tidal currents. From right to left, each spillover lobe is about 1 km wide.

landward from the shelf-edge reef and about 40 km along depositional strike. A diagrammatic cross-section and facies map illustrating the White Bank sand–wave complex are included in the memoir by Enos and Perkins (1977). Similar grainstone buildups occur behind the Guadalupian (Permian) reef in West Texas and New Mexico, and Wilson (1975) describes reef-related grainstone buildups on lower Cretaceous slope breaks in the Middle East (Thamama Group) and the ancestral Gulf of Mexico (Stuart City reef trend). In essence, typical slope-break facies consist of reefs and associated sand sheets or sand–wave complexes. On the Great Bahama Banks oolite sand waves occur on slope breaks without reefs (Figure 5.5). Sediments inboard from these slope-break grainstones are shallow subtidal wackestones to mudstones, or alternatively, platform interior wackestones to mudstones. Slope-break grainstones and reefs present three main kinds of depositional porosity: intergranular porosity in grainstones (including the detrital infill around reef masses), intraframe porosity within the entire skeletal or cementstone structure of the reef, and intraparticle porosity within individual skeletons in the reef mass. Of these pore types, intergranular porosity has the greatest potential to provide reservoir continuity and connectivity but may also have the highest potential for cementation by early marine diagenesis. Intraframe porosity and intraparticle porosity in fossil skeletons have a tendency to behave as separate vugs, thereby reducing their connectivity potential. Early diagenesis may drastically alter this pattern, especially if aragonitic skeletal components are abundant. Aragonite is metastable and quickly dissolves in undersaturated water to form moldic pores and may just as quickly be reprecipitated as pore-filling cement.

5.2.9 The Slope Environment

Bathymetric characteristics of slope environments vary depending on the mechanical slope stability, the slope angle, and the rigor of the physical oceanographic environment. Slope environments characterize both rimmed and open shelves but not ramps. With the possible exception of rare cases, slopes on distally steepened ramps are nothing more than changes in slope without accompanying differences in environmental characteristics or sedimentary rock properties. The absence of steep gradients in environmental processes above and below distal steepening on ramps indicates how monotonous sedimentation had to be for the distally steepened area to have rock properties no different from the surrounding seabed above and below it.

Carbonate slopes differ from siliciclastic slopes in that carbonate slopes are steeper (5°–15° as compared to 3°–6°), they tend to have a concave profile compared to the slightly convex or straight profile on siliciclastic slopes, they are prone to oversteepening as reef growth or cementation enhances slope steepness, and their slope angle increases with slope height (Coniglio and Dix, 1992). Physical processes that characterize slope environments vary depending on water depth, on the nature of the hydrologic regime, on slope characteristics, and on proximity to the slope.

Environmental processes on slopes are dominated by gravitational forces and pounding from waves. Upper slope zones in relatively shallow water may be subject to wind or storm waves, oceanic currents, and tides much the same as the slope-break environment. Steep slopes with abrupt and great changes in water depth are constantly subject to slope failure owing to steady gravitational pull on wave and storm-ridden upper slopes. Slopes are commonly sites for upwelling, initiation of density or turbidity currents, and initiation of slumps, rock falls, and debris flows triggered by slope failures. Middle slope and base of slope zones are typically below fair-weather wave base, below the influence of surface currents, and relatively less influenced by tidal currents than the upper slope zone, where the shorter water column is more vigorously moved during tidal exchange.

Deeper parts of slope zones are sites where rocks and sediments swept off the slope by shallow-water processes come to rest. Currents on the middle and outer slope are mainly geostrophic and density currents. Geostrophic currents may be active to any depth because they depend only on flow down a density gradient balanced by Coriolis force. These currents commonly flow parallel to the bathymetric contours along the toe-of-slope and are referred to as "contour currents." Density currents result when more dense water flows in the direction of less dense water. Typically, higher density results when salinity is high, or temperature is low, or when both factors operate simultaneously. Density currents driven by differences in the amount of suspended particulate material, or turbidity, in the water masses are known as turbidity currents and their deposits are called turbidites. Turbidites are a type of density current deposit but they differ from thermohaline density current deposits by having high matrix content.

Slope characteristics and proximity to the slope break probably have the greatest influence on what type of deposits will occur at the base of slope and farther beyond on the basin floor. Steep, unstable slopes are ideal places for rock falls, debrites, and coarse grain flows to be deposited. Gentle slopes are probably more likely to be associated with turbidites, rhythmites, and soft sediment slumps. Accretionary slopes that develop from reef growth and extensive sediment production and retention at the slope break are known as *depositional margins*. Slopes with inherited (antecedent) or eroded escarpment margins are known as *bypass margins* because their steepness facilitates slope failure and off-slope sediment shedding. The differences between depositional and bypass slopes were illustrated in simplified form by James and Mountjoy (1983). Steep slopes commonly fail and coarse debris is shed as "periplatform talus" or, farther away from the slope, as "debrites" and turbidites.

5.2.10 Depositional Rock Properties in the Slope and Slope-Toe Environments

Slope facies characteristics reflect environmental processes, slope characteristics, and proximity to the slope break above. Depending on the steepness of the slope, most sediment accumulation takes place near the base of the slope and just beyond. In other words, there is comparatively little deposition directly on steeply sloping surfaces. Depositional facies basinward from the toe of slope typically include slumps, slope talus, debris fans or debrites, grain flows, and turbidites. Gravity is the primary engine that drives sediment transport and the sediment source is material dislodged from the slope and slope break, material shed from the platform top, and admixed pelagic sediment. In situ sediment production is usually negligible because the slope environment is below the depth of production by the carbonate factory.

Slumps and debris flows consist of poorly sorted coarse particles in a finer matrix. In some cases slope failure yields blocks as large as automobiles and even entire segments of shelf margins can slump downslope (C. Kerans, 2006, personal communication). Debris flows, or debrites, commonly form as fan-shaped deposits at the toe of slope and beyond, depending on the competence and capacity of the transport mechanism. Examples from the Permian Basin of Texas are described in Saller et al. (1989) and Mazzullo (1994). Because debris flows at the base of slope and beyond usually have a high percentage of fine matrix, interparticle porosity is low. Intraparticle porosity formed before the debris was dislodged from the slope margin may be present, commonly as separate vugs. However, it may also be reduced by cement and matrix infill. In most slumps or debris flows reservoir porosity has either been created or enhanced by burial diagenesis (Mazzullo and Harris, 1992).

Grain flows and turbidites are finer grained than debrites. Grain flows are gravity-driven "rivers of sand" that pour down steep slopes after unstable accumulations of grainy sediment have been dislodged. The fundamental difference between grain flows and turbidites is that grain flows move in traction flow rather than in suspension and the flows are not driven by density contrasts induced by turbidity. Turbidites are deposits produced by a special form of density currents—turbidity currents—that result when fine sediment is put in suspension to form a water mass with greater density than its surroundings. Typically, turbidity is increased dramatically when storm surges pound sediments at shelf margins, during seismic tremors, and gravitational failure of unstable slope sediment. Turbidity currents can reach speeds of tens of kilometers per hour and may flow basinward for many kilometers. As with some grain flows and debris flows, proximal turbidites (those closest to the base of slope) may accumulate in fan or lobe-shaped deposits. Entrained grains may scour soft bottom sediments so that some turbidity currents excavate channels in the seabed. These channels are subsequently filled by channelized turbidites. Essentially all base-of-slope deposits are set in motion by an "event." Whether the event is a storm, a tremor, or a slope failure, the initial response is usually vigorous and intense sediment transport followed by declining transport velocity and subsequent deposition. Some deposits, particularly thermohaline current deposits and turbidites, reflect this waning energy by vertical changes in texture and sedimentary structures. Reservoirs in grain flow deposits and turbidites are subject to the same limitations on porosity as debris flows and slumps. These facies commonly include high matrix content that blocks pore throats and limits depositional porosity. In many cases, reservoir quality in deep slope deposits depends on burial diagenetic dissolution to create or open pores and pore throats.

5.2.11 Basinal Environments

The word "basinal" conjures images of cold, dark, forbidding depths. Images like those of the great ship *Titanic* lying miles beneath the North Atlantic Ocean or of mysterious sea creatures revealed in photographs taken in the abyss through windows of special submarines. From the practical viewpoint of geoscientists and engineers, the basinal environment is simply the end of the marine environmental spectrum that began at the strandline and ended at the deepest part of that particular sedimentary basin. There is no unique depth that identifies the "basinal" environment. There is not even a rigid definition of basinal environment or basinal facies. In fact, the greatest depth that exists in one basin may be the same measured depth as the shallow subtidal regime in another basin. For example, the maximum depth in the modern Persian Gulf is only about 200 m, but the maximum depth in the Mariana Trench is about 11 km. The 200-m depth, as we have mentioned earlier, is the arbitrary limit used by oceanographers to define the outer limit of the neritic (shallow subtidal) environment. It is commonly used to define the edge of the "continental shelf" or "continental terrace" as well, but in the Persian Gulf it represents basinal depth and the basinal environment. This lack of a unique and single definition for basinal environment and basinal facies also holds true for basinal deposits in the fossil record.

Epicontinental seas may not have been more than a few tens of meters deep on average, while the centers of the world oceans may have had average depths of several kilometers. For carbonate sediments to accumulate in the basinal environment, water depths must be shallower than the *carbonate compensation depth* (CCD) because below the CCD, carbonate particles dissolve. This depth varies depending mainly on water temperature and partial pressure of CO_2. It is also different for aragonite than for calcite because of their different solubilities. In the major world oceans today the CCD is at depths of several kilometers and it is shallower for aragonite than for calcite because aragonite is more soluble. We have already established that the optimum zone for carbonate sediment production, or the carbonate factory, extends from the surface to about 200 m. Below about 200 m, even in temperate environments with heterotrophic biota, there is little or no in situ sediment production; therefore deep-water deposits consist of imported sediment shed from platform tops, from slopes, and from the rain of pelagic organisms that live and die in the water column. Typical basinal sediments may include both carbonate and siliciclastic muds and fine sands, planktonic skeletal remains such as foraminifera, cocolithophoridae, discoasters, radiolarians, calpionellids, and tintinids, sponge spicules, and the odd coarser sediments carried in by density or turbidity currents.

Restricted basinal environments have limited water circulation. The Black Sea is a silled basin with such restricted circulation that the bottom sediments and interstitial waters are anoxic and reducing. The Permian sea in the Delaware Basin of West Texas had limited circulation and high salinity that allowed evaporite minerals to crystallize from interstitial waters on the basin floor. Evaporite crystal molds are easy to find on bedding planes in today's outcrops of these ancient basinal deposits. If basinal depths are greater than about 200 m or if water turbidity is high, little or no sunlight can penetrate to the seafloor. Depending on the amount and kind of current circulation in a given basin, bottom waters may be stagnant, anoxic, dark, cold, and above normal salinity. However, if the basin has open connections to the world oceans and circulation is vigorous, bottom waters will be in motion, oxygen will be more abundant, the Eh may be neutral to positive, and salinity may be closer to normal. Of course, if the depth is greater than the CCD, the preservation potential for fine carbonate sediments is low. Clearly, the number of environmental variables in basinal environments is great enough to prevent making rigid descriptions of basinal environmental characteristics.

Basinal environments, even those only a few hundreds of meters deep, are generally protected from wave action, surface currents, and ordinary tidal effects. But they are not immune from being swept by geostrophic and density currents. These deep-water currents can transport fine sediments and create sedimentary structures. Current ripples are common in basinal sediments, along with a variety of current-related markings such as flute casts and tool marks. In deeper basins, the environment is more hostile to life and only the best-adapted organisms colonize the seabed. Less hostile basinal settings may host a variety of benthic organisms that, in turn, produce many different kinds of tracks, trails, burrows, and other forms of *trace fossils*. Trace fossils have been studied extensively and a comprehensive literature exists about them. For purposes of reservoir studies, trace fossils are useful in identifying and subdividing basinal facies types when used in conjunction with physical stratigraphic and sedimentological criteria. In general, basinal settings can be characterized as cool, dark, high-pressure environments that may be dysoxic to anoxic, and that may have negative to neutral Eh. They usually have low taxonomic diversity, they may exhibit distinctive trace fossils, spiculiferous chert, and they may exhibit sedimentary structures produced by geostrophic and density currents.

5.2.12 Depositional Rock Properties in Basinal Environments

Slope facies overlap basinal facies. If slope-derived debris and grain flows are excluded, basinal deposits commonly consist of fine-grained carbonate and siliciclastic mud and sand, particulate organic matter, and pelagic microfossil remains. If the source of fine sediment is relatively constant and the basinal setting does not receive sediments from nearby slopes, the typical facies will be dark colored, thin bedded (millimeter scale), fine grained, sparsely fossiliferous, organic-rich deposits called *laminites*. Sets of millimeter-scale laminae are common and may be bundled in 5–30-cm beds alternating with centimeter-scale shale layers. Lamina sets are commonly graded from base to top. Coarser grained millimeter-scale laminae grade upward to finer ones. These graded beds are known as *rhythmites* because they may be cyclically repeated through many meters of vertical thickness. Occasionally turbidites may extend long distances from their source, typically shelf-slope areas, and become incorporated in the basinal facies melange. Ramps characteristically have low slopes that are not conducive to the initiation of turbidity currents and slope

Figure 5.6 Scanning electron micrograph of a Miocene chalk showing the characteristic microstructure of cocolith fragments. Scattered discoasters (star-shaped objects) are also present. The width of the micrograph is 22 μm. (From Scholle (1978). Reproduced with permission from the American Association of Petroleum Geologists.)

failures. Typical basinal facies associations include laminites that represent the ambient condition on the basinal floor of ramps and shelves, rhythmites that mark the distal parts of event bedding on any platform type, and distal turbidites that extended long distances from their source, mainly slopes and shelf margins. Depending on seawater and interstitial water chemistry, basinal facies may be rich in organic carbon, chert or chalcedony, or evaporites.

Chalks may have been deposited in basinal or shallow subtidal environments depending on the physiography of the basin. As we mentioned earlier, chalk deposition was particularly widespread during the Late Cretaceous and Early Tertiary in Europe, North America, and the Middle East. Most of the North American and Middle Eastern chalks are shallow subtidal deposits, but some of the European chalks were deposited in shallow basins. Chalk consists mainly of calcitic microfossils called cocoliths and discoasters along with various foraminifera (Figure 5.6). Shallow-water chalks also contain benthic organisms such as mollusks and echinoderms that add coarser grains to the chalk texture and aragonite to its mineralogical composition. Some European chalks contain abundant chert, or flint, nodules that are probably derived from siliceous sponge spicules. Metastable aragonite is more reactive to dissolution and reprecipitation than calcite; therefore chalks with large amounts of benthic megafossils may exhibit reduced porosity and permeability compared to purely pelagic chalks (Neugebauer, 1975). The main cause of porosity loss in chalks is compaction. A number of studies including those by Neugebauer (1973), Scholle (1977), Herrington et al. (1991), and Brasher and Vagle (1996) found that compaction and attendant porosity loss in chalk are related to high clay content, pelagic depositional mode, burial depths greater than 2 km, absence of overpressure, and late hydrocarbon migration. Brasher and Vagle (1996) devised a classification scheme for North Sea chalks to distinguish those deposited purely as pelagic chalks (Type I), those resedimented by slumping or turbidity currents (Type III), and intermediate types of chalks with some of both end-member characteristics (Type II). They found that the highest reservoir porosity, up to 40%, occurred in Type III chalks in overpressured formations that show evidence of early hydrocarbon migration (Figure 5.7). Most of the North American chalks, particularly the Austin Chalk of Texas, do not exhibit high porosity comparable to those in the Norwegian sector of the North Sea because they have undergone more compaction and cementation.

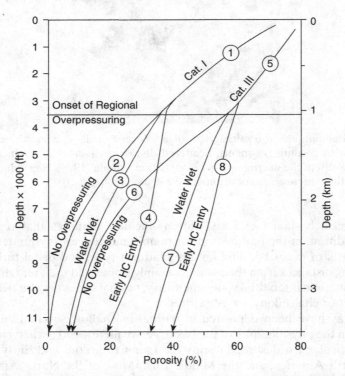

Figure 5.7 Porosity versus depth in chalks that have undergone different burial histories. Curves 1 and 2 represent normal porosity loss with depth. These curves follow the pattern of porosity decrease with depth illustrated in Schmoker and Halley (1982) mentioned in Chapter 2 of this book. Curves 3 and 4 represent porosity retention in spite of overburden pressure either because "overpressure" developed when pore water (3), or hydrocarbons (4) could not escape as overburden pressure increased. In both cases chemical compaction (pressure-solution) is retarded or prevented. Curve 5 represents porosity retention owing to spot-welding of loosely packed allochthonous chalks that resisted mechanical compaction to moderate depths, curve 6 represents the combined effects of mechanical and chemical compaction, and curve 7 illustrates porosity retention owing to trapped interstitial water under high overburden pressure that led to regional overpressuring. Curve 8 illustrates porosity preservation associated with early hydrocarbon entry and retention. (From Brasher and Vagle (1996).)

Scholle (1977) illustrated the difference between porosity in the Austin Chalk and that in a typical North Sea chalk. Although the Austin Chalk has low matrix porosity, it is a prolific producer from fractured reservoirs that extend across much of south central Texas. Fractured reservoirs are important in North Sea chalks that have diagenetically reduced matrix porosity. In some chalks with extensive flint beds, the more brittle flint exhibits more intense fracturing than the chalk.

5.2.13   Ideal Depositional Successions Illustrated

The ideal depositional successions and their typical locations on ramps, rimmed shelves, and open shelves are illustrated in Figures 5.8–5.14. All of the ideal successions are drawn to an arbitrary thickness scale of about 10 m, the length of a standard core barrel. Some additional comments about the drawings may be helpful (see Figure 5.15). Dunes are normally part of the beach succession, but in cases where sediment supply is limited dunes may not be present. The ideal depositional succession for dunes is therefore not always present. Attached beaches extend seaward from the nonmarine mainland shore but barrier island beaches extend seaward from an island shore with a lagoon separating the island from the mainland. Lagoonal successions may or may not contain storm washover deposits of poorly sorted,

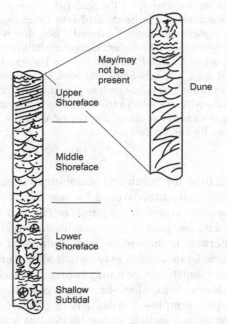

Figure 5.8   Ideal depositional succession of a beach–dune or barrier island environment. Barrier islands are simply beach–dune successions separated from the mainland shore by a lagoon. Each of the following successions is drawn in the form of an idealized borehole core that exhibits easily recognizable vertical sequences of sedimentary structures. Note that beach successions can be divided into lower, middle, and upper shoreface segments. Dunes may or may not be present at the top of the succession. If present, they exhibit meter-scale inclined or spillover beds that pass upward to smaller trough crossbeds and ripples. Plant roots (rhizocretions) or land animal tracks, trails, and burrows may mark the tops of dunes.

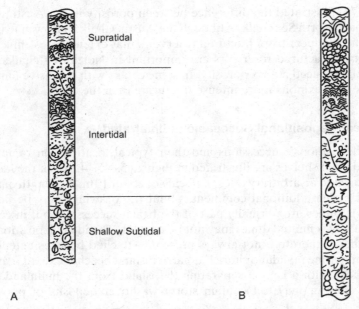

Supratidal

Intertidal

Shallow Subtidal

A   B

Figure 5.9   Ideal depositional succession for the tidal-flat environment in (a) humid and (b) arid climates. Tidal-flat successions can be divided into three main parts: the uppermost supratidal marsh, the alternately wet-and-dry intertidal belt that is usually marked by tidal channels and ponds, and the always-wet adjacent lagoon or shallow marine environment. The model for this succession is the Andros Island area of the Bahamas (see Figure 5.4), an isolated, low-energy, tropical shelf. The sebkha or arid climate tidal-flat succession is similar to the previous example except that evaporites such as gypsum and even anhydrite characterize the supratidal zone. Chicken-wire (nodular) and enterolithic (intestine-shaped) anhydrite are common in the upper parts of sebkhas along the Trucial Coast of the Arab Emirates (see Figure 5.3), the model area for this succession.

coarser material washed from the beach and nearshore marine environment. Typical shallow subtidal environments are covered by wackestones and mudstones but patch reefs may also be present, either on ramps or rimmed shelves. Exceptions to the general rule are grainstone–packstone accumulations on bathymetric highs that may punctuate the otherwise monotonous, mud-dominated subtidal environment. The shelf-edge (slope-break) succession may consist almost entirely of skeletal reefs with minor, interbedded grainstones, or it may consist of nearly equal parts of grain-stones and skeletal buildups. Some slope-break successions—some of those on the Great Bahama Banks, for example—consist only of grainstone buildups. Drawings of slope-toe successions do not include slump blocks but instead focus on debrites and turbidites. Distal turbidites, along with laminites and rhythmites, are the default basinal deposits that occur on both ramps and shelves.

5.3   PALEOTOPOGRAPHY AND DEPOSITIONAL FACIES

The seven idealized depositional successions and their variants portray typical deposits that characterize environmental cells across ramps and shelves. The under-

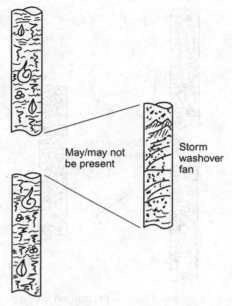

Figure 5.10 Ideal succession for the lagoonal environment. Lagoons are ponded behind barrier islands that characterize the nearshore zone on ramps, although they could exist on open shelves. Barrier islands and lagoons are not typical of rimmed shelves because rims act as baffles to incoming waves and currents causing the back-reef interior to be a "lower energy" environment. Lagoonal successions are characterized by mudstones and wacke-stones, low taxonomic diversity with specialized (euryhaline and eurythermic) biota, and usually by vertical, infaunal burrows. Storm washovers may or may not occur. They are so common along modern barrier island coastlines that they should be present in ancient examples also. They may not always be recognized even though they consist of coarser, unsorted material and open ocean fossils washed in from the seaward side of barriers.

lying assumption in assigning these model successions to specific environmental zones is that the bathymetry across the platform has low relief. That is, the seabed must have no more than a few meters of local relief in order for the hydrodynamic conditions to remain uniform, and hence for the deposit to be uniform over large areas along depositional strike. Topographic relief is rarely uniform and flat. If seabed topography includes prominences ("highs") and depressions ("lows"), then the patterns of sedimentation will reflect these local features as differences in thickness, differences in facies character, or both. In siliciclastic sedimentation, lows are always filled by sediment "thicks" that outline the lows. Bathymetric highs, on the other hand, are zones of limited deposition, nondeposition, or erosion. In the case of detrital carbonate sedimentation, the same rule applies. Filled lows become isopach thicks and antecedent highs are reflected as isopach thins. This fact enables us to use interval isopach maps to interpret platform *paleotopography*. For example, thickness patterns on interval isopach maps reveal the outlines of bathymetric lows and highs that existed during deposition of that stratigraphic interval. Ancient depressions are revealed as isopach thicks and paleo-highs are revealed as isopach thins. Presedimentation highs are zones of low sediment accumulation or erosion, assuming that there was enough wave or current action sweeping over the high to

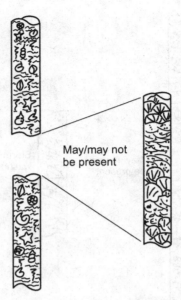

Figure 5.11   The shallow subtidal succession is present on all shelves and ramps. In tropical environments this succession may or may not include skeletal patch reefs with flanking skeletal grainstones and packstones depending on water clarity, nutrients, oxygenation, and suitable substrate to support reef buildups. The default subtidal succession on rimmed shelves and low-energy ramps is dominated by mudstones and wackestones. High-energy ramps and open shelves may exhibit more grain-dominated successions. Restricted interiors on rimmed shelves may have lower taxonomic diversity and muddier substrates than ramps and open shelves at the same latitudes because ramps and open shelves are subjected to ocean waves and currents with attendant nutrients, oxygen, salinity, and moderated temperature.

erode it. In some cases, the high may even be emergent, as in the case of islands.

Carbonates, unlike siliciclastics, are formed by biogenic and chemical processes as well as by detrital sedimentation. Abundant biological or chemical carbonate sedimentation may occur on highs with little or no comparable sedimentation in adjacent deeper water. This is particularly common in the case of reef and certain grainstone accumulations on tops and flanks of paleo-highs. Oolite grainstones are commonly found on the crests of salt-generated highs that marked the Jurassic seabed around the Gulf of Mexico rim during deposition of the Smackover and Cotton Valley Formations. Paleo-highs are favored locations for reefs, too. Reefs are commonly located on paleo-highs such as salt domes, horst blocks, or relict, erosional features. Bioclastic grainstones may accumulate preferentially on the crests of subtidal shoals because the skeletal producers favored the shallow crest of the high over the deeper, less favorable zones on the adjacent seabed. In these instances, relief on the bathymetric highs is increased by greater rates of deposition on the highs and comparatively lower rates of deposition in the adjacent deeper zones. In such cases, care must be taken to interpret interval isopach maps because they will exhibit thick deposits on antecedent highs and thins that outline coeval, antecedent lows—just the reverse of what one might expect for detrital sedimentation. The

Figure 5.12 Slope-break or shelf-edge successions are characteristic of rimmed shelf margins. Slope-break successions may include biogenic reefs and associated grainstones, floatstones, and rudstones. The proportions of skeletal reef, nonbedded detrital infill, and crossbedded grainstones that typify slope-break successions vary depending on relative sea-level history and location with respect to windward and leeward sides of the slope break. Similar successions may occur around local or regional slope breaks on islands or antecedent topographic highs, but they are easily recognized by their association with local topography rather than with continuous, laterally extensive shelf margins. Slope-break successions typically occur in tropical environments; they are rare or absent in temperate settings because skeletal reefs and oolitic grainstones are rare to absent in temperate environments.

take-home lesson is that biogenic and chemical carbonates may accumulate as local thicks on the crests of preexisting highs. This never occurs with siliciclastic sedimentation.

5.4   DIAGNOSIS AND MAPPING OF DEPOSITIONAL RESERVOIRS

To qualify as a depositional carbonate reservoir, reservoir porosity must be dominated by depositional rock properties such as texture, grain type (constituent composition), fabric, or sedimentary structures identified by direct observation of rock samples. Wireline logs, seismic profiles, and even borehole images do not provide enough information to identify and classify porosity. Although it is possible to identify and classify porosity by examining samples from one well, it is not possible to predict the size and shape of the reservoir at field scale without additional information on stratigraphy and structure. This is usually obtained as additional wells are drilled in a field. Predicting the size and shape of the reservoir body is the most

Debrite   Coarse Turbidite

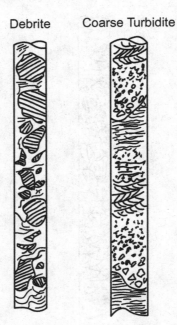

Figure 5.13 Slope-toe successions are characterized by debrites and coarse (proximal) turbidites or grain flows. Fine (distal) turbidites, rhythmites, and laminites occur farther from the slope and are more typical of basinal deposition. Debrites and coarse turbidites are generally absent on ramps owing to the absence of steep slopes that may be unstable and subject to collapse or shedding during storms, earthquakes, or other "events" that trigger slope failure.

basic task of the geoscientists and engineers who must decide where to drill the next well and how to develop a field in the most cost-effective manner. Identification and classification of porosity is the first step in determining the geological origin of a reservoir, predicting its 3D size and shape, and ultimately, in identifying and ranking individual flow units, baffles, and barriers.

The following checklist includes some of the basic methods for identifying and exploiting depositional carbonate reservoirs.

1. Determine the platform type and standard depositional succession or successions that make up the reservoir in order to predict depositional anatomy of the reservoir body. Next, identify and classify genetic pore types by examining rock samples from full-diameter or sidewall cores, or cuttings. Determine which pore types—depositional, diagenetic, fractures, or hybrids of those end members—dominate the reservoir interval. It they are depositional pores or hybrids with dominant depositional attributes, use facies maps as substitutes (proxies) for porosity. For hybrids with diagenetic attributes dominant, identify the diagenetic processes and timing of diagenetic changes in order to map the size and shape of flow units. For example, moldic porosity in a grainstone–packstone reservoir may or may not occupy the entire facies volume, in which case the portion of the facies that was affected by dissolution has to be determined by identifying stratigraphic characteristics such as unconformities, caliche surfaces, karst features, or other evidence of

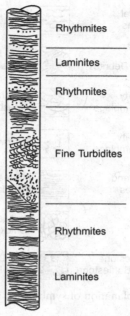

Figure 5.14 Basinal successions consist of fine (distal) turbidites, laminites, and rhythmites that are probably the default successions in the basinal environment on both ramps and shelves. Laminites are millimeter-scale, dark colored, sparsely fossiliferous, usually organic-rich beds that may occupy stratigraphic sections tens to hundreds of meters thick. They represent sedimentation in the cold, dark, dysoxic, deep water. Rhythmites are much like distal turbidites in that they fine-upward and may exhibit small ripples. They differ from turbidites by their small (millimeter-to-centimeter) scale. Typical rhythmites are rarely more than a few centimeters thick but they may be repeated cyclically over many meters in aggregate vertical thickness. The coarse–fine rhythms probably represent events: storms or small seismic events that stir shallow-water sediments to form density currents. Distal turbidites may exhibit typical "Bouma sequences."

dissolution diagenesis that may be interpreted from wireline log signatures or seismic clues.

2. From previous identification of the standard depositional succession represented in the reservoir, predict the 3D size and shape of the depositional body. Anticipate cyclical sedimentation and incomplete successions. Reservoir thickness depends on how many episodes (cycles) of sedimentation compose the reservoir and how many erosional or nondepositional events decreased total reservoir thickness. Expect variations in depositional thickness over long distances and variations in thickness from updip to downdip limits of the depositional body. For example, barrier island successions are thick if both beaches and dunes are present and "stacked" together; they are thin where dunes are absent and the beach was penetrated at its updip or downdip edge. Debrites and turbidites are thicker near the source and updip center of the sediment pile. They are thinner at the distal margins.

3. After several wells have been drilled, use the data to map average porosity in the reservoir interval and compare the porosity map with a structure map on

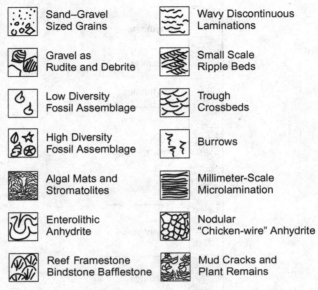

	Sand–Gravel Sized Grains		Wavy Discontinuous Laminations
	Gravel as Rudite and Debrite		Small Scale Ripple Beds
	Low Diversity Fossil Assemblage		Trough Crossbeds
	High Diversity Fossil Assemblage		Burrows
	Algal Mats and Stromatolites		Millimeter-Scale Microlamination
	Enterolithic Anhydrite		Nodular "Chicken-wire" Anhydrite
	Reef Framestone Bindstone Bafflestone		Mud Cracks and Plant Remains

Figure 5.15 Explanation of symbols for Figures 5.8–5.14.

the top of the reservoir horizon. If the maps have similar shapes it indicates that depositional trends follow present-day structure. If they do not match in shape, then go to step 4. If they do match, then map the average permeability of reservoir interval and compare porosity with permeability maps to isolate sectors with highest values of both porosity and permeability to identify flow units with potentially the highest reservoir quality.

4. If present structure and porosity maps are not similar, construct interval isopach maps of marker beds above the reservoir horizon. Isopach thicks and thins indicate lows and highs, respectively, on the surface of the top of the reservoir interval. Construct an isopach map of the reservoir horizon and compare the reservoir thickness with overlying bed isopach thicks and thins to distinguish between paleo-highs representing reservoir thicks and paleo-highs representing antecedent topography overlain only by thin veneers of reservoir rock. Recall that isopach thins over reefy horizons may indicate relief on reef buildups or relief on buildups plus elevation on antecedent highs beneath the buildups. Finally, compare average porosity and average permeability maps for selected reservoir intervals to determine where on paleostructure the best quality flow units occur.

5. When rock and reservoir characteristics have been determined, and flow units have been identified, compare wireline log and seismic data with rock and pore characteristics to determine whether a log or seismic signature can be used to map and correlate flow units. Caliper, neutron, density, sonic, NMR, and image logs are first choices. Gamma ray logs may help identify unconformities that may, in turn, be proximity indicators for porous zones. Once it is determined that certain log or seismic signatures "pick out" flow units, it should now be possible to correlate log or seismic signatures from well to well, making it possible to generate 3D maps of flow units, baffles, and barriers.

SUGGESTIONS FOR FURTHER READING

Extensive discussions of carbonate depositional environments from the strandline to the basin are presented in the AAPG Memoir 33, *Carbonate Depositional Environments* (Scholle et al., 1983). Individual papers are well illustrated with photos and diagrams, most of which are in color and extensive citations from the research literature are included. Good descriptions of carbonate depositional facies and vertical successions from a variety of environments and geographic locations are presented in Walker and James (1992), *Facies Models: Response to Sea Level Change.* A standard reference on depositional facies is by Reading (1996), *Sedimentary Environments: Processes Facies, and* Stratigraphy, 3rd edition. *Cool Water Carbonates* (James and Clarke, 1997) contains a variety of papers to illustrate the range of characteristics in temperate and cool-water carbonates. The most recent, comprehensive textbook on carbonate sedimentology is *Carbonate Sedimentology* by Tucker and Wright (1990). They provide detailed discussions on carbonate depositional environments, diagenesis, and geological trends in carbonate sedimentation through time. Wilson's 1975 book, *Carbonate Facies in Geologic History*, is an excellent review of carbonate depositional systems by geological age around the globe. Carbonate depositional systems are examined in sequence-stratigraphic context in *Carbonate Sequence Stratigraphy* (Loucks and Sarg, 1993).

REVIEW QUESTIONS

5.1. In purely depositional reservoirs, there are four pore types found in detrital carbonates. What are they?

5.2. Beach–barrier island successions consist of lower shoreface, middle shoreface, and upper shoreface segments. What physical oceanographic processes determine the depth at which the lower shoreface segment grades into the shallow subtidal environment?

5.3. What are some distinctive sedimentary structures in dunes that make them relatively easy to distinguish from beach successions?

5.4. We know that OOIP = $7758Ah\phi(1 - S_w)/B_{oi}$. Use a reservoir model based on the Pleistocene outcrops near Isla Cancun in Yucatán where dunes are about 30 m in average height, 100 m in average width, and extend about 7 km along the eastern rim of the island (Ward, 1975, 1997). Consider only a small segment of the total complex in which the hypothetical reservoir is 6 m (20 feet) thick above the oil–water contact. The reservoir is 4 miles long and 330 feet wide, or about 160 acres in map area. Average porosity is 15%, and S_w = 30%. Assume B_{oi} is 1.0.

(a) What is the estimated OOIP of our 160-acre field?

(b) In most reservoirs, the amount of recoverable oil is about 30% OOIP. Using an oil price of $75/barrel, what is the estimated dollar value of the oil in place?

(c) If we develop the field on 40-acre spacing, what is the average dollar value of each well?

5.5. Assuming an average porosity of 12% and $S_w = 25\%$, what is a reasonable estimate of OOIP for the Oaks Field? What is the average value of each of the 13 wells with $75 oil? Assuming a drilling cost of $95/foot for this area in 1979, what did it cost to reach the Smackover reservoir at 3.5-km depth? What errors might result in a geological cross section based on wireline logs from three wells, each in the center of the barriers and none on the flanks?

5.6. Tidal flat–lagoon successions may produce from diagenetically altered porosity, particularly in the transition zone between supratidal and upper intertidal zones, where fenestral porosity, intraclastic conglomerates, and scattered grainstone–packstone beds occur. One reservoir example is Cabin Creek Field, Williston Basin, Montana (Roehl and Choquette, 1985). This Silurian-aged reservoir is about 3 km × 14 km in total area, of which 8100 acres are productive. Average porosity and permeability are 15% and 5 md, respectively. Assume one pay zone about 12 feet thick, with $S_w = 35\%$. Forty-five percent of the porosity is fenestral with pore sizes ranging from 0.5 mm to 1 cm in diameter. The remaining 55% of porosity consists of sheet-like pores that connect most of the fenestrae, along with intergranular pores in the fine sand-sized peloidal matrix. The main rock types are wackestones with interbedded flat-pebble conglomerates and peloidal packstones, algal-laminated mudstones, and evaporites.

(a) What is the orientation of the reservoir body with respect to depositional dip and strike?

(b) What is the estimated OOIP?

(c) During field development, what potential problems may be encountered because of facies variability?

5.7. What part of the shallow subtidal zone should offer the greatest potential for depositional reservoir porosity? Why?

5.8. What is the anatomy and orientation of typical slope-break grainstone successions? These grainstone complexes consist of overlapping lime sand lobes shaped by different episodes of vigorous tidal currents.

(a) What potential problems could this anatomical variability pose for geologists attempting to map the size and shape of flow units?

(b) Referring to Chapter 2, what kinds of reef characteristics might appear in slope-break buildups of Mesozoic, Cenozoic, and Paleozoic age?

(c) What classification scheme can be used to help identify the degree and kind of variability in reef porosity?

(d) What information in the Lucia classification of carbonate porosity might indicate the need for caution in calculating recoverable reserves in reef reservoirs?

5.9. What exploration strategies would you use to find rocks with the greatest depositional porosity in slope-toe deposits?

(a) What distinguishes between turbidites or grain flows and ordinary basinal bottom sediments?

(b) What types of slope-toe deposits probably have the greatest depositional porosity and permeability?

(c) What methods might be used to predict the size and shape of slope-toe reservoirs?

(d) What uncertainties exist in calculations of OOIP for slope deposits?

5.10. Laminites, rhythmites, and distal turbidites are not usually good reservoir rocks. Why?

(a) What kind of basinal deposits might have enough depositional porosity to be commercially important?

(b) What strategy would you use to predict the three-dimensional form of the reservoir body?

(c) Would you expect compartmentalization or laterally extensive flow units in such rocks? Why?

CHAPTER SIX

DIAGENETIC CARBONATE RESERVOIRS

6.1 DIAGENESIS AND DIAGENETIC PROCESSES

This chapter covers processes and products of diagenesis and how they affect carbonate reservoirs. Emphasis is on recognition of diagenetic and highly altered hybrid pore types and on methods for analyzing and exploiting diagenetic reservoirs. While diagenetic porosity and permeability are created or modified by postdepositional processes, depositional porosity and permeability are formed at the time of deposition. The key to understanding and exploiting diagenetic reservoirs lies in the ability to identify and correctly classify pore types by origin. A further challenge is to determine whether the pore systems are purely diagenetic or some hybrid of diagenetic plus depositional or diagenetic plus fracture porosity. The economic bottom line is to assess the extent to which diagenesis has modified depositional porosity and permeability and how those modifications have influenced reservoir performance. Purely diagenetic porosity such as intercrystalline porosity in dolostones requires exploration and development strategies built around a geological concept that enables one to map and predict the spatial distribution of the porous and permeable dolostone. That is, one has to understand the origin and spatial distribution of diagenetic changes—not the distribution of depositional facies. As with depositional reservoirs, the main materials needed to analyze diagenetic reservoirs are borehole cores, cuttings, and subsurface cross sections and maps. Depositional facies maps are commonly used as proxies for maps for porosity in reservoirs with purely depositional or lightly altered hybrid pore types. Most wireline logs and seismic records are not useful for identifying diagenetic porosity, but that information is helpful in identifying structural or stratigraphic trends that may have influenced diagenetic patterns. Borehole logs and other petrophysical characteristics

Geology of Carbonate Reservoirs: The Identification, Description, and Characterization of Hydrocarbon Reservoirs in Carbonate Rocks
By Wayne M. Ahr   Copyright © 2008 John Wiley & Sons, Inc.

such as capillary pressure and NMR measurements are very important for identifying flow units, baffles, and barriers within reservoirs and for establishing a quality ranking system for reservoir flow units, but logs alone cannot distinguish between genetic pore types.

When diagenetic porosity and permeability are intimately related to depositional rock properties, reservoir boundaries conform to depositional facies boundaries. If diagenetic changes follow fracture or joint patterns, then determining the size and shape of the reservoir may be a job of interpreting the fracture distribution pattern rather than one of interpreting patterns of diagenesis. In purely diagenetic pore systems that do not conform to fracture or depositional trends, the techniques for analyzing reservoir performance are defined by the type and extent of diagenesis that created the porosity. Reservoir size and shape may depend on the mechanism of diagenesis, the environment of diagenesis, and the size and shape of the zones that were exposed to diagenesis. For example, reservoirs formed by replacement diagenesis—as in replacement dolostones after limestones—could have boundaries roughly related to the size and shape of ancient coastal salinas, where evaporative brines reacted with metastable carbonates to form new minerals, fabrics, pores, and pore throats. Diagenetic reservoir architecture may correspond to the geometry of an ancient water table, to an unconformity or exposure surface, to a paleosol horizon, or to a karst surface. In short, diagenetic porosity may or may not correspond to depositional or structural trends. The challenge lies in doing the geological detective work to determine relationships between diagenetic porosity and other geological attributes of reservoir zones.

6.1.1 Definition of Diagenesis

The term diagenesis derives from ancient Greek *dia*—across or through—and *genesis*—origin or generation. In today's literature, diagenesis is generally interpreted to mean "across generations" in the sense that diagenetic changes cut across (modify) different generations of minerals or rock properties. Most current reference books more or less have the same definition for diagenesis; namely, it is *all of the changes that happen to sedimentary rocks after deposition and before metamorphism*. All changes in size, shape, volume, chemical composition, or crystalline structure of a sedimentary rock after its detrital, biogenic, or crystalline constituents have been deposited. It is easy to imagine changes happening to sedimentary particles but the boundary between diagenesis and metamorphism is not as easy to recognize. Because that is a gradational boundary and it is not easy to identify one side as purely sedimentary and the other side as purely metamorphic, most modern workers accept that the boundary is gradational and focus instead on the key rock properties that can be identified as low-temperature and low-pressure changes that occurred in the burial diagenetic domain. For those wanting more precise definitions of this boundary region, sedimentary geochemists have devised a scheme to measure "organic metamorphism" based on the degree to which kerogen and other organic constituents are heated during burial. This scheme is used to describe the maturity of kerogen-rich source rocks that yield petroleum hydrocarbons when heated to high enough temperatures. Kerogen releases petroleum hydrocarbons at burial depths characterized by temperatures ranging between about 65 and 150 °C (Selley, 1985). This temperature range is sometimes called the "oil window" if the principal hydrocarbons liberated from kerogen are oil rather than gas. Traditional metamorphic minerals and facies do not form at temperatures and pressures that accompany organic metamorphism. A good example of hydrothermal mineralization in the metamorphic domain rather than the sedimentary diagenetic domain is the greenschist metamorphic mineralization that occurred at temperatures of about 390 °C in subsurface sandstones around the Salton Basin geothermal area (Helgeson, 1968; Hulen et al., 2003). Metamorphic petrologists suggest that the onset of metamorphism is marked by the appearance of muscovite, epidote, albite, lawsonite, laumontite, or pyrophyllite (Ehlers and Blatt, 1982). Although all of these minerals do not form at the same temperatures and pressures in the subsurface, they are reliable indicators of metamorphism and would not be expected to occur together with oil or gas. That said, one would probably never find these minerals in sedimentary carbonates, particularly in reservoir rocks. In practice, deep-burial diagenesis in carbonate reservoir rocks never reaches the level of conventionally defined metamorphism even though minerals such as metallic sulfides, fluorite, and saddle dolomite may be formed during the invasion of deep-burial fluids into carbonate reservoirs. Mineralization of this type is called MVT (Mississippi Valley type) diagenesis. It is a form of mineralization that commonly occurs in association with the influx of hydrocarbons and their associated mineralizing fluids during migration from source rock to reservoir.

6.1.2 Diagenetic Processes

The mechanisms of diagenesis may be mechanical, biological, or chemical, or several of them in some combination. An example of mechanical diagenesis is volume reduction by compaction during burial. In fine-grained rocks such as lime mudstones, compaction is usually accompanied by expulsion of interstitial water, reduction in thickness of beds or laminae, and an increase in bulk density. In granular rocks or some reef rocks, compaction can modify grain packing; it can cause penetrative grain-to-grain contacts (fitted contacts) to form or may cause grain breakage. Stylolites (Figure 6.1) are produced by the combination of compaction and dissolution and are common in all carbonate rock textures. After chemical diagenesis, mechanical compaction has the greatest influence on carbonate reservoir porosity.

Biological diagenesis consists mainly of bioerosion, or the grinding, rasping, boring, and otherwise eroding of rock surfaces by plants and animals. Some organisms may produce secretions that dissolve rock. Bioerosion is particularly visible in some modern tropical carbonate coasts, where notches and overhangs have been formed at water level along steep, rocky shorelines by the combination of bioerosion and breaking waves. Bioerosion is relatively unimportant as an agent of change in carbonate reservoir porosity but it may produce a large percentage of the mud fraction in sediments from some depositional settings.

Chemical diagenesis is the most important agent of change in carbonate reservoirs. It is a process of rock-and-water interaction that proceeds in rates and directions dictated by the equilibrium between the rocks and the water. The main processes in chemical diagenesis are dissolution, cementation (precipitation), recrystallization, and replacement. Dissolution occurs when the rock–water system is out of equilibrium. In such a case, the water is undersaturated with respect to $CaCO_3$. For example, meteoric water dissolves $CaCO_3$ until saturation equilibrium is reached

Figure 6.1   The stylolite just below the 1-cent coin is typical of these features that are formed by the combination of pressure and dissolution. Most of them are filled with insoluble material, commonly including or limited to bituminous residue and most stylolites are baffles or barriers to flow. A few may remain open, or become opened during later dissolution, in which case they may provide access for fluid migration (see the sketch in Figure 6.6). This stylolite, like most, resembles a trace on a seismograph or an electrocardiogram. Note the small natural fractures and dissolution vugs also in this core segment from a Mississippian-aged dolomitic reservoir in Texas.

between rocks and water. Generally, as long as a continued supply of undersaturated water is available to react with the rock, dissolution continues. Extensive dissolution creates karst features, such as caverns and sinkholes, and enlarges pores to form molds and vugs. Water moving down aquifers dissolves as it migrates so long as it is undersaturated in $CaCO_3$. If the aquifer water becomes saturated, dissolution stops. If excess calcium carbonate is in solution and a trigger exists for precipitation, then cementation will take place in the water-filled pores. Similarly, fresh water may percolate from the surface downward through rock or soil, dissolving as it goes until saturation equilibrium is reached or exceeded. As with aquifers, cementation can occur when carbonate saturation is high and a trigger mechanism causes precipitation. Stagnant aquifers are those in which water does not move or moves very little; consequently, there is comparatively low potential for either dissolution or cementation (Longman, 1980).

Carbonate cementation binds loose grains and fills pores with one or more carbonate crystal growth forms. Cement crystals may occur in a variety of individual shapes and forms depending on the crystallographic system in which the mineral forms and on the environment in which crystallization occurs. For example, calcite may crystallize as pointed scalenohedra ("dogtooth spar") or as flat, bladed rhombohedra, depending on environmental conditions. Calcite varieties can also form in crystal bundles such as radiaxial fibrous and fascicular optical calcite. Some calcite cements are micritic or peloidal. Peloids are typically spheroidal in shape and are a few micrometers in diameter. The spheroids consist of crystallites arranged radially around grain centers, or nuclei. Most peloidal cements are interpreted to result from microbial or biochemical processes because the peloids occur in cavities, sometimes as geopetal fillings, within reefs and "mud mounds." Microbial origin for some carbonate deposits may be difficult to document; thus there is commonly some controversy over their origin. The muddy interstices in these reef and mound environments must have been fertile culture media for microbial life and there is usually little or no evidence of animal life that could have produced peloids as fecal pellets. There is no evidence of inorganic precipitation such as a continuous horizon of peloidal grains indicating a sediment–water chemical interface. There is also no evidence of mechanical abrasion to produce uniform, spheroidal, silt-sized grains that typically fill sheltered cavities. Mechanically abraded grains, or intraclasts, are generally not uniform in size and shape. Aragonite commonly crystallizes as needle-like orthorhombic prisms only a few micrometers in size that can form grain-coating, isopachous crusts, or rim cements (Figure 6.2a). These pore-lining aragonite cements indicate a marine phreatic origin. Botryoidal masses of aragonite crystals up to tens of centimeters in length are commonly found in modern and ancient reefs (Figure 6.2b), and some aragonitic cement is present in cave dripstones. Dolomite (Figure 6.2c) typically crystallizes as simple rhombohedra in most surface and near-surface environments but it may grow into a baroque form called saddle dolomite (Figure 6.2d), particularly in the burial environment where water temperatures are 60 °C and higher (Radke and Mathis, 1980). A great deal of attention has been given to "hydrothermal dolomite" in recent times. Some disagreement exists about the best definition of "hydrothermal" and about the source of dolomitizing fluids that created saddle dolomite. It is commonly associated with but not limited to fractured reservoirs in carbonates (Ahr, 1982). Saddle dolomite may also form as a by-product of thermochemical sulfate reduction in nonfractured reservoirs. Some of the early discussion on "TSR"-formed saddle dolomite was given by Machel (1987b).

Recrystallization is the process by which crystal morphology is changed without major changes in the mineral composition. Sometimes this process is called "neomorphism," a term coined by R. L. Folk (1965), who defined neomorphism to include both true recrystallization and mineralogical inversion. True recrystallization is a change in crystal form without changes in mineralogical composition: for example, micrometer-sized crystals of calcite micrite going through transformation to millimeter-sized blocky calcite crystals (Figure 6.3) during what Folk (1965) called aggradational neomorphism. Mineralogical inversion is not strictly recrystallization; rather, as an example, it is the process by which a metastable mineral such as aragonite or Mg-calcite undergoes both crystallographic and compositional change to become ordinary calcite. Even though the basic composition of aragonite is $CaCO_3$, it may have a different trace element and isotopic composition than its postneomorphism daughter product just as Mg-calcite will lose Mg to become ordinary calcite. Aggradational neomorphism in microporous micrites can produce close-fitting crystal mosaics that have lower porosity and permeability than their precursor. Finally, micritization, or grain diminution, is a form of degradational neomorphism. One form of micritization, or at least a form of grain-size reduction, is sometimes

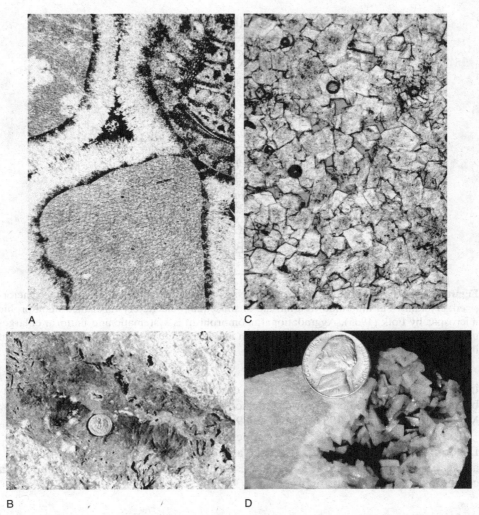

Figure 6.2   (a) Photomicrograph of a modern skeletal grainstone with isopachous aragonite rim cement typical of the marine phreatic environment. The grains are foraminiferal and red algal fragments. The width of the photo is 2.0 mm. (b) An outcrop photo of botryoidal aragonite cement filling a cavity in the Guadalupian (Permian) reef of West Texas. The dark brown cement is presently calcite but it retains the original aragonite crystal structure visible as fan-shaped crystal arrays, two of which are to the right of the coin and one is to the left. The bright (white) marks are hammer scars left by careless students. (c) Rhombohedral dolomite crystals typically indicative of early diagenetic replacement. The width of this photo is 2 mm. (d) Saddle dolomite crystals filling a large vug in a Mississippian limestone. Saddle dolomite is typical of late burial replacement and is formed at temperatures of 60 °C and above. (Photo (c) courtesy of David Kopaska-Merkel; the others are by the author.)

Neomorphism (Recrystrallization)

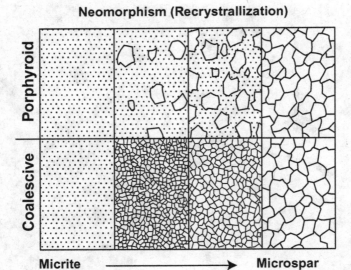

Figure 6.3 Diagram illustrating neomorphism—the diagenetic change from micrometer-sized, crystallites in lime mud to a mosaic of subsequant calcite crystals called neomorphic microspar by Folk (1965). Aggradational neomorphism is systematic and uniform enlargement of crystallites, porphyritic neomorphism is enlargement in patches. (Adapted from an illustration in Folk (1965).)

incorrectly called "chalkification." In this case, rather than true chalk, a microrhombic microcrystalline calcitic fabric is formed by diagenetic alteration of aragonite or Mg-calcite ooids (Ahr, 1989; Dravis, 1989; Moshier, 1989).

Replacement is the process of complete change from one mineral to another. Dolomite commonly replaces calcite and aragonite, although replacement by minerals such as anhydrite and SiO_2 are almost as common and both calcite and silica commonly replace evaporite precursors (Ulmer-Scholle and Scholle, 1994).

6.2 DIAGENETIC POROSITY

Diagenetic porosity can be formed by dissolution—the enlargement of existing pores (e.g., solution-enlarged intergranular pores) or the creation of new ones (e.g., karst features). It can be formed by replacement with successor minerals that ultimately occur in some fashion that is associated with greater porosity than the parent rock (dolomitization), and it can be formed by dissolution or stabilization followed by reprecipitation that yields a porous, neomorphic product (e.g., microporosity). When porosity is created or enhanced by these processes, the previous texture, fabric, and sometimes the mineralogy of the host rock may be radically changed. Purely diagenetic porosity is not as common as diagenetically altered depositional porosity. However, some purely diagenetic alterations exist that may not correspond with depositional characteristics. Intercrystalline porosity in dolomite (replacement) may be unrelated to depositional facies boundaries or fabrics, as we learned in carbonate porosity classifications. Vuggy or cavernous porosity and porosity associated with karst processes (dissolution) may correspond with paleotopography and positions of ancient water tables instead of depositional fabric or facies, or it may correspond to the distribution of metastable precursor minerals such as aragonite or Mg-calcite. In such cases, selective dissolution of metastable constituents can produce patchworks of moldic or vuggy pores that mark the previous position of aragonitic or magnesium-calcite components, or simply the position of ancient water tables. Neomorphism of limestone resulting in a net porosity increase is uncommon because limestone neomorphism usually produces close-fitting crystal mosaics in which compromise crystal boundaries (close fit between crystal faces at angles of 120°) are abundant. Intercrystalline porosity is low in this case and the resulting sheet-like pore throats are too small (micrometer-sized) to allow easy liquid flow, although microporous gas reservoirs are common.

Porosity created by diagenesis, not simply modified by it, can be grouped into four end-member categories: (1) intercrystalline, (2) moldic, (3) vuggy, and (4) cavernous pores. A variety of combinations can exist. Some examples include solution-enhanced versions of inter- and intragranular pores, intercrystalline pores, solution-enlarged fractures or joints, and complex porosity in solution-collapse or karst-related breccias. Stylolites may act as pores, but they are usually flow barriers because they are typically plugged with insoluble residue. Intercrystalline porosity is common in dolomite replacements of limestone. It is rare in neomorphosed limestones with one noteworthy exception: microporous, microrhombic low-Mg calcite that forms from metastable precursors, probably Mg-calcite, by a stabilization process that is not well understood (Ahr, 1989; Dravis, 1989; Moshier, 1989). This form of "chalky" porosity is the primary reservoir rock in many natural gas fields, including many Cotton Valley Limestone (Jurassic) fields in the East Texas basin.

Intercrystalline porosity, literally pore spaces between crystals, requires neomorphism, replacement, and perhaps enlarging grain boundaries by dissolution. Replacement of limestone by dolomite has been cited as a mechanism by which, in theory, a solid volume decrease of up to 13% can be accomplished (Weyl, 1960). Porosity measurements on synthetic dolomite in laboratory studies by Bubb and Perry (1968) were within the limits of the theoretical predictions, but in practice, the theory has not been shown to work. Weyl's paper gained wide acceptance because many productive reservoirs with intercrystalline porosity are in dolostones. However, for such a volume change to occur requires substitution of Mg^{2+} into the $CaCO_3$ lattice along with removal of Ca^{2+}. If Mg^{2+} and CO_3^{2-} are supplied simultaneously to the dolomitizing reaction, there is a volume increase of 75–88% (Morrow, 1990). Replacement of limestone by dolomite does not automatically produce a 13% increase in total reservoir porosity (Lucia, 2000). Why then do most dolomite reservoirs with a given porosity have greater permeability than limestone reservoirs with the same porosity (Wardlaw, 1979)? Intercrystalline porosity decreases with initial increase in the dolomite/calcite ratio in ancient carbonates up to about 50% replacement by dolomite (Murray, 1960). But Murray found that as dolomite content increased beyond 50%, intercrystalline porosity increased proportionately, leading him to conclude that high percentages of intercrystalline porosity are the result of dissolution of the last remaining calcite host rock. His observations are purely empirical and have not been duplicated in the laboratory. His interpretation that dolomite was concentrated by subtle dissolution that removed the last remaining calcite from between the replacement dolomite crystals is compelling. This process would produce high dolomite/calcite ratios, enhance porosity and permeability, and generally create good reservoir rocks. This, along with the tetrahedral geometry of dolomite intercrystalline pore throats that we discussed in Chapter 3, is probably the main reason why dolostones have higher permeability than limestones with the same porosity. Recall from Chapter 3 that the expression for capillary pressure can be written differently for tetrahedral porosity in dolostones as compared to interparticle porosity in limestones and sandstones. Petrographic examination shows that crystals in porous dolostones are generally uniform in size and euhedral to subhedral in form. Dolomite crystals do not usually exhibit compromise boundaries, but calcite crystals almost always do. That is, dolomite crystals appear to have stopped growing when initial contact was made with adjacent crystals—a phenomenon described as *contact inhibition* by Wardlaw (1979). It is probably the idiomorphic (well-formed) shape, uniform crystal size, and absence of compromise boundaries between dolomite crystals that give rise to the sucrosic texture described by Archie (1952) and that make this rock type a more permeable reservoir than limestones with comparable porosity (Wardlaw, 1979). Replacement of limestone by dolomite does not a priori produce a net increase in reservoir porosity, however. This is particularly true if the dolostone/limestone ratio is small. Moreover, it is difficult to prove that calcite was removed from dolomitic rocks by dissolution because most dolomite crystals in reservoir rocks show little evidence of corrosion or dissolution. Instead, they are typically euhedral to subhedral and exhibit sharp crystal edges. Perhaps the volume of dissolved calcite was small compared to the volume of dolomite remaining, and because dolomite is less soluble than calcite it was comparatively unaffected. The complex association between dolomite and reservoir porosity is reviewed by Sun (1995), who concludes that porosity evolution in dolomites involves both syndolomitization dissolution and postdolomitization modification such as karstification, fracturing, and burial corrosion. This line of investigation was pursued by Saller and Henderson (1998), who described porosity increasing basinward in dolostone reservoirs on the Central Basin Platform in West Texas. Those authors proposed that dolomite was more abundant on the updip portion of the platform, where evaporation had concentrated the seawater, and that such abundance allowed "excess" dolomitization to occur in the form of replacement dolomite followed by dolomite cementation. This scenario is similar to that described by Lucia and Major (1994) for the Plio-Pleistocene carbonates of Bonaire, Netherland Antilles. Lucia and Major emphasize that dolomitization of limestones does not lead a priori to increased porosity and may, in fact, lead to decreased porosity after continued emplacement of diagenetic dolomite cements. In their study of dolomitized rocks in the Permian Basin, Saller and Henderson (1998) argued that as diagenetic fluids migrate basinward, they could become less saturated with respect to dolomite after much of it had been "used-up" by replacement and cementation. From this point basinward, the migrating brines could trigger simultaneous dolomitization and dissolution of $CaCO_3$ with the end result being more porous and permeable dolostones near the shelf margin, where excess dolomitization as cement did not occur. Morrow (2001) argued that the process described by Saller and Henderson (1998) could have occurred in two stages: a massive episode of dolomitization on the updip platform followed by downdip reflux of brines capable of forming dolomite but undersaturated with respect to $CaCO_3$ in order to dissolve any nonreplaced limestone. The take-home message for readers is this: as Wardlaw (1979) observed, dolostones are in general more permeable reservoir rocks than limestones with the same porosity. There are no unique or universally applicable answers to the questions of how, at what rate, and under what controlling circumstances dolomitization occurred, but the job of finding and developing dolostone reservoirs is very real. As one is challenged to predict the spatial distribution of porosity and permeability in dolomitized reservoirs, the questions listed above must be answered. The answers will be found only by studying the rocks first-hand. Seismology and wireline logs will offer little or no help.

Moldic, vuggy, and cavernous pores are formed when carbonate rock matrix and pore walls are dissolved when fluids passing through the pore system are undersaturated with respect to the reservoir rock. These pore types have a common origin but they differ greatly in size and shape. The dissolution that affected them may occur in shallow- or deep-burial environments. Molds represent dissolved former grains or crystals, vugs represent dissolved spaces larger than the surrounding grains or crystals, and caverns include very large dissolution pores. Carlsbad Caverns in New Mexico can be thought of as a gigantic pore system. Moldic and vuggy porosity may be related to depositional fabric or depositional facies in some cases. Moldic pores typically indicate differential solubility between more and less soluble carbonate constituents in the rock fabric. In some cases, the more soluble constituent may be a metastable mineral such as Mg-calcite or aragonite. Preferential dissolution may be due to a difference in grain surface area/volume ratio (Walter, 1985). Smaller grains with large surface area/volume ratio can be dissolved more easily than larger ones. Some moldic pores owe their existence to dissolution of large grains within a mud matrix after the mud, but not the grains, has undergone replacement or neomorphism to become less soluble. Subsequent exposure to leaching removed the soluble grains but the transformed matrix remained intact. Some vugs are simply moldic pores that were dissolved to sizes greater than the preexisting grain or crystal boundaries.

6.3 DIAGENETIC ENVIRONMENTS AND FACIES

The principal diagenetic environments are the vadose, meteoric phreatic, mixing-zone, marine phreatic, and subsurface or burial environments (Figure 6.4). The vadose environment lies above the water table and all pores in this domain are filled with both air and water. Water resides only temporarily in vadose pores, depending on the quantity and frequency of precipitation. Water moves through the vadose zone leaving only surface-tension films on grain surfaces and meniscus films across pore throats. The depth where all pores are filled with water is called the water table. That surface represents the top of the phreatic zone. Below that surface, where all pores are filled with fresh water, is the meteoric or freshwater phreatic zone. The depth to the water table varies depending on subsurface geology, topography, capillarity, and climate. Near the marine environment, fresh water mixes with seawater in the mixed phreatic or mixing zone and aquifers saturated with seawater make up the marine phreatic environment. Subsurface or burial diagenetic environments are those in which water chemistry is unlike either the meteoric or marine phreatic zones, and temperature and pressure become increasingly important. Subsurface water chemistry is different because it reflects rock–water interactions and water

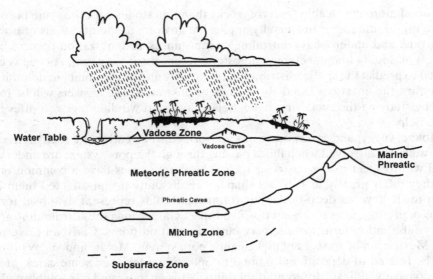

Figure 6.4   A sketch showing the principal diagenetic environments from the surface vadose environment through freshwater and marine phreatic, to the mixing zone, to the shallow- and deep-burial environments. (Adapted from an illustration in McIlreath and Morrow (1990).)

expulsion during compaction, among other possible sources that can create hybrid water composition.

Diagenetic environments are classified mainly on the basis of water chemistry and location with respect to the earth's surface. Fresh water, brackish water, seawater, and expulsion brines from the subsurface are the main kinds of diagenetic fluids that come in contact with carbonate rocks during their diagenetic histories. True connate water (original water of deposition trapped in buried rock) probably does not exist because extensive studies on subsurface brine compositions show that those brines have dramatically different compositions than do surface waters in marine and terrestrial settings. Instead, subsurface waters are mixtures of trapped surface waters, expulsion fluids from mechanical compaction during burial, and diagenetic fluids from rock–water interactions during burial. Shallow- and deep-burial environments are distinguished not by prescribed depth zones but by presence or absence of clues to elevated temperature and pressure along with evidence of exotic water chemistry typical of basinal brines. Other clues include diagenetic fabric, mineralogy and cement characteristics, and geochemical indicators of environmental parameters. Some fabric characteristics that indicate pressure include penetrative grain contacts ("overcompaction") and stylolites. Mineralogical indications of deeper burial include the presence of high-temperature minerals such as saddle dolomite, high levels of kerogen metamorphism, and the amount, crystallographic character, and isotopic composition of pore-filling cements. Carbon and oxygen isotopic compositions in combination with fluid inclusion geochemistry are commonly used to identify the type or source of carbon and the temperature of crystallization, respectively. There are many more clues and techniques that can be used to identify the kinds of diagenetic alteration and the environments in which they formed. Some of them will be mentioned in this chapter, but no attempt is made to present a comprehensive review of carbonate diagenesis.

6.3.1   Diagenetic Facies

Depositional facies were defined earlier as laterally continuous, mappable horizons with well-defined associations of fundamental rock properties and, in some cases, fossils. Depositional facies are usually described as sedimentary blankets of relatively local extent and limited thickness. Diagenetic facies are also laterally continuous zones that display mappable patterns of diagenetic alteration. For example, rocks exposed to freshwater dissolution at an unconformity or at an ancient water table could be grouped into diagenetic facies on the basis of the amount and kind of diagenetic changes they exhibit. These diagenetic patterns can be treated as "facies" to make the task of mapping somewhat similar to the mapping of depositional facies. In this case, the diagenetic characteristics are mapped instead of depositional rock properties. Mapping diagenetic facies requires information about the vertical and lateral extent of the stratigraphic horizon that is associated with the diagenesis. Care must be taken to ensure that the diagenetic facies include only those rocks that were associated with that specific stratigraphic horizon during one diagenetic episode because it is possible that more than one episode may have occurred at different times during burial. Stratigraphic markers are usually necessary to establish datums for mapping diagenetic facies. Such markers include hardgrounds, unconformities, paleosols, or other features typically associated with subareal exposure or ancient water tables.

Diagenetic facies are different from depositional facies and the methods used to predict their spatial distribution are correspondingly different. Diagenetic alterations may or may not have followed depositional facies boundaries and they may not be blanket-like in form. Sinkholes in karst topography are good examples of one diagenetic terrane that cuts across both lateral and vertical depositional boundaries. Instead of following traditional depositional boundaries, diagenetic facies may reflect the outlines of ancient water tables, or zones where evaporative concentration formed enriched brines that percolated through the shallow subsurface and selectively replaced metastable carbonates. Diagenetic facies may have followed ancient topography, where highs were exposed to weathering and erosion while nearby submerged low areas remained unaffected. They may mark subsurface regions where rock–water interactions were part of circulation systems driven by temperature or concentration gradients as in Kohout circulation, a system of platform-wide circulation cells associated with regional diagenetic alteration (Kohout, 1967). Zones of extensive compaction and pressure solution may occur in halos around tectonic features where stresses were concentrated. Away from stress concentrations, beds of the same lithology and the same age will not show the same degree of mechanical alteration. Because nonselective diagenesis cuts across depositional boundaries, it is important to identify the distinguishing characteristics of each type of alteration.

Diagenetic facies maps can be constructed using these distinguishing characteristics even though it may be difficult at first to identify the size and shape of the altered zones in the subsurface. Of course, reservoir rocks have almost always been affected by more than one episode of diagenesis during their burial history; therefore it is important to record each type of alteration, temporal relationships between episodes of alteration, and correspondence between diagenesis and tectonosedimentary history. Because diagenetic histories can be complex, a practical approach to identifying and mapping diagenetic facies involves using the genetic classification scheme for carbonate porosity presented in Chapter 2. In essence, reservoir pore types are mapped on the basis of their depositional and diagenetic characteristics so that *pore facies* maps can be constructed to pinpoint the spatial distribution of the pore types with highest corresponding permeability and lowest resistance to fluid flow and that can be correlated stratigraphically at field scale. Each episode of diagenesis leaves distinctive traces that can be compared with other microscopic traces to reveal the timing of each event. This tracing of diagenetic events is done by identifying *cross-cutting relationships* in thin sections. The latest or last event cuts across the previous one, and so on, until the first diagenetic event can be isolated. In this way, episodes of dissolution, cementation, compaction, or other forms of diagenesis can be identified and placed in chronological sequence to reveal the burial history and the geological cause–effect system that modified the reservoir rocks.

6.4   DIAGENETICALLY ENHANCED POROSITY

Diagenetic trends may or may not cut across depositional facies boundaries and they may include a variety of pore types that represent more than one diagenetic event. Knowing this, the traditional approach for many geoscientists is to focus on case histories—catalogs of reservoir examples—in which cross-cutting relationships, genetic categories of pore types, or chronologies of pore formation and modification are compiled. Case histories are usually used as models or analogs to help understand other reservoirs by providing a similar appearance, a "look-alike" or a template. However useful they may be as references for comparison, analogs cannot explain how different reservoirs formed (cause–effect relationships) or the relative timing of porosity formation, and analogs cannot provide reliable means with which to predict the spatial distribution of reservoir properties. Analogs are commonly used without doing additional studies of depositional and diagenetic histories that influenced the formation of pore types, flow units, baffles, and barriers. Because analogs and case histories have limited application in critical analysis of reservoir origin and distribution, the focus of the following discussions is on *how to recognize and map diagenetic pore systems*. Diagenetically enhanced porosity is considered first. Diagenetically reduced porosity is discussed later. Once diagenetically enhanced porosity is identified, geoscientists and engineers can focus directly on individual pore types, their relationship to fundamental rock and reservoir properties, and their times of origin relative to mappable depositional and diagenetic events. For example, it is relatively common nowadays for specialists to identify the relative times of origin and cause–effect relationships linked to specific diagenetic events during burial history. Individual pore types can be distinguished even though multiple episodes of diagenesis may be present in the rock. With this information in place, it is usually possible to identify reservoir flow units based on petrographic and petrophysical attributes of pore types and on the cross-cutting relationships they display within the rock framework. Finally, the spatial distribution of flow units can be determined by correlating pore "facies" based on pore attributes and pore genesis from zone to zone and well to well. This process allows mapping of flow units and flow barriers at field scale.

Diagenetic processes that create or enhance porosity are dissolution, replacement, and recrystallization. Dissolution of solid rock in undersaturated waters produces molds, vugs, caverns, and channels with or without collapse features (commonly associated with caves and karst features), and solution-enhanced interparticle pores. These pore types share a common origin and differ in degree rather than kind. Common origin suggests similar geological settings in which the pores formed and helps geologists eliminate settings that were unlikely to have been sites for dissolution. Geological settings where dissolution is common include the top of the meteoric phreatic zone, the mixing zone, and parts of the vadose zone. Dissolution can also occur in the subsurface when rocks and water are out of chemical equilibrium. Rock–water interaction where undersaturated fluids are present may be involved in stabilization reactions that alter metastable carbonates to stable ones (recrystallization, also called neomorphism). Weathering and soil-forming processes at unconformities can involve a combination of diagenetic processes including dissolution, precipitation, biological activity, and neomorphism. Soils and lightly weathered zones are not usually important as reservoir rocks because matrix pore sizes in carbonate soils are small and capillary pressures are correspondingly high.

Dissolution requires undersaturated water, usually meteoric or mixing-zone water, to interact with carbonate country rock. Caves are formed in both vadose and meteoric phreatic environments as illustrated in Figure 6.4. An idealized cave system with both vadose and phreatic characteristics is shown in Figure 6.5. Dissolution is common as corrosion and pore enlargement in deep-burial settings, too. Deep-burial dissolution is also called *mesogenetic dissolution*, following the terminology in the Choquette and Pray (1970) carbonate porosity classification (Mazzullo

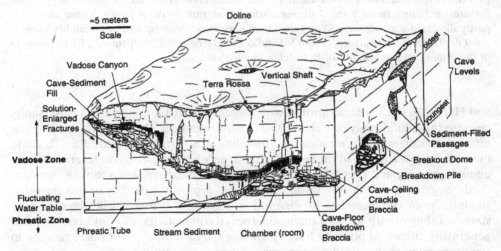

Figure 6.5 Sketch of an idealized cave system showing both vadose and phreatic diagenetic characteristics. Note the brecciated material on the cave floors and the "crackle breccia" on the ceiling. Although buried paleocave reservoirs may have large open caverns, they are more likely to be filled with collapse breccias and other forms of internal sediment. (From Loucks (1999). Reprinted with permission from the American Association of Petroleum Geologists.)

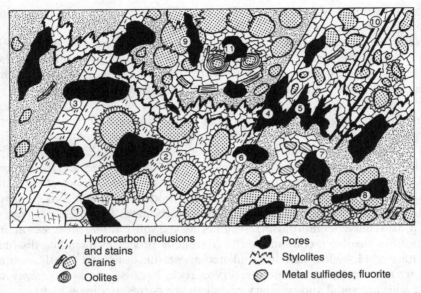

Figure 6.6 A sketch of a hypothetical thin section showing associations of diagenetic attributes characteristic of deeper burial (mesogenetic) diagenesis. The numbers refer to late burial dissolution events that cross-cut the following numbered features: (1) saddle dolomite; (2) coarse, burial calcite cements with hydrocarbon inclusions; (3) cemented fractures with hydrocarbon inclusions in the cement; (4) along fractures that postdate stylolitization (note the jagged stylolites that cut left to right across most of the drawing); (5) along stylolites themselves; (6) along hydrocarbon stained fractures; (7) of cements associated with metallic sulfides or fluorite; (8) of compacted grains; (9) of pressure-solution compacted grains; and (10) of postcompaction cements. In short, late burial dissolution has cut across all previous features including those formed during earlier moderate to deep burial. These examples partly illustrate why carbonate reservoirs may have good porosity and permeability even at great depths in the subsurface. (From Mazzullo and Harris (1992). Reprinted with permission of the American Association of Petroleum Geologists.)

and Harris, 1992). Undersaturation with respect to $CaCO_3$ in burial fluids is usually because these fluids may be rich in CO_2, H_2S, or organic acids. A sketch showing ten different occurrences of mesogenetic dissolution is shown in Figure 6.6. This late burial dissolution is one of the reasons that deeply buried carbonate reservoirs have enough porosity and permeability to produce hydrocarbons, even though the standard "dogma" is that deeply buried carbonates typically have 5% or less porosity. In short, pore enhancement by dissolution diagenesis can create a range of pore sizes and shapes with widely ranging degrees of connectivity. A brief review of diagenetically enhanced porosity follows, tracing the development of pores ranging in size from micrometer-sized micropores to karst caves and caverns.

6.4.1 Enhancement by Recrystallization

Decades ago, all diagenetic changes that affected texture, fabric, and even mineralogy in carbonates were included in the umbrella term *recrystallization*. As discussed earlier in this book, the term neomorphism originated with Folk (1965), who defined it to include both true recrystallization and mineralogical inversion. True recrystallization occurs when very small crystals of calcite dissolve and reprecipitate as larger, neomorphic spar crystals. Inversion is the process by which metastable minerals change to stable minerals, generally undergoing some compositional as well as crystallographic changes such as the inversion of strontium-rich aragonite or Mg-calcite to low-Mg calcite. Recrystallization is also used to describe the processes of crystal reformation such as those that take place during metamorphism of limestone to marble or as new crystals form during application of great stresses. This type of recrystallization can be accompanied by elimination of crystal imperfections and fluid inclusions to produce nearly perfect lattice structures.

Examples of porosity enhancement associated with stabilizing neomorphism are described from Cotton Valley Formation (Jurassic) reservoirs in the East Texas Basin (Ahr, 1989; Dravis, 1989; Moshier, 1989). One of the most striking examples of neomorphic microporosity is found at Overton Field, Texas. There, individual ooid grainstone deposits (Figure 6.7a) exhibit micrometer-sized, microrhombic microporosity (Figure 6.7b) interpreted to have resulted from neomorphic stabilization of metastable ooid mineralogy to present-day low-Mg calcite. The mechanism of the

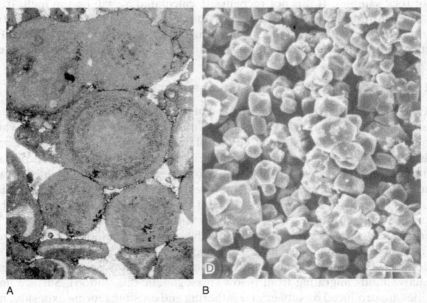

A   B

Figure 6.7 (a) Thin section photomicrograph of an altered oolite grainstone in the Jurassic Cotton Valley Formation of East Texas. The photomicrograph shows the degree of degradational diagenesis that has almost obliterated the original microstructure of the ooid. Only traces of concentric structure of the original ooid cortices are visible in the now-altered, "chalky" ooid that makes up a large part of the natural gas reservoir rock in Exxon-Mobil's Overton Field in Texas. The width of the photo is 2.5 mm. (b) An SEM micrograph of the microstructure in the large ooid in the center of the photo in (a). This microporous, microrhombic microcalcite fabric resulted from a type of recrystallization that is not well understood but is widespread and makes up many gas reservoir rocks around the world. The bar scale faintly visible in the lower right-hand corner of the image is 5 μm long. (From Ahr (1989).)

neomorphic change is not understood, but it appears to occur preferentially in grainstones linked to paleo-highs. This association led to the interpretation that the neomorphism was an early burial phenomenon produced by invasion of waters that were out of equilibrium with the original, probably Mg-calcite, ooids (Ahr, 1989). Other evidence suggests that some of the disequilibrium fluids were introduced late in the burial history of the Cotton Valley reservoir (Dravis, 1989; Fretwell, 1994). Regardless of the time of formation of the microporosity, it is a common and widespread type of diagenetic porosity that serves as the principal pore type in many gas reservoirs in carbonate rocks. It is not a good candidate pore type for oil reservoirs because the pore dimensions are in the micrometer range with pore throats in the submicrometer size range. Capillary pressures associated with such miniscule pore apertures are great enough to block oil movement.

Neomorphic microporosity is diagenetic but it lacks the characteristic vugs, molds, caverns, channels, and other solution-related features that typify dissolution porosity; therefore its origin is attributed to stabilization reactions rather than macroscale dissolution. It is important to recognize that these micropores can store natural gas and they can store formation water. A common example of bimodal porosity in carbonates is large pores, such as molds and vugs, together with micrometer-sized micropores. Bimodal porosity of this type has to be recognized by direct examination of rock samples. If it is not recognized, calculated S_w will be too high. In such cases, the saturation values are S_{wt}, or total saturation, rather than S_{we}, effective saturation. Asquith and Jacka (1992) emphasized the difference between total and effective saturation and the significance it has on estimating hydrocarbon volume.

6.4.2 Enhancement by Solution Enlargement

Solution-enlarged porosity includes (1) enlarged interparticle pores, (2) moldic pores, (3) vugs, (4) channels (including solution-enlarged fractures), and various combinations of (1)–(4) that may be completely or partially formed (e.g., partial molds with some original grain material remaining). Solution-enlarged pores are small-scale (millimeter range) features. Large-scale (centimeter to meter range) phenomena such as caves, caverns, sinkholes, and collapse features are discussed later. Reservoirs with solution-enlarged and karst porosity may be associated with (1) surface unconformities, (2) present or ancient topographic highs that have undergone exposure to vadose or phreatic dissolution in meteoric or mixing zones, (3) fracture zones that may have been exposed to waters undersaturated with respect to $CaCO_3$, and (4) subsurface permeability pathways that allowed passage of reactive fluids migrating from below (mesogenetic dissolution). Surface unconformities are produced by subaerial weathering and erosion. During exposure, meteoric water percolates through porous rock beneath the unconformity. Pore types and pore volumes created or influenced by this exposure depend to a great extent on the duration of exposure, solubility of rock constituents, and the degree of disequilibrium between water and rock. Saller et al. (1999) found that short (a few thousand years) duration of exposure to dissolution had little effect on depositional porosity, but that intermediate duration (50,000–130,000 years) modified the pore system by forming micropores and channelized porosity. They found that long duration time correlated with extensive porosity reduction by cementation, although some channelized pores were enlarged. In sum, those authors argue that intermediate times of duration seem to have had the greatest positive effect on porosity enhancement by dissolution. Intuitively, one would expect long duration, comparatively soluble rock, and persistent strong disequilibrium to produce karst features with cavernous porosity. Mild disequilibrium, lower solubility, and shorter duration may produce only minor enlargement of interparticle pores. Intermediate situations include connected moldic and vuggy pores or solution-enlarged fracture porosity.

Dissolution commonly occurs during late burial diagenesis as well as at exposure surfaces and during early burial. There is no single result from exposure to vadose and meteoric water dissolution at unconformities or from reactive fluids in deep-burial environments. Each situation must be evaluated on its characteristics in order to diagnose the extent, type, and timing of alteration. It is relatively easy to identify solution-enlarged pores in cuttings, cores, and especially in thin sections because the boundaries of grains, cements, and matrix show truncation, corrosion, or otherwise irregular and transgressive edges. In short, the boundaries of the enlarged pore "transgress" original outlines of grains, cement crystals, and muds.

The gamma ray log can be a useful tool for indirect identification of dissolution at exposure surfaces. Many carbonates contain insoluble residues rich in clays and organic matter that emit natural radioactivity. As the carbonates are dissolved on unconformities and the insoluble constituents are concentrated along the dissolution surface, the amount of natural radioactivity associated with the surface is also increased. Simple contouring of API units from gamma ray logs on wells that penetrated an unconformity may indicate high, intermediate, and low concentrations of insolubles. It is common to find higher effective porosity in zones just below the main surface of dissolution, because the unconformity surface per se may be plugged with insolubles or cements. An example of this type of low to moderate dissolution-enhanced porosity is characteristic of the reservoir at Lisbon Field, Utah. Fouret (1996) found that high API gamma values correspond to zones with the greatest amount of dissolution diagenesis found in cores. The greatest dissolution in the Lisbon Field cores was found in wells that penetrated dissolution surfaces on the crest of a large anticlinal structure that is clearly revealed on interval isopach maps. Pore enlargement by dissolution during deep burial has been known for decades. Moore and Druckman (1981) described deep-burial solution-enlarged pores in the Jurassic Smackover Formation in Louisiana, as did Ahr and Hull (1983) along with many later workers. Deep-burial dissolution is commonly followed by partial or complete pore filling by minerals such as saddle dolomite, fluorite, sphalerite, and a variety of metallic sulfides. These minerals provide clear evidence of burial diagenesis because they do not exist in ordinary sedimentary environments. Saddle dolomite is a particularly useful mineral to aid in recognition of deep-burial diagenesis because it can readily be identified in cuttings and it commonly occurs as vug or fracture fillings that may in turn be correlated with dramatic changes on such logs as the acoustic log (rapid scale change or "cycle skipping"), the caliper log (washouts at fracture and vuggy zones), the density–porosity log (very high porosity), and the drilling time log (drilling breaks).

6.4.3 Large-Scale Dissolution-Related Porosity

Porosity at Lisbon Field is mainly the result of grain-scale dissolution on an unconformity surface. Solution-enlarged pores and molds generally did not undergo dramatic changes in shape or volume. By comparison, karst features such as caves, caverns, sinkholes, towers, and pinnacles represent massive dissolution of carbonate strata that produces landforms so large and distinctive that they can be recognized on seismic profiles (Loucks, 1999; Purdy and Waltham, 1999). Towers, pinnacles, and cones, the striking landforms seen in South China, Sarawak, and Java, respectively, are interpreted to be the result of karst development following regional uplift attended by lowering of the base level of erosion (Purdy and Waltham, 1999). On the other hand, sinkholes, caves, and caverns are in general more commonly associated with passive sea-level lowering without tectonic uplift.

Loucks (1999) focused attention on karst-related paleocave systems as hydrocarbon reservoirs. He listed as examples of giant oilfields in karst reservoirs the Lower Ordovician Puckett Field in West Texas, the Permian Yates Field in West Texas, and the Lower Cretaceous Golden Lane Field in eastern Mexico. Loucks (1999) also formulated an idealized developmental history of pore types in paleocave reservoirs. According to his scheme, cave-related pore networks change character with increasing depth of burial. He points out that large dissolution pores may persist with burial depth to several thousand meters, "but eventually (the pores) collapse forming smaller interbreccia pores and fractures associated with crackle and mosaic breccias. Coarse-interbreccia pores between large clasts are reduced by rotation of clasts to more stable positions and by rebrecciation of clasts to smaller fragments. As passages and large interbreccia pores in the cave system collapse, fine-interbreccia porosity first increases and then decreases, whereas fracture pore types become more abundant." More recently, it has been argued that coastal-zone caves like those in northeastern Yucatán, Mexico may be more realistic than continental caves as modern analogs for paleocave reservoirs (Smart and Whitaker, 2003).

Purdy and Waltham (1999) argue that positive karst features such as towers and cones are associated with uplift, which is in turn commonly associated with tectonic activity and attendant fracturing. Passive sea-level lowering, in contrast, is typically associated with more prolonged runoff and sustained dissolution that produces connected vug and channel systems with a high degree of connectivity. In their words with minor omissions: "Uplifted grain-dominated carbonates seem to be particularly well lithified and prone to develop fault and fracture avenues of solution that leave tower and cone karst as intervening residuals. The net effect is to minimize matrix porosity in the grain-dominated lithologies. Carbonates with this history essentially behave as fractured reservoirs. In the case of nonfractured karst residuals, the slope of the uplifted carbonate surface controls runoff direction and resulting karst morphology. In this case, solution conduits and caves provide large-scale permeability avenues. In both cases, however, recovery efficiency is geared to matrix porosity: the higher the matrix porosity, the greater the rate of deliverability of matrix oil to that produced from fracture and karst conduits. Recovery efficiency is generally higher in these reservoirs than in their uplifted counterparts. In contrast, a passive drop in sea level militates against development of positive karst relief. The slopes of the subaerially exposed surfaces generally are the lesser ones of deposition rather than uplift and, therefore, the probability of developing runoff directions with more than a few degrees of slope and resulting karst relief is correspondingly reduced. Additionally, these subaerially exposed surfaces have reduced potential for the occurrence of coeval fracturing and faulting. Extensive moldic and vuggy porosity dominates, and permeability is provided largely by interconnected solution channels." In both the paleocave reservoirs described by Loucks (1999) and the systems described by Purdy and Waltham (1999), it is matrix porosity that ultimately determines the volume of producible hydrocarbons. Knowing the array of pore characteristics that may appear with increasing depth of burial in collapsed caves, it is probably equally important to distinguish between caves in uplifted karst terranes, those developed during passive sea-level lowering, and those developed as coastal, anchialine (mixing-zone) caves.

6.4.4   Porosity Enhancement by Replacement

The most common replacement mineral in carbonate reservoirs is dolomite. Replacement of limestone by dolomite is discussed at the beginning of this chapter. In general, replacement of limestone by dolomite can actually reduce porosity rather than enhance it (Lucia, 2000). Permeability is increased only in cases where limestone has been replaced totally by dolomite and intercrystalline porosity is preserved without "excess" dolomite cement. Porosity enhancement in dolostone replacements after limestones may have been accomplished largely by dissolution of limestone simultaneously with dolomite replacement leaving a residual, porous network of dolomite crystals. Other common replacement minerals include anhydrite, silica as chert and flint, and sulfide minerals such as MVT ore minerals. MVT ore emplacement is not generally associated with porosity enhancement or with live hydrocarbons, although MVT ores are commonly associated with bitumen.

6.4.5   Recognizing Enhanced Porosity

Enhanced porosity is easy to recognize. It is not as easy to determine the cause or causes of diagenetic enhancement, how many times the pores were altered, and to what geographic extent the alterations modified reservoir zones. Porosity enhancement by dissolution is arguably the easiest to identify and sort out. Caverns, vugs, enlarged inter- and intragranular pores, and molds obviously represent enlargement by dissolution. Other types of diagenetically enhanced porosity may not be as easy to recognize but they still represent pore enlargement, improved storage capacity, and perhaps improved flow capacity. Recognition of reservoirs with purely depositional porosity is, as we have seen previously, a matter of identifying pore characteristics that correspond closely with depositional successions and facies. Purely diagenetic porosity and extensively altered hybrids (diagenetic attributes dominant) have few characteristics in common with depositional facies. Methods used to determine reservoir boundaries in those cases depend on the kinds of diagenesis that have influenced porosity, in which diagenetic environment the changes occurred, and the relative timing of the alterations. Dissolution is, as previously discussed, commonly associated with exposure surfaces, unconformities, soil zones, and karst surfaces.

Replacement by dolomite is commonly associated with lagoonal or tidal-flat evaporites such that stratigraphic models of evaporite facies may predict the occurrence of seepage-reflux dolomite replacements. Replacement by evaporites, silicates, sulfides, or other minerals usually reduces depositional porosity instead of enhancing it. Recrystallization is commonly associated with neomorphic microspar, or a tightly interlocked crystal mosaic that results from the dissolution and reprecipitation of metastable carbonates such as aragonite or Mg-calcite with the end result being porosity reduction instead of enhancement. One example of recrystallization that does enhance porosity is the microcrystalline microporosity interpreted by Ahr (1989) to be the result of neomorphic stabilization of aragonite or Mg-calcite.

Diagenetically enhanced porosity may correspond to the outline of a paleoaquifer, some topographic feature that influenced paleohydrology, or to fractures and joints. This may mean that the enhanced porosity conforms to the outline of a present or paleostructural high, to the slope of a paleoaquifer, to the shape of an evaporite lagoon, or to the boundaries of some other kind of system in which chemically reactive waters came in contact with carbonate rocks—good examples of which are the extensive coastal-zone caves of Mexico.

The job is to trace the "diagenetic pathway," pinpointing the processes that created reservoir porosity and determining the relative timing of diagenetic events during burial history. This is not always simple because several events that occurred at different times during burial may have shaped the final pore system much like multiple exposures influence the final image on a photograph. Fortunately, some simplifying generalizations can help solve the problem. Virtually all diagenetic processes require reactive fluids to pass through connected pores in the host rock, reacting with the rock as the fluids migrate. Large-scale removal of rock as in the formation of caves, caverns, towers, and other karst features requires enormous amounts of undersaturated water. Such great volumes of water are virtually impossible to obtain except at the earth's surface or in shallow aquifers. It follows that large-scale dissolution diagenesis is a priori an early event in burial history. The same is generally true for extensive networks of vugs, molds, and channels. Dolomitization reactions are commonly, but not exclusively, associated with evaporites and shallowing-upward cycles in carbonate strata. Pore enhancement by burial diagenesis must follow some pattern of preserved porosity and permeability, usually a depositional fabric or metastable original mineralogical composition inherited from the original rock. The conduit through which burial fluids migrated may not match exactly the size and shape of the original depositional facies; rather, burial alteration may correspond to paleotopography or present structure where ancient or present structural contours intersect facies surfaces, creating an interesting coincidence of three-dimensional geometry and geology.

6.5 POROSITY REDUCTION BY DIAGENESIS

Diagenetic processes that reduce porosity include some of the same ones that enhance porosity—recrystallization and replacement—but cementation and compaction stand alone as mechanisms that only reduce porosity. Cementation happens when crystalline precipitates fill pore spaces. This can happen just after deposition and at any time during burial. Because cementation can occur immediately after deposition, it is tempting to infer that most porosity reduction is accomplished that way, but cemented rocks may be compacted to cause even further porosity reduction.

6.5.1 Pore Reduction by Compaction

Mechanical compaction results from overburden stress during burial and from tectonism. Compaction effects are sometimes accentuated or exaggerated when dissolution and compaction act together to form stylolites by pressure acting concurrently with dissolution. Graphs showing porosity decrease with increasing depth of burial (Schmoker and Halley, 1982) suggest that compaction may be more influential than cementation in reducing porosity. Of course, cementation takes place during burial, so porosity reduction with depth is not a function of one variable. The relative importance of compaction and cementation in reduction of porosity can be ascertained by counting the number and kind of grain contacts in samples from different burial depths to estimate the extent to which compaction has reduced original intergranular porosity. Rocks with flattened or penetrative grain contacts have significantly more grain contacts per area and lower porosity than uncompacted rocks where individual grain contacts are tangential and sparse, especially when counted in 2D thin section views. Pore reduction by cementation can be crudely estimated by measuring the 2D volume of cement in pore spaces along several transects across a thin section. If compaction had a greater influence on porosity reduction, there will be successively more contacts per grain with depth and the contacts will progress from being tangential contacts at shallow depths to penetrative and stylolitic contacts at depth. As compaction continues with combined pressure and dissolution, stylolites are formed. Generally, stylolites are more common in mud-supported rocks (Dickson and Saller, 1995) than in grainstones and packstones, and in general, they reduce porosity and permeability (Nelson, 1981). The literature is replete with references to stylolites in Middle Eastern carbonate reservoirs and how they form permeability barriers that can be used to zone or map reservoir flow units, baffles, and barriers. However, poststylolite diagenesis can create porosity and permeability in previously tight rocks (Dawson, 1988). The take-home message is that one has to look at the rocks to isolate the cause of reduced or enhanced porosity and permeability. Wireline logs and seismic data cannot yet distinguish between cementation, compaction, recrystallization, dissolution, and replacement.

6.5.2 Pore Reduction by Recrystallization

Neomorphic stabilization can enhance porosity as discussed earlier in reference to the Overton gas field, Texas. However, neomorphism as a type of recrystallization (inversion of aragonite or Mg-calcite to calcite) usually reduces porosity. Folk (1965) described a form of coalescive neomorphism in which highly porous, "felted" networks of acicular microcrystals were recrystallized to produce larger crystals of neomorphic microspar. Formation of microspar involves neomorphic ("new form") crystal growth at the expense of smaller, precursor micrite particles. The process forms a crystalline mosaic with abundant compromise boundaries and virtually no intercrystalline porosity. Neomorphic microspar is the common form of calcite seen in almost all stabilized, recrystallized lime muds, whether as muddy patches in grainy rocks or in mudstones and wackestones. In carbonate reservoirs neomorphic microspars are usually candidates for seals rather than reservoirs. In general, recrystallization reduces porosity and permeability because the original sedimentary constituents were composed of micrometer-sized crystals of metastable aragonite or

Mg-calcite. In short, highly porous parent sediment composed of metastable aragonite, Mg-calcite, or small calcite particles with high surface area to volume ratio will
undergo coalescive or porphyroid neomorphism with attendant loss of interparticle
porosity.

6.5.3 Pore Reduction by Replacement

Replacement of precursor carbonate minerals by silica, anhydrite, sulfide minerals,
and dolomite can reduce original porosity by replacing both the original mineral and
the original rock fabric at the expense of depositional or pre-replacement diagenetic
porosity. Anhydrite, saddle dolomite, and silica are common replacement minerals
in carbonate reservoirs. Anhydrite is the stable form of calcium sulfate at depths
below about 3000 feet in the subsurface (Hardie, 1967), where it may be a burial
transformation of gypsum. Anhydrite may occur as massive beds, nodules, and pore
fillings, and as replacements that transgress grain boundaries and pore walls. Bedded
or massive anhydrite forms during deposition and may mark specific parts of stratigraphic successions such as the tops of shallowing-upward successions or parasequences. Massive anhydrite layers may be identifiable by distinct signatures on
wireline logs or may otherwise be predicted to occur as cycle capping beds in parasequences identified by their sequence-stratigraphic stacking patterns. This type of
anhydrite is typically dewatered gypsum that has been altered during burial. Pore-
filling and replacement anhydrite usually form during early burial diagenesis as
sulfate-rich brines percolate downward through porous and permeable carbonates.
Pore fillings result from precipitation of gypsum or anhydrite from the migrating
brines. As brine migration continues, more and more pore spaces are filled and reactions between the sulfate-rich water and the carbonate rock result in replacement
of carbonate by anhydrite. This type of anhydrite may occur in beds up to several
meters below the tops of shallowing-upward cycles, where the anhydrite derives from
interstitial brines that drained downward from exposure surfaces, restricted lagoons,
or tidal flats. Pore-filling and replacement anhydrite can dramatically reduce depositional porosity. On the positive side, anhydrite plugging may form seals to prevent
hydrocarbon leakage from reservoir-quality rocks below. It is relatively common in
shallowing-upward successions to find the best porosity and permeability in beds
beneath cycle-capping, anhydrite-cemented grainstones. Anhydrite may also come
from the deep-burial environment and form as a late burial cement or replacement
derived from upward-migrating fluids that invaded the reservoir from below.

Saddle dolomite, silica, and sulfide replacements are not common in the shallow
burial domain; instead, they usually indicate deeper-burial diagenetic replacements.
Saddle dolomite occurs in two different settings: (1) fractured reservoirs (see Figure
8.18) in which deep subsurface fluids migrate up the fracture systems and into reservoir pores, and (2) nonfractured reservoirs (Figure 6.8) in which thermochemical
sulfate reduction (TSR) is interpreted to be a major factor influencing the formation
of saddle dolomite (Machel, 1987b). Silica replacement as chalcedony is common
in deep-burial replacements and may develop in association with saddle dolomite,
fluorite, sulfide minerals, and hydrocarbons. Hydrothermal silicates also are known
as replacements for evaporite minerals (Ulmer-Scholle et al., 1993); consequently,
early burial sulfates such as anhydrite may be found to have been replaced by chalcedony or other varieties of quartz during later burial alteration.

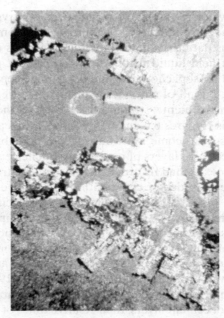

Figure 6.8 Photomicrograph of saddle dolomite replacing micritized ooids in the Jurassic Cotton Valley Formation at Teague Townsite Field, Texas. The dark stains at the margins of the replacive saddle dolomite are residual hydrocarbons, suggesting that the dolomite-forming liquid was associated with the hydrocarbon migration "front." The width of the photo is 1.5 mm.

6.5.4   Pore Reduction by Cementation

Cementation can occur several times in the diagenetic history of a carbonate rock beginning with cementation in the marine environment just after deposition and continuing through vadose, shallow, intermediate, and deep-burial environments. The literature on early marine cementation is too extensive to summarize here, but one of the early and dramatic examples of marine cementation is in a paper by Shinn (1969). He described a marine cement crust on a soft drink bottle found on the shallow seabed of the Persian Gulf. At the time, this discovery was particularly noteworthy because a controversy in carbonate sedimentological circles during the 1950s centered on whether submarine cementation could actually happen. Marine phreatic cements usually coat the entire perimeter of constituent particles, giving rise to the name "isopachous" marine cement. Original mineralogy of marine cements in today's oceans consists primarily of Mg-calcite and aragonite in various crystal habits ranging from micrite to needle crystals to botryoidal masses.

The mineralogy and crystal forms of carbonate cements change as water chemistry and diagenetic environments change from marine phreatic to meteoric phreatic to shallow and deep subsurface waters. Folk (1974) was one of the first to emphasize the importance of Mg content in interstitial water, salinity of the water, and environment of diagenesis (vadose, phreatic, or subsurface) as controls on the mineralogy and crystal form of the resulting cements. Marine phreatic cements in today's oceans are Mg-calcite or aragonite because abundant Mg favors precipitation of aragonite and Mg-calcite. Freshwater vadose and phreatic cements crystallize as calcite with blocky crystals and bladed rhomb crystal habits. Vadose cements typically exhibit the distinctive meniscus patterns where cementation occurs only at grain contacts where a meniscus film held liquid from which the cement precipitated. Phreatic cements typically occur as isopachous rim and pore-lining cements. As more micro-layers of cement form, a kind of chronology or microstratigraphy of cementation can be recognized. This cement "stratigraphy" displays the sequence of events during burial diagenesis (Figures 6.6 and 6.9). Some of the early work that defined cement stratigraphy, as it has come to be known, was done by Meyers (1974), who recognized that the successive layers of pore-filling calcite cements in Mississippian limestones from New Mexico had recorded a kind of microstratigraphy of cementa-tion events. Successive episodes of cementation are most easily recognized with cathode luminescence (CL). This technique involves placing thin sections in a vacuum and bombarding them with electrons to induce luminescence in some min-erals. The principle is somewhat similar to the way a television CRT works. If car-bonate minerals contain the right mixture of trace elements, they luminesce in different colors and intensities. Trace element compositions and their effects on CL have been described by many authors, notably by Machel (1987a). Generally, each cement microlayer luminesces (or does not) uniformly, enabling the petrographer to identify the number of cement layers, their relative times of origin, and something about their trace element compositions.

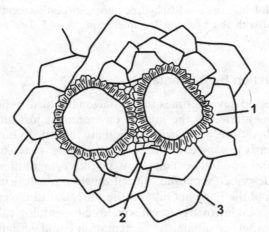

1.   **MARINE PHREATIC** – ISOPACHOUS RIM CEMENT

2.   **VADOSE** – MENISCUS CEMENT

3.   **BLOCKY CALCITE** – LATE BURIAL CEMENT

Figure 6.9   A sketch illustrating "cement stratigraphy" that can reveal different times and styles of cementation before and after burial. The first stage marine phreatic cement is in the form of an acicular, isopachous rim around the grains. The next stage cement is meniscus cement that forms only in the vadose diagenetic environment, indicating that the marine-cemented grain was exposed to the vadose environment after initial cementation. Finally, two stages of blocky, burial calcite cement mark the last of the "cement stratigraphic" episodes. (Adapted from an illustration in McIlreath and Morrow (1990).)

As burial proceeds and interstitial water composition changes, cement mineralogy and crystal form change accordingly. Water composition changes during burial as upward-migrating expulsion waters mingle with in situ interstitial water, which may have already been through a variety of rock–water reactions. Calcite, dolomite, and other mineral cements may form, depending on water composition and chemical equilibrium for each mineral species. If burial calcite forms, it commonly takes the form of large, clear crystals that fill remaining pores and may encompass pore spaces around several constituent grains. These multipore, large crystals are called *poikilotopic crystals*, a name borrowed from igneous petrology. Because Mg is depleted in subsurface waters as compared to the marine phreatic environment, burial calcites are low-Mg calcites, like freshwater varieties. Burial calcites may include relatively higher iron content than marine or freshwater calcites, depending on availability of iron and the oxidation–reduction state of the cementing environment. Higher iron content in the calcite lattice is characteristic of reducing burial environments and is easily recognized by applying a potassium ferricyanide stain to thin sections. Ferroan calcites (high iron content) are stained with a blue tint in this solution. In summary, cementation can occur in marine, fresh, or subsurface (burial) waters. Different episodes of cementation can be recognized with CL and ordinary petrography enabling the petrographer to establish a cement stratigraphy for a given rock specimen. This method has been refined and automated with computer-assisted image capturing techniques to enable petrographers to reconstruct the cementation history of carbonate reservoir flow units and barriers (Mowers and Budd, 1996; Witkowski et al., 2000).

Noncarbonate cements are less abundant in carbonate reservoirs, but when they are present they can greatly reduce effective porosity. One of the most common cements found in carbonate reservoirs is anhydrite. It is usually associated with evaporite deposits that commonly cap shallowing-upward sequences, but it can be introduced as late-stage burial cements as well. Dolomite, including saddle dolomite, may occur as cement as well as a replacement mineral. Saddle dolomite is common as a vug and fracture lining cement in many reservoirs. Chalcedony, a variety of silica cement, is moderately common as pore-reducing cement, especially in association with late-stage burial cements such as saddle dolomite, sphalerite, and fluorite. These minerals are commonly associated with residual heavy hydrocarbons, suggesting that the exotic cementation took place in association with hydrocarbon migration to form the MVT mineral association discussed previously in this chapter.

Porosity reduction by cementation can usually be identified by cross-cutting relationships in cement stratigraphy. Isopachous cements that rim the framework grains may be the first cementation event that occurred in the marine phreatic zone, for example. Coarser, centripetal cements that further reduce porosity may follow these early cements. Finally, remaining effective porosity might be plugged with poikilotopic cements formed in the deeper-burial environment. All of the cements might be cross-cut by mineral-filled fractures or by exotic deep-burial cements. Exotic mineral cements and mineralized fractures are good indications that new permeability avenues were created, allowing upward migration of mineralizing fluids. Such exotic fluids are commonly associated with migrating hydrocarbons and the new permeability avenues are usually fractures that postdate other diagenetic events.

6.5.5   Recognizing Diagenetically Reduced Porosity

Porosity reduction by cementation, compaction, replacement, or recrystallization is usually easy to recognize but it may not be as easy to determine how many episodes of diagenesis have occurred to reach the final levels of porosity and permeability in the reservoir rocks. Multiple episodes of diagenesis involving several types of alteration over a long time, especially when early episodes are cross-cut by later ones, make it challenging to interpret the origin and spatial distribution of the final poroperm characteristics in altered reservoirs. Diagenetically reduced porosity is analyzed using most of the same methods that are employed to analyze diagenetically enhanced porosity. That is, the first item on a checklist is to determine if porosity is purely diagenetic or a hybrid of diagenetic and depositional or fracture porosity. Hybrids of diagenetic and depositional pore types can be identified by finding correspondence between types of diagenesis (cementation, compaction, replacement, or recrystallization) and depositional facies. It should be relatively easy to recognize porosity loss due to diagenetic changes. It is not always easy to determine which of the diagenetic events caused the greatest reduction in porosity and when that event, or events, happened during burial. In hybrid pore types where depositional attributes dominate, petrophysical characteristics reflect more of the depositional character than of the diagenetic alteration; therefore it is not as critical to trace zones that have been cemented or compacted, or where recrystallization or replacement have reduced depositional porosity. On the other hand, in hybrid porosity dominated by diagenetic attributes, it is critically important to identify the type of diagenesis, the diagenetic environment in which it occurred, and the time sequence of diagenetic events as each event modified the original depositional pore geometry.

When diagenetic attributes dominate, porosity does not follow facies boundaries. Instead, it may follow certain phases in stratigraphic cycles, it may follow the shape of ancient water tables, it may follow unconformities, or it may follow paleotopography. Ruppel (1992) described a compartmentalized reservoir of Leonardian (Permian) age in West Texas, where stratigraphic cyclicity (mainly shallowing-upward cycles) and paleotopography played dominant roles in shaping the final pore geometry. Grainy facies tended to be associated with upper parts of shallowing-upward cycles and with paleotopographic highs. The highs were also associated with dissolution diagenesis that enhanced intergranular pores in grain-supported rocks. Low porosity tended to be associated with tidal-flat facies with poorly connected fenestral porosity, pore-filling anhydrite, and fine siliciclastics. Pore-reducing diagenesis may correspond to paleotopographic or paleohydrological boundaries. Kopaska-Merkel and Mann (1993) described stratigraphic cyclicity in Jurassic Smackover facies from Alabama, where shallowing-upward cycles ended with low-porosity horizons including paleosols, evaporites, and sebkha deposits. It is relatively common in Permian carbonate reservoirs of the southwestern United States to find pores plugged by anhydrite that capped shallowing-upward stratigraphic cycles. Calcium sulfate from the tops of cycles was dissolved, percolated downward as a dense brine, and reprecipitated as pore-filling cement in rocks immediately below cycle capping grainstones (Amare, 1996). Pore-filling carbonate cements may occur in certain stratigraphic horizons below unconformities or simply in progressively deeper positions along a paleoaquifer. In either case, cementation below a zone of exposure and dissolution commonly coincides with the outline of a paleotopographic feature (unconformity surface or upper part of a paleoaquifer). Exposed highs subjected to dissolution can provide $CaCO_3$ that migrates down the hydraulic gradient to be precipitated as pore-filling cement. In some cases it may be possible to identify where along the paleowater table certain cements were precipitated by examining their geochemistry, petrography, and cathode luminescence (Grover and Read, 1983). Similarly, the timing, associated paleohydrological environments, and controls on replacement by dolomite, and dolomite stabilization (neomorphism) were investigated by Barnaby and Read (1992) and Montañez and Read (1992), respectively. Meyers' (1974) pioneering work on "cement stratigraphy" in carbonate burial environments showed that it is possible to identify and correlate individual calcite cement types over several square miles by their petrographic and geochemical characteristics.

Patterns of compaction and pressure solution can be somewhat more difficult to identify because they are the result of mechanical and chemical processes acting simultaneously. Meyers (1980) was able to determine the timing of compaction in Mississippian limestones from New Mexico on the basis of their physical appearance in chert replacements for which relatively precise age dates were available. That is, he found compacted and broken constituent grains within chert replacements of a specific age but not in rocks of other ages. He found that depositional porosity was reduced by at least 50% and up to as much as 75% in the compacted rocks. Furthermore, he was able to determine that burial compaction took place between Mississippian deposition and burial until Permian times. The methods used by Meyers (1980) can be employed to determine the amount of porosity reduction by mechanical and chemical compaction, the timing of compaction events, and perhaps the spatial extent of compaction-reduced porosity at reservoir scale. Because many diagenetic processes are linked to subaerial exposure surfaces or paleo water tables, clues to diagenetic porosity reduction can be found when those features are recognized.

6.6 DIAGNOSING AND MAPPING DIAGENETIC RESERVOIRS

Diagenesis usually follows more than one pathway in creating, enhancing, or reducing porosity. Hybrid pore types are probably the rule rather than the exception. Original (depositional) porosity may be reduced, destroyed, or enlarged (enhanced). Purely diagenetic reservoirs such as blanket dolomitization of limestone parent rocks cannot be traced directly back to depositional rock properties. Hybrid pore systems where depositional attributes are dominant will exhibit either enlarged or reduced intergranular, intragranular, fenestral, shelter vug, keystone vug, and "reef" pores. Diagnostic procedures to evaluate this type of hybrid reservoirs are essentially the same as those used to identify and map depositional reservoirs. In short, reservoir porosity is facies selective but diagenetically altered. Maps of depositional facies are still proxies for porosity.

If diagenesis does not follow facies, lithostratigraphic, or chronostratigraphic boundaries, then some other cause–effect relationships must be examined. The less obvious boundaries of diagenetic reservoirs can be identified by examining paleo-structure, paleohydrology, and relative timing of diagenetic events. Further diagnostic methods include recognition, classification, and thin section measurement of pore geometry. Wireline logs, seismic profiles, and even borehole images do not provide enough information to identify and classify porosity; therefore it is absolutely necessary to examine rock samples, preferably full-diameter cores. Although it is possible to identify and classify porosity in samples from a single well, it is not possible to predict the size and shape of the three-dimensional porous reservoir without additional information about how the one well fits in the field-scale stratigraphic and structural architecture. This information is usually obtained as additional wells are drilled. Predicting the size and shape of the reservoir body is the most basic task of the geoscientists and engineers who must decide where to drill the next well and how to develop a field in the most cost-effective manner. Identification and classification of genetic pore types is the first step in determining the geological origin of a reservoir, predicting its spatial dimensions, and, ultimately, in delineating individual flow units, baffles, and barriers.

The following checklist includes procedures for identifying and exploiting diagenetic reservoirs.

1. Classify the genetic pore types—purely diagenetic, hybrid type 1A (depositional attributes dominant), and hybrid type 1B (diagenetic attributes dominant)—to determine which has the most influence on reservoir quality. Identify the type of diagenesis—cementation, compaction, dissolution, replacement, or recrystallization—associated with the pore types in flow units, baffles, and barriers.

2. Identify the diagenetic environment or environments in which pore alteration took place, how many episodes of change took place, and in what order of occurrence the changes took place. The point is to determine which diagenetic events caused changes that have the greatest influence on today's porosity and permeability, in which environments those changes took place, when during burial history they took place, and how to use that information to predict the spatial distribution of flow units, baffles, and barriers.

3. Search for evidence of paleoaquifers, exposure surfaces, stratigraphic cycles that may include evaporites, paleosols, or karst features by comparing lithological logs from cores or cuttings with subsurface geological data (structure and stratigraphy), seismic profiles and seismic attributes, biostratigraphic data, and geochemical data. If joints or fractures are part of the pore systems, fracture geometry usually occurs in predictable patterns on faults and folds.

4. Diagenetic porosity may correspond to present structure, paleostructure, proximity to unconformities, proximity to facies such as evaporites or lacustrine deposits, or proximity to fracture systems that were conduits for migrating fluids. Compare structure maps with interval porosity maps to test for contour shape similarity. Strong correspondence between structural highs and high porosity values indicates that structural form has remained basically unchanged since diagenesis took place. It also suggests that paleotopographic highs were more affected by diagenesis than paleo-lows.

5. If porosity and present structure maps do not have similar shapes, test for correspondence between porosity and paleostructure. Make interval isopach maps of the thinnest interval possible that overlies the reservoir. Isopach thins indicate antecedent highs on the reservoir surface; thicks indicate lows. If

porosity values are high on isopach thins, it indicates that ancient structural highs and lows influenced diagenesis and resulting pore—pore throat attributes. Be aware when doing volumetric calculations or planning future drilling locations that paleostructurally influenced reservoir geometry does not follow present structure.

6. Purely diagenetic porosity (e.g., dolostones in which all traces of original limestone properties have been replaced) may correspond to position in stratigraphic cycles (e.g., top, middle, or bottom of the cycle) or to horizons where evaporites, ordinary dolomites, silicates and sulfides, or saddle dolomites occur together. Dolomite forms in a wide variety of geological settings including deep subsurface environments where saddle dolomite is typical. Although saddle dolomite is found in fractures and solution-enlarged vugs, it is not always an indicator of enhanced porosity. On the negative side, replacement by exotic minerals and dolomite may reduce porosity instead of enhancing it. Use the same methods that are used to find flow units (this checklist) to avoid baffles and barriers.

7. Develop geological concepts to explain reservoir characteristics by incorporating all available data on rock and petrophysical characteristics. Compare wireline log and seismic data with rock and pore characteristics observed in full-diameter cores to determine whether borehole log or seismic "signatures" can be used to identify porous zones in the borehole. Caliper logs, porosity logs, image logs, and NMR logs are first choices. Gamma ray logs may help identify concentrations of insoluble organic matter at unconformities, which in turn may be proximity indicators to porous zones. Once characteristic signatures are found that correspond to porous zones, the locations of genetic pore types that govern reservoir quality can be pinpointed and correlated from well to well.

Mapping flow units in diagenetically altered pore systems can be done as with depositional pore systems, but instead of focusing on depositional rock characteristics, cycles, and sequences, mapping flow units with diagenetic porosity focuses almost entirely on classification of pore types, determining their modes and times of origin, and their spatial distribution within the reservoir. If diagenesis has not removed all traces of depositional influence and porosity still conforms closely to depositional textures and fabrics, then identification of flow units is simply the identification of depositional rock properties that have the highest combined values of porosity and permeability. In such cases, individual flow units can be ranked by identifying pore categories that exhibit highest poroperm paired values (highest porosity paired with highest permeability for any given stratigraphic horizon within the reservoir), the intermediate poroperm values, and the lowest poroperm values.

Mapping flow units in purely diagenetic pore systems that show no relationship to depositional rock properties requires identifying the following sequence of events: (1) the types of diagenesis that influenced porosity; (2) the direction of that influence (pore creation, pore enhancement, or pore reduction); (3) the relative timing of the different diagenetic events with respect to each other; and finally, (4) the geological conditions or events that caused the different episodes of diagenesis to occur. Distinguishing different types of diagenesis enables one to isolate reservoir zones according to diagenetic origins. Purely diagenetic porosity in reservoir zones can be traced with more certainty by following the pathways that diagenesis took during burial history, such as pore reduction by cementation or pore enhancement by dissolution. Then, identifying times during which each diagenetic event occurred enables one to establish the sequence of burial diagenetic events that shaped the final pore system. Finally, reservoir flow units can be identified, mapped, and ranked at field scale if the geological conditions that existed during key diagenetic episodes can be identified. Did pore enhancement by dissolution influence reservoir porosity, for example? Are the dissolution zones linked to some kind of marker like an unconformity in the stratigraphic section? Careful observation will point the way to identifying more of the dissolution porosity if the stratigraphic and geographic coordinates of the unconformity, or other type of marker, are known. The geological setting that existed in order for the marker event to happen in the first place may have been tectonism associated with local or regional structural features, fluctuations of relative sea level, or dramatic climate change. Knowing the major causes of geological change that set the stage for specific diagenetic environments and kinds of diagenesis to occur provides the large-scale spatial reference to use in predicting the locations of "good" and "bad" diagenetic changes at field or regional scale.

SUGGESTIONS FOR FURTHER READING

A good starting point for nonspecialists is the book by McIlreath and Morrow (1990), *Diagenesis*. This book is a compilation of papers on diagenesis in both carbonates and siliciclastics. It is well illustrated and has an extensive list of references for that time period. Many of the illustrations are excellent line drawings that illustrate diagenetic features more vividly than photographs. A must-read reference is the book by Tucker and Wright (1990), *Carbonate Sedimentology*. Chapters 6–9 provide excellent discussions, illustrations, and references for carbonate mineralogy and chemistry (Chapter 6), diagenetic processes, products, and environments (Chapter 7), dolomites and dolomitization models (Chapter 8), and carbonate rocks throughout geological time (Chapter 9). The geological history of carbonate rocks through time includes discussions about temporal and secular changes in carbonate mineralogy and chemistry over time as well as discussions on variations in depositional styles and facies through time. Those interested in the more academic aspects of carbonate diagenesis will find a thorough review of carbonate geochemistry in Morse and Mackenzie (1990), *Geochemistry of Sedimentary Carbonates*. An older but useful reference on carbonate sedimentology and diagenesis is Bathurst (1975), *Carbonate Sediments and Their Diagenesis*, 2nd edition. A recent and comprehensive review of topics on burial diagenesis is by Machel (2005), "Investigations of Burial Diagenesis in Carbonate Hydrocarbon Reservoir Rocks." The most common techniques, other than standard petrography, for identifying the types of diagenesis and the environmental conditions under which the changes took place involve one or more of (1) stable isotope geochemistry, (2) cathode luminescence microscopy, (3) analyses of fluid inclusions, or (4) one of several types of microprobe analyses to identify trace element concentrations. These techniques require extensive explanations with examples—something beyond the scope and purpose of this book. Those interested in a brief summary of cathode luminescence applied to carbonates should read Machel (1987a). An extensive reference on fluid inclusion analyses is Roedder (1984), "Fluid Inclusions: An Introduction to Studies of All Types of Fluid Inclusions, Gas, Liquid, or Melt, Trapped in Materials from Earth and Space, and Their Application to the Understanding of Geologic Processes. A recently updated book on stable isotope geochemistry is by Hoefs (2004), *Stable Isotope Geochemistry*. A good reference on electron microprobe analysis is by Newbury et al. (2003) *Scanning Electron Microscopy and X-Ray Microanalysis.*

REVIEW QUESTIONS

6.1. Is diagenesis the same thing as metamorphism? If not, how are they different?

6.2. How would you rank the mechanisms of diagenesis in their order of influence, or impact, on reservoir quality?

6.3. What is the difference between replacement and recrystallization?

6.4. What are good indicators to distinguish between hybrid 1A (depositional attributes dominant) and 1B (diagenetic attributes dominant) pore types?

6.5. What examples can you give of type 1A and 1B hybrid pores?

6.6. Explain how recrystallization can either enhance or reduce porosity.

6.7. What types of diagenesis are associated exclusively with pore enhancement and with pore reduction?

6.8. What value is the gamma ray log as an aid in searching for erosional unconformities (Type I sequence boundaries) that may be related to diagenetically altered pore systems?

6.9. Dolomitization is always associated with enhanced porosity, ideally with about a 12% increase over the original amount of porosity in the parent limestone. True or false? Explain.

6.10. The most rapid decrease in porosity with burial occurs in the upper 1000 m of depth. True or false? Explain.

6.11. Can wireline logs, the seismograph, or other indirect methods of measurement identify genetic pore types in carbonate rocks? If yes, explain which methods and how they work. If no, can anything be done to identify genetic pore types?

6.12. What is the point of comparing structure maps and interval isopach maps with porosity maps in diagenetic reservoirs?

CHAPTER SEVEN

FRACTURED RESERVOIRS

7.1 FRACTURES AND FRACTURED RESERVOIRS

This chapter covers fractured carbonate reservoirs—how they form, where and why they form, what characteristics make them different from depositional and diagenetic reservoirs, and finally the basic methods for identifying, mapping, and exploiting them.

Working with fractured reservoirs is very different from working with depositional and diagenetic reservoirs and it requires an adjustment in ways of thinking. Fractures are generally unrelated to depositional or diagenetic rock properties because fractures are produced by mechanical stresses after the rocks have been lithified. Depositional or diagenetic processes are long-since over by the time the fractures form, although the fractures could be affected by later dissolution or cementation. It is also possible for fractures to be linked with depositional or diagenetic rock properties to form hybrid pore types. Diagenetic-fracture hybrids can be confused with purely fractured reservoirs unless cores or image logs are examined. Vuggy, cavernous, channeled, or otherwise connected dissolution-diagenetic porosity may exhibit petrophysical and flow characteristics that mimic fracture porosity and permeability, making it more difficult to separate hybrids from purely fractured reservoirs without direct observation of borehole cores or image logs. Above all, finding fractures requires knowledge of how they formed so workers can narrow the search to areas where fracture-forming processes are known or suspected to exist. Fractures may be associated with tectonic features such as folds, faults, regional flexures, or localized zones of differential compaction around buried shelf margins, large reefs and mounds, or other prominent and resistant bodies. More than one generation of fractures may be present in a single reservoir, although most

Geology of Carbonate Reservoirs: The Identification, Description, and Characterization of Hydrocarbon Reservoirs in Carbonate Rocks
By Wayne M. Ahr   Copyright © 2008 John Wiley & Sons, Inc.

of the fractures may be closed by cements, gouge, or compaction. In such cases, one fracture set—the one oriented parallel to maximum stress under present-day conditions—will remain open. The open set will, of course, have the greatest influence on reservoir behavior. Field development in fractured reservoirs requires knowledge of the extent to which fracture porosity and permeability influence the reservoir. At one end of the spectrum, fractures contribute the bulk of reservoir porosity and permeability; while at the other end, they may be impediments to flow because they have been filled with gouge or mineral cements.

7.1.1   Definition of Fractures

Fractures are defined as naturally occurring macroscopic planar discontinuities in rock due to deformation or physical diagenesis (Nelson, 2001). Fractures may form as the result of brittle failure under differential stress but ductile fractures also occur. Most fractured reservoirs, especially in carbonates, are brittle fractures. Brittle failure means that rocks failed by rupture after the application of differential stresses exceeded their elastic limit. Rocks behave as brittle materials as long as they are subjected to deformation within the *brittle domain*, which is characterized by comparatively low temperatures and confining pressures such as those at the earth's surface and at relatively shallow burial depths. Silica-cemented sandstones, dolostones, limestones, some chalks, and cherts are brittle at earth surface temperatures and pressures. At higher temperatures and confining pressures ductile behavior dominates and buckling or flowage may result instead of fracturing. Ductile and plastic behavior (high strain under low to moderate stress) can occur at low temperatures and confining pressures in rocks such as shale, mudrocks, salt, evaporites, marls, and clay-rich chalks. An example of ductile behavior can be observed by putting soft metal wire such as copper or gold under extension stress. The wire gradually lengthens, stretches (necks-down), and finally fails. Ductile fractures occur under higher confining pressures or high temperatures and they commonly occur as bands of deformed rock rather than distinct, sharply defined "cracks" that we usually visualize in association with the word fracture. Finally, plastic behavior is characterized by large amounts of strain produced by limited stress. An example of this behavior is the flowage of "silly putty" on a tabletop. Plastic behavior is not associated with brittle failure or fracturing, but rather it is the style of deformation that characterizes similar folds and material "flowage."

Fractured reservoirs are, according to Nelson (2001), "reservoirs in which natural fractures have, or are predicted to have, a significant effect on fluid flow either in the form of increased reservoir permeability and/or reserves or increased permeability anisotropy." He emphasizes that the words "predicted to" are especially important because the data needed to conduct quantitative analyses on fractured reservoirs must be collected early in the life of the reservoir in order to take advantage of and manage the "significant effects" of the fractures.

7.1.2   Types of Fractures

Two main variables determine which types of fractures will form: stress characteristics and material behavior. *Stress* is defined as force acting on an area and *strain* is deformation due to stress. In the case of fractures, strain is failure by rupture

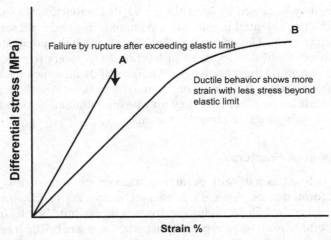

Figure 7.1 Stress–strain curves illustrate material behavior under stress. (Brittle failure (A) and ductile deformation (B) curves are not to scale.) This example shows typical curves for brittle and ductile behavior. Note that brittle behavior is indicated by a linear stress–strain curve up to the point at which the elastic limit of the material is reached and brittle failure (fracture) occurs. Think of a brittle glass plate that can be bent and released before breaking. It returns to its original shape because its response is elastic under that amount of stress. Exceed that level of stress and the glass snaps in brittle failure. Ductile behavior is indicated by a stress–strain curve that shows a major increase in strain with little increase in stress after a certain point is reached. This behavior is similar to the process of "necking-down" a copper rod or wire under extension stress. At some point the center of the wire becomes thinner and thinner until it finally fails. Ductile behavior is typical in folding; brittle behavior is typical in faulting.

(Figure 7.1). Stresses are represented as vectors with magnitude and direction. In the usual convention, the three principal stresses are identified by their magnitudes—maximum, intermediate, and minimum—and they are represented by the symbols σ_1, σ_2, and σ_3, respectively. In brittle behavior, different fracture types can result depending on whether compression, extension, or shear stresses caused failure. Laboratory experiments illustrate how extension and shear fractures are produced in a specimen subjected to compressive stress (Figure 7.2). Conjugate shear fractures are produced at an acute angle to the maximum principal stress σ_1, and a single extension fracture is oriented in a plane parallel to σ_2. Extension fractures are always oriented parallel to σ_1 and σ_2 and perpendicular to σ_3 and only when principal stresses are compressive. They can occur in all "low mean stress" subsurface conditions, according to Nelson (2001). Tension fractures have the same spatial orientation but occur only when σ_3 is negative. Tension fractures only occur in the near subsurface environment and are much less common than extension fractures (Nelson, 2001).

7.1.3 Genetic Classification of Fractures

Nelson's (2001) genetic classification of natural fractures identifies (1) tectonic fractures, (2) regional fractures, (3) contractional fractures, and (4) surface-related

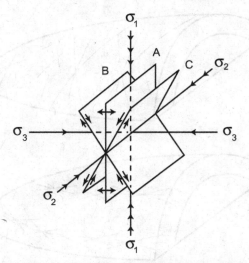

Figure 7.2 The typical orientation of conjugate shear and extension fractures with respect to the axes of maximum principal stress. When the maximum principal stress (σ_1) is vertical, fractures typically occur in pairs called conjugate shear sets. The acute angle between the fractures is bisected by an imaginary line representing σ_1. If σ_1 is horizontal, the acute angle between the conjugate fractures will also be horizontal. Intermediate principal stresses are always in the plane of the shear fractures and the minimum principal stresses are always perpendicular to the plane of the fractures.

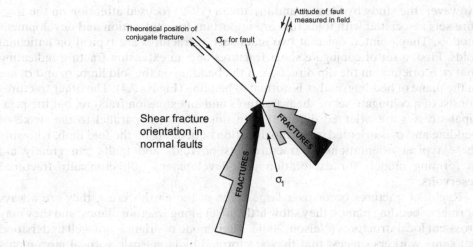

Figure 7.3 Fracture orientation with respect to maximum principal stress (σ_1) in normal and reverse faults. (From Stearns (1968).)

fractures. Tectonic fractures typically form with and have spatial relationships dictated by faults and folds. Fracture orientation with respect to σ_1 in normal faults is illustrated in Figure 7.3. Stearns (1968) illustrated the relationships between the attitudes of normal faults in the field, maximum principal stresses for faulting, and the orientation of fractures produced by the faulting.

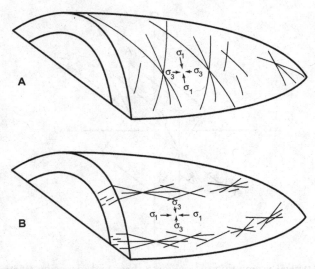

Figure 7.4 Typical fracture patterns in folds where maximum principal stress is oriented vertically and horizontally: (a) type 1 pattern with maximum principal stress vertical; (b) type 2 pattern with maximum principal stress horizontal. Fracture patterns follow bedding rather than fold geometry. (From Stearns and Friedman (1972).)

Tectonic fractures associated with folding have been studied by many workers. However, the study by Stearns and Friedman (1972) focused attention on the fracture sets associated with folds that are important for exploration and development models. They pointed out that two main sets of fractures are typical on anticlinal folds. First, a set of conjugate shear fractures and an extension fracture indicating that σ_1 is oriented in the dip direction of the bedding on the fold limb, σ_1 and σ_3 are in the plane of bedding, and σ_2 is normal to bedding (Figure 7.4). The other fractures consist of a conjugate set of shear fractures and an extension fracture, but the principal stresses are oriented differently. In this case, σ_1 is parallel to the strike of bedding and σ_3 is oriented in the dip direction of bedding on the fold limb. Knowing these typical orientations of fracture sets on folds and faults can greatly aid in forming models for exploration and development of tectonically fractured reservoirs.

Regional fractures occur over large areas of the earth's crust, they are always normal to bedding planes, they show little offset along fracture planes, and they may cross-cut local structures (Nelson, 2001). Their mode of origin is not well understood but some workers suggest that they are formed by large-scale vertical movements in the earth's crust. Nelson describes regional fractures around the Colorado Plateau, for example. He further notes that excellent reservoirs can result when regional fractures are cut by later tectonic fractures.

Contractional fractures are extensional and tensional fractures formed as the result of bulk volume reduction in the parent rock. Volume reduction can be produced by desiccation, syneresis, mineral phase changes, and temperature change. Contractional fractures are not as important as tectonic and regional fractures in hydrocarbon reservoirs.

Figure 7.5   Photo of a microbial bioherm in the Wilberns Formation (Cambrian) of Central Texas showing differential compaction that has caused fracturing in bedded rocks immediately above and below the bioherm. This type of fracturing is commonly associated with large reef and mound buildups. (Photo by Richard Rezak in Ahr (1971).)

Surface-related fractures are created by unloading stresses. Quarrying and weathering, for example, may remove stabilizing masses and create instability that leads to collapse of quarry walls. Surface-related fractures are not considered to be important in hydrocarbon reservoirs.

Fractures can also be produced by differential compaction at the local scale, such as compaction of strata above buried topography. Antecedent features such as reefs and mounds, shelf edges, erosional outliers, or horst blocks could cause overlying beds to drape, extend, and fracture in patterns related to the size and shape of the antecedent topographic feature. For example, brittle fractures have been documented by this author in the basal 5–20 meters of Mississippian "mud mounds" in the Williston Basin of North Dakota, where thin (1–15 cm thick) mudstone and cementstone beds appear to have been fractured by the overburden load of the mature mound (up to 150 meters thick). Differential compaction of carbonate strata above and below resistant mounds and reefs is common. An excellent example is illustrated by a Cambrian microbialite mound in Central Texas (Figure 7.5), where the beds above have draped (extension) over the mound and the beds below have been compressed (compaction) by the concentrated overburden. Larger-scale differential compaction fractures can occur along buried shelf edges, large reefs, erosional remnants, and fault blocks.

7.1.4   Fracture Morphology

In fractured reservoirs, total porosity and permeability values consist of both matrix and fracture components. In order to assess the importance of fracture porosity and permeability to total reservoir performance, it is necessary to determine the relative contributions of matrix porosity and permeability as compared to fracture porosity and permeability. Nelson (2001) ranks the four most useful petrophysical determinations to make on fractured reservoirs (in order of increasing difficulty of calculation) as (1) fracture permeability, (2) fracture porosity, (3) fluid saturations within fractures, and (4) fluid recovery factor expected from the fracture system. The kinds of data necessary for these determinations can be obtained from well tests and, according to Nelson, from whole core analyses. He points out that calculations from wireline log data do not provide accurate information for the evaluation of fracture contributions to reservoir performance. Imaging logs such as the Schlumberger FMI® (formation microresistivity imaging log) and FMS® (formation microscanning log) are, however, very useful for the identification of fractures and for determining their orientation (Figure 7.6). Fracture porosity and permeability are greatly influenced by fracture morphology. Nelson (2001) has identified four categories of fracture morphology: (1) open fractures, (2) deformed fractures, (3) mineral-filled fractures, and (4) vuggy fractures. The following paragraphs are based on material in Nelson (2001).

Open fractures are those that have not been deformed and that are not plugged with tectonic gouge or mineral precipitates. Permeability in open fractures is determined by the original fracture width, roughness, and the effective stress component that is oriented perpendicular to the plane of the fracture. Fracture width, roughness, and the contact area along fracture walls are usually determined by the texture of the host rock, that is, by its constituent particle size. Open fractures greatly increase permeability in the direction of the fracture plane but there is little change in permeability in the direction perpendicular to the fracture system.

Deformed fractures are produced either as ductile shear zones or as once-open fractures later altered by shear stresses. Two different fracture morphologies may result: gouge-filled deformation bands or slickensided fractures. Gouge is pulverized rock produced by the grinding and sliding motion of fracture walls against each other. Gouge partially or completely fills fractures and reduces both fracture porosity and permeability. In general, reduction in permeability due to gouge fill is greater in the direction perpendicular to the sense of sliding, or shear motion. Because gouge is typically fine grained, it may have high S_w that reduces relative permeability to hydrocarbons. Slickensides are simply gouge material that has been melted during cataclasis or grinding and pulverization that occurs during shear motion on fault surfaces. Slickensides generally occupy less volume than gouge because they consist of melted residue; consequently, the deformed rock volume around the fracture is lower than in the case of gouge filling. Slickensides can be thought of as glassy coatings that line fracture walls and that vary in thickness from less than 1 mm to a few millimeters. Gouge, on the other hand, may be several centimeters thick depending on the material involved, the mechanics of gouge formation, and the duration of the process. In slickensided fractures permeability is decreased perpendicular to the direction of the slip surface, but some increase in permeability may occur if there is a "mismatch" between slickensided textures on opposing fracture walls. In effect, the slickensides can prop-open fractures in some instances. Evaluating gouge and slickensided fractures involves knowing the mechanical system that produced the fractures and identifying the rock types that were more susceptible to cataclasis and production of gouge.

Mineral-filled fractures have been partially or completely filled by diagenetic precipitates after fracturing. As discussed in Chapter 6, diagenesis can take place at the surface or at great depth, it may occur once or many times, and it may decrease or increase porosity. Mineral fillings decrease original fracture porosity but mineral

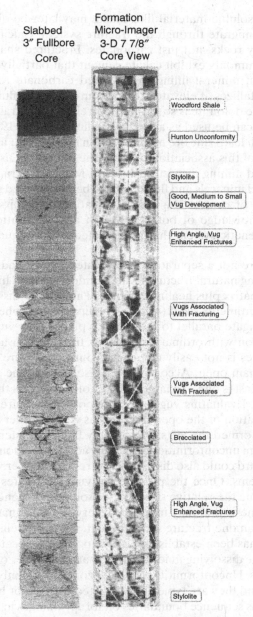

Slabbed
3″ Fullbore
Core

Formation
Micro-Imager
3-D 7 7/8″
Core View

Woodford Shale

Hunton Unconformity

Stylolite

Good, Medium to Small
Vug Development

High Angle, Vug
Enhanced Fractures

Vugs Associated
With Fracturing

Vugs Associated
With Fractures

Brecciated

High Angle, Vug
Enhanced Fractures

Stylolite

Figure 7.6 Imaging logs such as this FMI® log by Schlumberger are invaluable aids to determine fracture location, orientation, and spacing. Note the photo of the borehole core to the left of the image log. Some of the fractures and vugs are visible on the core photo and even more clearly on the image log. Most of the fractures are in conjugate sets that are highlighted by lines drawn on the image. Imaging may be based on differences in electrical resistivity, as the FMI log, or on acoustic responses. Extensive discussion on imaging logs can be found in Rider (1996) and Asquith and Krygowski (2004). (Illustration from Garber (2001), courtesy of the author.)

fillings composed of soluble material like calcite may later be dissolved by under-saturated fluids that migrate through the fracture system. Calcite healed fractures are common in many rocks, not just carbonates. Fractured shale, mudrocks, and other siliciclastics commonly exhibit calcite cement that partially or completely fills fractures. A common mineral filling in fractured carbonate reservoirs is saddle dolomite, which crystallizes at elevated temperature in burial fluids. In many cases, the presence of saddle dolomite is so strongly correlated with fractures that its presence in well cuttings can be used as a fracture indicator (Ahr, 1982). It is also commonly associated with lead–zinc ore mineralization in veins that are essentially filled fractures. Examples of this association occur in the Tri-State District of the United States and in the lead mining district of the Irish Midlands. Fracture filling by sulfides, particularly lead minerals and fluorite, have been exploited since Roman times in the Peak District of central England. Evaluating reservoirs with mineralized fractures requires knowledge of both the origin and orientation of the fracture system and the diagenetic system that precipitated or subsequently dissolved the mineral fillings.

Vuggy fractures are not a separate fracture category; instead, they are enlargements within and along natural fractures where undersaturated fluids have removed the rock matrix, or matrix plus healing cements. This type of dissolution diagenesis typically produces centimeter-sized (or larger) ellipsoidal to spherical or subspherical vugs that are elongate parallel to the fracture planes. The vugs greatly increase porosity in comparison with ordinary, nonvuggy fractures. In addition, the vuggy nature of the fractures is not easily compacted during pressure drawdown, so the fractures tend to remain open. According to Nelson (2001) the supergiant Asmari Field in the Middle East is one of a large number of giant fields that produce mainly from vuggy fractures. Evaluating vuggy fracture porosity requires understanding of the origin and orientation of the open fractures as well as understanding the diagenetic processes that formed the vugs. Most vuggy fracture systems are enlarged by water percolating from unconformities or shallow aquifers, although some expulsion fluids migrating upward could also dissolve fracture walls. Other vuggy fractures are related to karst systems. Once the presence of vuggy fractures is confirmed, it is necessary to determine the relative timing of dissolution diagenesis with respect to other forms of diagenesis that may influence reservoir performance. Timing is also important if more than one fracture set is present. Once the relative timing of vug-forming dissolution has been established, it is important to locate possible sources, or entry points, for the dissolving fluids because that will be the zone with the highest concentration of vugs. Unconformities and karst zones commonly show up as spikes on gamma ray logs and they are typically associated with major breaks in the stratigraphic record such as sequence boundaries or abrupt changes in bedding character. Some unconformities and karst surfaces may also be visible on image logs.

7.1.5 Where Do Fractures Occur?

Natural fractures are present in carbonates in settings that vary from undeformed, near horizontal beds to fold and thrust belts (Lorenz et al., 1997). Those writers describe fractured carbonate reservoirs in mildly, moderately, and severely deformed strata. Fractures in mildly deformed beds include regional fractures. There has been controversy over the origin of regional fractures, but Rhett (2001) offers a convincing argument in support of the extension fracture mechanism involving high pore pressure described by Lorenz et al. (1991). On fractured carbonates in general, Lorenz et al. (1997) conclude that extension fractures are common to virtually all examples in their studies and that lithology exerts great control on fracture characteristics. Dolomitic rocks fracture more easily than limestones, fine-grained rocks fracture more easily than coarse-grained ones, thin beds are more prone to fracturing than thick beds, and structural position has a great influence on fracture character. More fractures form along hinges of folds (areas of stress concentration) than along limbs, for example. Importantly, they also note that multiple generations of fracturing may occur in which older fractures are overprinted by younger ones, especially in highly deformed rocks. The younger fracture sets may control reservoir behavior because they are more likely to remain open and to be oriented parallel to in situ stresses. Lorenz et al. (1997) frame their discussion of fractures in terms of *mechanical stratigraphic units* to differentiate between lithostratigraphy and mechanical stratigraphy. Sonnenfeld, in Lorenz et al. (1997), described one type of mechanical stratigraphic unit in the following terms: "[a] *hierarchy of stratigraphic cycles that controls the distribution of small-scale evaporite collapse breccias, which in turn create a fracture-prone mechanical stratigraphy* ... [with] *the following intrinsic factors: 1) thin bedding, 2) ... dolomitic lithologies, 3) fine grain sizes, and 4) accentuated bedding-plane slip due to mechanical contrasts between ... evaporite collapse breccias and intervening competent ... dolomite beds.*" Although regional fractures are common in mildly deformed strata, tectonic fractures dominate in moderately and severely deformed rocks.

Tectonic fractures commonly occur in predictable patterns determined by the geometry of the associated faults or folds. Corbett et al. (1991) mapped fractures in the Cretaceous Austin Chalk of Texas and established that four different structural configurations had specific fracture patterns depending on the nature of the structural feature. The four structural types included anticlinal folds, monoclinal flexures, listric normal faults, and graben-in-graben normal faults (Figure 7.7). Stearns and Friedman (1972) demonstrated that extension fractures occur along and parallel to fold crests while conjugate sets of shear fractures typically occur along fold limbs of anticlinal folds. The orientation of the respective fracture sets differs according to the orientation of the maximum, intermediate, and minimum principal stresses. Fractures associated with faults are parallel to fault slip planes and fracture density is commonly higher on hanging walls of normal faults than on footwalls (Friedman and Wiltschko, 1992). Fractures that result from mechanical compaction, or physical diagenesis, also occur in predictable patterns along stratigraphic hinge lines such as the inflection zones on monoclinal flexures. Fractures may form radial or concentric patterns peripheral to and above or below buried reefs and mounds, where the beds overlying the buildups have undergone differential compaction with respect to the surrounding strata. Radial and tangential faults and fractures are commonly formed as salt domes grew upward. Much of the fracturing at Ekofisk Field in the North Sea was formed by salt tectonics (Farrell, in Lorenz et al., 1997). Differential compaction fractures around buried reefs are important contributors to total production in the Mississippian mound reservoirs of the Williston Basin (Young et al., 1998). The orientation of compaction fractures in the subsurface is not always obvious or easy to deduce; consequently, image logs and borehole cores are essential for determination of fracture orientation, spacing, and density.

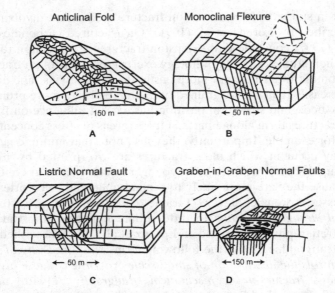

Figure 7.7   Diagram illustrating the most common types of fractures mapped in outcrops of the Austin Chalk (Cretaceous) by Corbett et al. (1991). Note fracture patterns on folds, on regional-scale monoclinal flexures, in half-grabens, and in normal faults. Fracture intensity is greatest on the hanging wall of faults and where stresses were concentrated along the crests of folds.

7.2   FRACTURE PERMEABILITY, POROSITY, AND S_W

Before discussing ways to evaluate the relative contribution of fractures to total reservoir permeability and porosity, it is necessary to define fracture permeability and porosity as compared to matrix permeability and porosity. The two major factors that distinguish fracture permeability and porosity from matrix pore systems are *fracture width e* and *fracture spacing D*. We considered Darcy's equation for permeability in a homogeneous, porous medium under single-phase, Newtonian flow conditions as

$$Q = KA\frac{dh}{dl}$$

where K = Hydraulic conductivity
    A = Cross sectional area of the porous medium
    dh/dl = Gradient in hydraulic head

Hubbert (1940) determined that $K = \rho g/\mu$ and that $k = Nd^2$. In this case, k is intrinsic permeability with dimensions of L^2, ρ is fluid density, g is acceleration due to gravity, μ is fluid viscosity, N is a dimensionless coefficient characteristic of the porous medium, and d is the average diameter of constituent grains in the rock (a condition most applicable to siliciclastic sandstones and less so to carbonates). As noted in the next paragraph, values for Hubbert's Nd^2 could not be reliably defined and other methods had to be developed to model flow in fractures.

7.2.1 Fracture Permeability

Nelson (2001) notes that the value of Nd^2 could not be reliably defined when attempts were made to model fluid flow in fractures. In order to model fluid flow in fractures, the parallel plate theory for fluid flow was developed. The essence of this theory is the assumption that flow occurs between two smooth parallel plates a distance e apart. The following equation represents flow that should pass through the two plates:

$$\frac{Q}{A} = \frac{e^3}{12D} \cdot \frac{dh}{dl} \cdot \frac{\rho g}{\mu}$$

where Q is discharge in volume per time, A is area of the slot between the plates, D is plate spacing, or the average distance between regularly spaced parallel plates, and e is slot width. This is the expression for laminar flow in parallel fractures with only nominal variation in fracture width.

The Darcy equation deals with matrix permeability or, as Nelson (2001) calls it, the "intact portion of the rock," whereas the equation for flow in parallel fractures deals only with the theory of flow between parallel plates. An expression to deal with flow through both matrix and parallel fractures was developed by Parsons (1966). The combined flow expression is

$$k_{fr} = k_r + \frac{e^3 \cos^2 \alpha}{12D}$$

and flow through fractures only is given by

$$k_f = \frac{e^2}{12} \cdot \frac{\rho g}{\mu}$$

where k_{fr} = Permeability of matrix (intact rock) plus fracture

  k_f = Fracture permeability

  k_r = Matrix permeability

  α = Angle between the fracture planes and the axis of the subsurface pressure gradient

Of course, the underlying assumptions for all of these equations are that flow is laminar, the plates are smooth and do not move, and fracture width and spacing are constant. The equations provide a method for estimating fracture permeability knowing some of the fracture properties. If the assumed conditions are not met, these equations will not provide realistic results.

It is also important to understand that fracture permeability (k_f) and fracture width (e) decrease exponentially with depth and confining pressure. According to

Nelson (2001), fracture widths of "crystalline carbonates" and chalks at simulated depths of 10,000 feet are 10^{-1} and 10^{-1} to 10^{-6} cm, respectively. If k_r and k_f are known at a given confining pressure, the earlier equations will provide results that are more realistic for subsurface reservoir conditions. Fracture spacing, D, also influences fracture permeability; however, determining reliable values for D is not easy, especially in the subsurface where observations are usually limited to core samples. Nelson (2001) uses the definition of Parsons (1966) for fracture spacing; that is, "*the average distance between regularly spaced fractures measured perpendicular to a parallel set of fractures of a given orientation.*" Fracture spacing can be calculated with statistical and geometrical methods from measurements made on cores or outcrops, but those methods are beyond the scope of this book.

7.2.2 Fracture Porosity

Calculating fracture porosity is similar to calculating matrix porosity in that ϕ_f is the percentage of total rock volume made up of fracture pores. Whereas matrix porosity is represented by $\phi_r = (V_p/V_t) \times 100$, fracture porosity is expressed in terms of fracture width and spacing, or $\phi_f = [e/(D + e)] \times 100$

where ϕ_r = Matrix porosity in percent
   V_p = Pore volume in intact (matrix) rock
   V_t = Total rock volume
   ϕ_f = Fracture porosity
   e = Average effective fracture width
   D = Average spacing between parallel fractures

Note that ϕ_f for a constant e varies as a function of distance between fractures, indicating that fracture porosity is *scale dependent*. Matrix porosity is not. If fracture spacing is measured over a small area (e.g., cm^2) and only one fracture is counted, the resulting fracture porosity value will be high because the proportion of porosity contributed by the one fracture in a small area is a large part of the total porosity. If the measurement area is increased in size, matrix porosity remains fixed, and more fractures are counted in the larger area, then ϕ_f will be proportionately smaller because the contribution of matrix porosity is greater over the larger area and the proportion of fracture porosity to matrix porosity becomes smaller. For this reason, it is important to utilize a large enough measurement area to encounter several fractures. It also means that obtaining statistically valid sampling is more difficult when it is not possible to measure fracture spacing directly as on outcrops.

Measurement of fracture porosity also requires larger samples. Conventional 1-inch diameter "perm plugs" that are used for routine core analyses do not have enough volume to adequately sample both matrix and fracture porosity components; therefore full-diameter core segments are necessary. It is also important to have experienced geologists examine core samples before measuring porosity in order to distinguish between induced fractures and natural fractures. And it is helpful to identify which natural fracture sets are likely to be the principal contributors to fracture porosity, especially if several fracture sets are present but only one

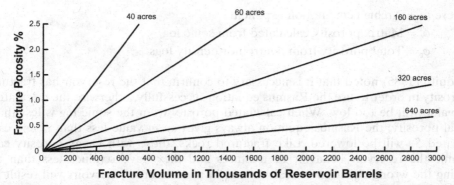

Figure 7.8 Diagram illustrating the relationship between fracture porosity, fracture volume, and reservoir drainage area. Intuitively, smaller drainage areas have smaller fracture volume. (Adapted from an illustration in Nelson (2001).)

or two contribute to reservoir performance. Such situations exist when several fracture sets are present but only the ones oriented in the direction of in situ principal stress are open to flow.

Determination of fracture porosity in subsurface reservoirs is difficult. Nelson (2001) notes that if certain conditions are met and fracture permeability values have been determined from reservoir flow tests, then fracture porosity can be calculated from the empirical relationship between fracture permeability and porosity. The conditions that must be met are: (1) a flow test permeability calculation must be made from a zone in which a core has been pulled; (2) core analysis must show that matrix porosity and permeability contributed negligible flow to the flow test; and (3) a good estimate of fracture spacing must be obtainable from core examination. It must be recognized that fracture porosity is more compressible than matrix porosity, especially in brittle rocks; therefore fracture porosity and permeability are more susceptible to reduction due to confining pressure than are matrix porosity and permeability.

Fracture porosity is generally only a small percentage of total reservoir porosity, but because the fractures are connected, the small fracture volume can contribute enormously to total permeability. If fracture porosity amounts only to about 1% in a thick and aerially extensive reservoir, fracture volume can be very large, justifying well spacing of hundreds to 1000 acres, according to Nelson (2001). A relationship between fracture porosity, fracture volume, and reservoir drainage area is shown in Figure 7.8.

7.2.3 S_w in Fractured Reservoirs

Determining S_w in fractured reservoirs using the Archie equation is complicated because the cementation exponent, m, may be as low as 1.0 according to Asquith (1985), who cites an equation by Rasmus (1983) for calculating m:

$$m = \frac{\log[\phi_s^3 + \phi_s^2(1 - \phi_t) + (\phi_t - \phi_s)]}{\log \phi_t}$$

where m = Archie cementation exponent

   ϕ_s = Matrix porosity calculated from sonic log

   ϕ_t = Total porosity from neutron or density logs

Asquith (1985) noted that it is necessary to confirm that the reservoir has fracture porosity in order to use the Rasmus equation successfully, otherwise the calculated m value will be too low. When calculated porosity from the sonic log is less than total porosity, the Rasmus equation always gives an m value less than 2, and calculated S_w will be lowered. Like fractured rocks, those with vuggy porosity can mimic the reservoir behavior of fractures and they have ϕ_s values less than ϕ_t. Using the wrong equations to calculate m in nonfractured reservoirs will result in serious errors in calculated S_w. Asquith (1985) notes that m values for rocks with vuggy porosity should be greater than 2.0—not less than 2.0 as when the Rasmus equation is used. He presents an example of the errors that result from misapplying the Rasmus equation to reservoirs with vuggy porosity. He gives the following values for an example calculation: R_w = 0.04, R_t = 20, ϕ_s = 0.05, and ϕ_t = 0.15. It is assumed that the reservoir is vuggy and that the Nugent (1984) equation, $m \geq (2\log\phi_s)/(\log\phi_t)$ is used to compute m. The results are m = 3.16 and S_w = 89.6%. If the Rasmus equation is used instead, assuming that the reservoir has both fracture and vuggy porosity, the results are m = 1.20 and S_w = 13.9%. If the S_w = 89.6% value is used, the reservoir is water-wet and nonproductive. If the 13.9% value is used, the reservoir is productive. This dramatic difference in outcomes illustrates the importance of recognizing pore type and rock properties from direct observation of core samples before undertaking evaluation of carbonate reservoirs.

7.3 CLASSIFICATION OF FRACTURED RESERVOIRS

Fractured reservoirs can be divided into four types, following Nelson (2001):

   Type I. Fractures provide essential reservoir porosity and permeability.

   Type II. Fractures provide essential permeability.

   Type III. Fractures assist permeability in an already producible reservoir.

   Type IV. Fractures provide no additional porosity or permeability but they do impose significant anisotropies such as barriers to flow.

Examples of type I reservoirs include Amal Field in Libya, with reserves of about 1700 MMbbls and several Ellenburger dolomite fields in Texas. Normally reservoirs of this type do not have large reserves because matrix porosity, the principal storage volume in reservoirs, is insignificant. The large reserves at Amal Field exist because the reservoir is 800 feet thick and extends over 100,000 acres. Fracture porosity in Amal Field is only about 1.7%, assuming no contribution from the Cambrian quartzite matrix (Nelson, 2001).

Type II reservoirs have greater storage capacity because matrix porosity is important. Because of this, type II reservoirs typically have greater reserve volumes than type I reservoirs. Examples of type II reservoirs include the Agha Jari Field in Iran with 9500-MMbbl reserves, the Haft Kel Field in Iran (2660 MMbbls), and the Spraberry trend in Texas (447 MMbbls) according to Nelson (2001).

Type III reservoirs are confirmed to occur in some of the largest fields in the world where both matrix and fractures are capable of significant flow, although fractures provide only a permeability assist. Among fields with type III reservoirs are Kirkuk Field in Iran (15,000 MMbbls), Hassi Messaoud, Algeria (6000 MMbbls), and Dukhan, Qatar (4570 MMbbls). Kirkuk Field had initial flow rates of up to 100,000 BOPD on one of the first wells drilled into the fractured Asmari Limestone. Because the pressure differential on initial flow was very low, the reservoir was recognized as a fractured system immediately and was managed as such from that time onward (Nelson, 2001). It is not always easy to recognize fractured reservoirs and many secondary recovery efforts fail because injected fluids follow unknown fracture systems instead of sweeping the expected matrix porosity systems.

7.4 DETECTING FRACTURED RESERVOIRS

There are a variety of methods for finding fractures in the subsurface: (1) seismology, (2) subsurface geological methods, (3) mapping fractures on outcrops and extrapolating fracture patterns to the nearby subsurface, and (4) exploratory drilling of structures that are likely to have tectonic fractures. Extensively fractured rock can have a distinctive seismic signature that is sometimes recognizable directly on seismic records or less directly with the aid of *seismic attributes* (seismic wave characteristics such as amplitude, frequency, polarity, spatial extent, shear wave character, and amplitude versus offset—AVO). AVO analysis is a data-processing technique used on prestack seismic data and is widely used in exploring for gas sands. Although much of the literature on AVO processing focuses on gas sand reservoirs, the method should also be useful in carbonates under certain circumstances. Ideally, both P and S wave velocities would be recorded along with offset and bulk density of the target rocks. Differences in return velocities of shear and compressional waves and P wave polarity, which changes as the wave passes from water-wet sands to gas sands, can indicate the presence of gas. AVO methods are also useful for detecting fractured reservoirs because the bulk density, returned P–S wave velocities, and possibly the fluid characteristics in fractured zones differ from those in nonfractured zones. Most seismic records do not include S wave data, but P wave data at different offsets are routinely recorded. That information can be processed to record a component of S wave data. A detailed explanation of the AVO technique is beyond the scope of this book. Suggestions for further reading are included at the end of this chapter.

Subsurface geological methods utilize data from previous drilling, coring, logging, completion, and production. Image logs and cores are direct proof that fractures exist. Drilling time logs, mud circulation logs, caliper logs, and acoustic logs are usually helpful indicators of fractures, but they have limited use in distinguishing between cavernous dissolution porosity and fractures. More about borehole logs and their application appears later in the discussion on indirect methods for detecting fractures (Section 7.4.2).

Exploratory drilling of structures is based on the concept that folds and faults have specific fracture patterns that correspond to structural geometry. As we saw earlier, fractures associated with faults are generally parallel to fault slip planes and fracture density is higher on hanging walls of normal faults than on footwalls (Friedman and Wiltschko, 1992). In addition, specific fracture types and geometrical arrangements occur on anticlines. It is riskier to drill structural geometry in search of fractures than it is to drill prospects based on seismological data or zones where subsurface geology indicates fractures. In the best situations, drilling should be based on information from all methods: seismology, subsurface geology, and structural geometry. Once fractures have been detected in the subsurface, the next step is to identify or predict the location of zones with highest fracture intensity (closest spacing) to ensure the most economical and efficient field development. That can be accomplished by either direct or indirect observation.

7.4.1   Direct Observation of Fractures in the Borehole

Direct observations include core examination, viewing borehole walls with downhole TV cameras, and making borehole impressions with inflatable packers. Core examination is the best method for direct determination of fracture dip, fracture intensity, rock strength, rock fabric, and what type of fractured reservoir (of the four categories discussed above) exists. Knowing the extent to which fracture and matrix porosity influence reservoir performance will greatly improve the odds for improved management strategies and development outcomes. Downhole cameras can only be used in dry, gas-filled, or clear, water-filled boreholes, but they can provide information about the presence of fractures and their orientation, and some information about fracture intensity. Inflatable impression packers have a coating that enables an impression of the borehole wall to be made when the packers are set in the borehole and inflated. They are deflated and removed to examine the impressions. The method is not sensitive enough to detect small natural fractures, it is ineffective in mud-caked holes, and inflated packers commonly fail in large, jagged washouts, a common characteristic of some extensively fractured zones.

7.4.2   Indirect Methods to Detect Fractures in the Borehole

The main methods for indirect detection of fractures in the borehole are (1) well log evaluation, (2) flow test evaluation, and (3) mathematical and graphical manipulation of reservoir rock data. Nelson (2001) discusses nine different logging tools and their use in indirect detection of fractures. He notes that log responses have been used with varying degrees of success to detect fractures and fracture intensity, but not fracture spacing. And because the log responses are not unique to fractures, the log analyst must have detailed knowledge of each tool and its response to a wide variety of rock properties, some of which could cause responses in unfractured rock that resemble those in fractured reservoirs.

One of the logging tools commonly used to determine the presence of fractures is the *sonic amplitude log*. Compressional wave amplitudes are reduced dramatically when they encounter fluid-filled fractures and shear wave amplitudes virtually disappear. The *acoustic log* (Sonic Log® Schlumberger registered trademark) can be used as a qualitative check for possible fractured zones, especially when it is used in conjunction with caliper, drilling time, and mud logs, including the lithology log from cuttings. When fractured zones with dramatically increased interval acoustic transit time are encountered in the borehole, the log scale changes as many as 2 to 4 times almost instantly, depending on the difference in acoustic transit time between fractured and unfractured rock. Rapid scale changes, increased drilling rate, and the presence of saddle dolomite in cuttings were correctly interpreted as fracture indicators in Mississippian "mud-mound" reservoirs in Texas (Ahr and Ross, 1982). *Caliper logs* are useful for locating washouts and enlarged boreholes that may have been susceptible to washing out because the fractured borehole wall was highly friable and tended to cave in. *Drilling time logs* are useful to compare drilling rate with borehole diameter, lithology log, and records of mud circulation. Fractured rock is more easily drilled than unaffected matrix rock; therefore drilling time will decrease, sometimes dramatically. Fractured zones commonly cause lost mud circulation or problems with maintaining mud circulation, and these events coincide with drilling "breaks"—or high rates of penetration by the bit. These changes in mud circulation patterns are reflected on the *mud log*. *Lithology logs* prepared by qualified and observant persons may show traces of large crystals or veins of fracture-filling minerals. Calcite or exotic minerals such as saddle dolomite cements are commonly found in cuttings as mineral fillings from subsurface fractured rock. Saddle dolomite is particularly useful in this regard (Ahr, 1982). *Borehole imaging logs* have become the principal tool for fracture detection in today's industry. Both acoustic and electrical imaging systems are used. Acoustic imagers such as Schlumberger's UBI (ultrasonic borehole imager) and FMI (formation micro imager) are typical examples. The acoustic tool is used mainly in wells drilled with oil-based muds in which the resistivity tool does not function well. The imaging logs produce an unrolled "picture" of the borehole (Figure 7.6). With digital data and workstation interpretation software, the dip and strike of all features such as fractures and sedimentary structures can be calculated automatically. The *dipmeter* is another device commonly used to detect natural fractures. The four-pad dipmeter can function as a caliper log by comparing borehole enlargement in two directions at 90° from each other, and it can function as a four-pad resistivity device that measures vertical displacements in resistivity response to the four pads. The resistivity responses are assumed to be generated by fluid-filled fractures.

Some well testing procedures are widely used by engineers as indirect methods for detecting fractured reservoirs. One of the most commonly used ones is the pressure buildup test because it can be performed at any time in the life of a well. The only requirement is that the well be shut in. The shut-in flow rate (zero) is easier to control than a constant rate flow test; therefore buildup tests are the preferred method of pressure-transient testing (Lee, 1992). Fractures are indicated by the longer-term pressure behavior of the well in the manner illustrated in Figure 7.9. Borehole effects are dominant during the early stages of the pressure-transient test and are shown as a straight-line segment on the semilog plot. This segment reflects reservoir behavior influenced by fractures only. The response is immediate and does not yet reflect pressure contribution from matrix permeability. The second straight-line segment reflects the transient flow from the matrix until it stabilizes with flow from fractures. At that point, the third straight-line segment represents combined flow from fractures and matrix. This characteristic semilog plot is known as the

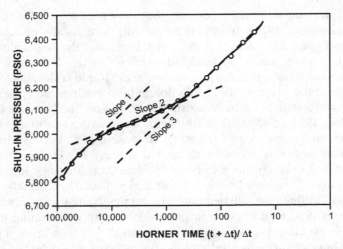

Figure 7.9 Reservoir pressure behavior over time, depicted here as a "Horner plot," is a useful technique to determine whether or not fractures contribute to reservoir pressure behavior. The open circles are data points on a sigmoidal curve that has two inflection points and three slope segments. The first slope segment of the curve shows the influence of flow from fractures, the second slope segment shows the transition between flow purely from fractures and flow from the unfractured "matrix." The final slope represents combined flow from the matrix and from fractures. (Adapted from an illustration in Tiab and Donaldson (2004).)

Horner plot for the person who first introduced the technique in the petroleum literature. The y axis of the plot is well pressure in psi. The abscissa of the plot is "Horner time," or the ratio of producing time, t_p, plus shut-in time Δt, to shut-in time.

Some of the more common methods for mathematical and graphical manipulation of reservoir data include crossplotting log k versus ϕ from measured values in cores (core analyses), k_v and k_h from whole core analyses, core permeability plotted against flow-test permeability, and core porosity plotted against calculated porosity from neutron logs. Permeability–porosity crossplots can indicate unusual relationships such as low porosity with high permeability. In such a case, fractures are a likely cause of the high permeability values, although connected vugs and channelized porosity can produce similar results. Vertical and horizontal permeability from whole core analyses can indicate fractures if the k_v component is large because most layered sedimentary rocks tend to have higher k_h values owing to preferential flow parallel to bedding. Core permeability and flow-test permeability are usually compared to determine a range of reasonable values for k. However, if flow-test values are significantly greater than values from corresponding core analyses, it is usually taken as an indication of fractures. Core porosity compared with calculated porosity from neutron logs can suggest fractures if the values differ widely. Core porosity is generally taken to represent matrix porosity and neutron porosity is taken to represent total porosity. If core porosity plots very low compared to the neutron porosity value, the core porosity is interpreted to represent matrix in a fractured zone. Nelson (2001) notes, however, that the value for fracture porosity is usually too small to observe numerically on the neutron log.

7.5 PREDICTING RESERVOIR FRACTURE SPACING AND INTENSITY

Fracture spacing and intensity result from a complex interplay between stress concentration, orientation, and magnitude along with material characteristics. Because it is easier to observe and measure rock properties (material characteristics) than to deduce stress trajectories and magnitudes, this section focuses on rock properties and their relationship to fracture spacing and intensity.

7.5.1 Factors that Influence Fracture Spacing and Intensity

Geological factors that influence fracture spacing and intensity include carbonate rock composition, bulk rock porosity, bed thickness, and position of the fractured zone with respect to the geometry of faults and folds. In general, as noted by Stearns and Friedman (1972), rocks composed mainly of brittle constituents exhibit closer spaced fractures than those without brittle constituents. Typical brittle constituents common in carbonate reservoirs are dolomite, chert, and, in some instances, calcite.

Porosity influences fracture spacing because rock strength decreases as porosity increases. If composition and fabric do not vary, rocks with lower porosity will have closer spaced fractures and greater fracture intensity, although this situation may be uncommon because significant changes in carbonate porosity are generally accompanied by variations in fabric or texture.

Bed thickness influences fracture spacing and intensity in that, all else being equal, thin beds will have closer spaced fractures than thick ones. Nelson's (2001) work suggests that when bed thickness and structural geometry are known, fracture spacing can be predicted. Laubach et al. (2002) have had success in using observations of microfractures to predict the character of large-scale fractures in some sandstone reservoirs.

Finally, the influence of structural geometry on fracture spacing and intensity is based on the premise that higher fracture intensity and closer spacing correspond to positions on structures where the greatest rate of change in dip or bed curvature exists per unit of horizontal distance across the structures. This premise depends on three assumptions: (1) the rocks involved behave as brittle material and fail by brittle fracture, (2) increases in bed curvature correspond to increases in strain, and (3) increased strain causes higher fracture intensity.

7.6 IDENTIFYING AND DEVELOPING FRACTURED RESERVOIRS

The following checklist offers some helpful hints for use in working fractured carbonate reservoirs.

1. Evaluate seismic data, subsurface geology, or projections of structural geometry from outcrop to subsurface to maximize the likelihood of finding structures with tectonic fractures. Look for the fracture patterns predicted by Stearns and Friedman (1972) if the structures are folds. In the case of faults, expect higher fracture intensity on the hanging walls. Regional fractures may be present in areas that have been subjected to regional stresses, such as

monoclinal flexures over buried shelf edges or buried reef trends. Seismic attributes may provide indications of regional fractures in low-dipping beds. On a smaller scale, differential compaction fractures tend to be localized around antecedent highs such as horsts, buried mounds, or erosional outliers.

2. After the presence of fractures has been confirmed, determine whether they are related to present structure or paleostructure. Using image logs or cores, if available, determine fracture orientation, spacing, and intensity with respect to structural geometry. Determine whether fracture spacing and intensity have any relationships to rock and stratigraphic properties such as mineralogical composition, grain or crystal size, bed thickness, or bulk-rock porosity.

3. Evaluate well test and production data to determine the extent to which fractures influence reservoir porosity and permeability so you can classify the fracture system according to Nelson's (2001) four categories. Develop strategies that incorporate the positive and negative characteristics of each of the four fractured reservoir categories. Examples of some positive attributes and common problems associated with the different types of fractured reservoirs are listed in Tables 7.1 and 7.2.

4. Using data from at least one well, refine the geological concept. By this stage, the types of fractures have been classified, their relationship to paleostructural or present structural geometry is known, and the rock and stratigraphic characteristics of the fractured zones are known. Utilize observations from cores or image logs to refine estimates of fracture spacing, intensity, width, and orientation. Incorporate these refined estimates in volumetric calculations and future development planning.

TABLE 7.1   Positive Attributes of Fractured Reservoirs Types I–III

Type I Fractures: Fractures Provide Essential Porosity and Permeability	Type II Fractures: Fractures Provide Essential Permeability	Type III Fractures: Fractures Provide Permeability Assist
Large drainage areas per well	Can develop in low-permeability rocks	Reserves are dominated by matrix properties
Few wells needed for field development	May have unexpectedly high well rates	Reserve distribution may be relatively homogeneous
Good correlation between well flow rates and well reservoirs	Hydrocarbon charge is commonly facilitated by fractures	High sustained well rates
Best wells are commonly early		Excellent reservoir continuity
High initial potential (IP) common		
Can produce from nonreservoir quality and nonstandard rocks		

Source: Adapted from Nelson (2001).

TABLE 7.2 Commonly Encountered Problems Involved in Fractured Reservoir Types I–IV

Type I Fractures Fractures Provide Essential Porosity and Permeability	Type II Fractures Fractures Provide Essential Permeability	Type III Fractures Fractures Provide Permeability Assist	Type IV Fractures Fractures Create Barriers to Flow
Often have rapid decline curves	Poor matrix-to-fracture communication leading to poor matrix recovery efficiency and disastrous secondary recovery results	Highly anisotropic permeability	Reservoirs commonly compartmentalized
Possible early water encroachment—requires controlled production rates	Possible early water encroachment—requires controlled production rates	Commonly can have unexpected response to secondary recovery	Wells underperform in comparison to matrix flow capabilities
Difficult to determine drainage area size and shape	Fracture intensity and dip angles critical	Drainage areas commonly elliptical	Recovery factor highly variable across field
Difficult to constrain reserve calculations	Development pattern must be tailored to reservoir	Interconnected reservoirs common	Permeability anisotropy commonly unlike that in adjacent fractured reservoirs with different fracture styles
Development wells add production rate but not additional reserves	Recovery factor difficult to determine and may vary widely	Commonly have poor correlation between log response, core analyses, and well test or performance data	
	Fracture closure may occur in overpressured reservoirs		

Source: Adapted from Nelson (2001).

5. Expand the scale of investigation to field size by adjusting the geological concept (fracture model) to assume that fractures behave as ideal fracture types (standard models in the literature). Using those ideal fracture models, predict the spatial distribution of fracture types, intensity, and spacing at field scale using mathematical and geometrical manipulation of subsurface data as outlined in Nelson (2001).

6. Refine the model for fluid flow in the types of fractures and matrix rock known to exist in the reservoir based on well test or production data. Estimate proportions of matrix and fracture porosity and permeability to develop a reservoir model for optimum and sustained recovery.

Full-diameter cores offer the best source of information about fracture and matrix properties. Use that information to develop a more accurate model for total reservoir characteristics. Keep in mind that fracture porosity is calculated differently than matrix porosity and that crossflow (matrix flow interacting with fracture flow) can occur in bimodal porosity systems. Pressure testing is especially useful to detect crossflow.

SUGGESTIONS FOR FURTHER READING

The most comprehensive discussions on fractured reservoirs are in the first and second editions of *Geological Analysis of Naturally Fractured Reservoirs* by Nelson (1985, 2001). Examples of fractured reservoirs in mildly to severely deformed strata are described and illustrated in Lorenz et al. (1997), in a chapter in *Carbonate Seismology*. Reservoir engineering aspects of fractured reservoirs are discussed in detail in Chapter 8 of Tiab and Donaldson's (2004) text, *Petrophysics*, 2nd edition. A concise and clear discussion of AVO analysis in seismic exploration for hydrocarbon reservoirs is given by Davies et al. (2003).

REVIEW QUESTIONS

7.1. Define stress and strain.

7.2. What are three types of stress and how would structures produced by them be different?

7.3. What is the typical material behavior associated with faults? With folds?

7.4. According to the work by Stearns and Friedman (1972), what types of fractures are typically found on limbs of anticlinal folds? What effect does the orientation of maximum principal stress have on those fractures?

7.5. What types of fractures would you expect to find on the crests of anticlines or synclines?

7.6. Fracture porosity is described by Nelson (2001) as "scale-dependent." What does that mean?

7.7. How might horizontal boreholes (directional drilling) be more effective than vertical ones in exploiting fractured reservoirs along normal faults?

7.8. How could you determine whether a fractured reservoir is a Type I, II, III, or IV variety?

7.9. What clues might indicate the presence of fractures during the drilling process?

7.10. How could you confirm the presence of fractures in the borehole after drilling?

7.11. The genetic porosity classification in this book indicates that depositional and diagenetic attributes influence fracture porosity. What influence can diagenetic or depositional characteristics have on fracture intensity?

7.12. What is meant by "mechanical stratigraphy"?

7.13. What problems might you encounter in calculating fluid saturation for fractured reservoirs? How would you determine the appropriate "*m*" value for the Archie equation in fractured reservoirs?

7.14. Of Nelson's four principal types of fractured reservoirs, which has the greatest storage capacity? For the fractured reservoir type you identified, is capacity to store fluids equivalent to capacity to flow? Explain.

CHAPTER EIGHT

SUMMARY: GEOLOGY OF CARBONATE RESERVOIRS

The first three chapters in this book introduce carbonate rock and reservoir properties to set the stage for understanding the relationships between rocks and the fluids that move through them. A genetic classification of carbonate reservoir porosity—depositional, diagenetic, and fracture pore types—is emphasized because finding and developing carbonate reservoirs is a job of determining where porous and permeable rocks occur in the subsurface. Correlating flow units within fields then requires understanding the geological causes that formed porosity and permeability. Carbonate sedimentology and stratigraphy, discussed in Chapter 4, put rocks in temporal and spatial context to illustrate how reservoir bodies—reservoir architecture—can be visualized. Chapters 5–7 deal specifically with depositional, diagenetic, and fractured reservoirs.

This chapter is a summary of the discussions in the first seven chapters, it includes some comments about methods for exploration and development in carbonate reservoirs based on the author's personal experience, and it includes selected field examples of the three genetic reservoir types. The purposes of this summary are to highlight the similarities and differences between carbonate and siliciclastic reservoirs and between reservoir types in carbonates, and to focus on practical ways to identify, describe, and develop carbonate reservoirs.

The first rule in working carbonate reservoirs is "look at the rocks." The importance of direct observation of rocks, cores, cuttings, or outcrops cannot be overemphasized. It is virtually impossible with present technology to identify carbonate facies, genetic pore types, or reservoir categories based on those pore types without direct examination of rocks. Seismic and logging methods, including data from imaging logs and seismic attributes, cannot discriminate between carbonate pore types, depositional facies and diagenetic facies, nor can they make 100% reliable

Geology of Carbonate Reservoirs: The Identification, Description, and Characterization of Hydrocarbon Reservoirs in Carbonate Rocks
By Wayne M. Ahr   Copyright © 2008 John Wiley & Sons, Inc.

and accurate determinations of mineralogical composition in multicomponent reservoir rocks. One might wonder why mineralogical composition is important in carbonate reservoirs because they consist mainly of calcite or dolomite. The answer is that porosity and permeability in some reservoirs is highly dependent on mineralogical composition, such as in those that produce exclusively from intercrystalline porosity in dolostones. In those reservoirs small percentages of "accessory minerals" may significantly alter reservoir quality. Calcite, anhydrite, clay minerals, quartz, or other minerals generally have a negative influence on reservoir quality in dolostone reservoirs because those minerals usually plug pore spaces that would otherwise be open. Without direct observation to confirm mineralogical composition, the risk of error increases in direct proportion to the number of different minerals that may be in the rocks and to the reliability of the logging or seismic methods to estimate mineralogical composition.

8.1   ROCK PROPERTIES AND DIAGNOSTIC METHODS

Carbonates and siliciclastics have very different origins. That fact is important because the ways in which rock properties are interpreted to identify carbonate depositional and diagenetic environments and facies and how those characteristics determine reservoir behavior depend on understanding those differences. Three of the most obvious differences were recognized by Ham and Pray (1962): (1) carbonates form locally, within the basin of deposition; (2) carbonates are almost exclusively composed of biological constituents; and (3) carbonates are highly susceptible to diagenesis. The consequences of those differences are not often discussed in the literature, yet they have profound impact on the final carbonate reservoir rock composition, texture, and reservoir properties. Carbonate rock composition, at least in reservoir rocks, is almost exclusively calcite (limestone) or dolomite (dolostones) or a mixture of the two minerals. Carbonates are not classified on the basis of mineral content as are siliciclastics, consequently, it is not possible to assign a tectonosedimentary history to a carbonate rock based on its mineral composition. Because carbonate depositional textures reflect biological (skeletal) characteristics, chemically precipitated crystalline fabrics, or alteration of porosity by diagenetic effects, they do not indicate a history of transportation, abrasion, and size sorting as do siliciclastics and it is risky to interpret depositional environments in carbonates on the basis of textural characteristics alone. That said, it is common practice to assume that mud-rich rocks reflect low-energy environments and grain-rich rocks reflect high-energy environments as reflected in the Dunham classification for detrital carbonates. Rock fabrics in carbonates can be created or influenced by chemical, mechanical, biological, or depositional processes. Terrigenous sandstone fabrics and porosity are almost exclusively the result of detrital sedimentation with only limited alteration by mechanical or chemical diagenesis during burial. Carbonate rocks may behave as brittle or ductile material under stress, but they behave quite differently than most siliciclastic rocks because they generally have crystalline fabrics while siliciclastics have granular fabrics. Microscale deformation of crystalline material involves crystal lattice structures such as twin planes or dislocation zones, along with crystal boundary patterns. Deformation of many granular rocks involves grain rotation and translation without deformation of crystal lattices. Under research conditions, variations in seismic velocity through carbonates have been shown to depend on pore characteristics (Wang, 1997), but electrofacies mapping methods using gamma ray and resistivity log data that are widely used on siliciclastic reservoirs are not useful in carbonates because carbonates do not ordinarily contain radioactive minerals and carbonate pores are not simply indicators of depositional textures. With enough well control in an established field, log signatures of carbonate rock and pore characteristics can be indicators of facies characteristics, especially if the reservoir is facies selective and relatively free of complications or partitioning by diagenesis. Perhaps density and neutron logs may be included with gamma ray and resistivity logs to enable a kind of electrofacies mapping to be done with reasonable success. Density–neutron crossplots are commonly done to identify rock types in carbonates. The Schlumberger M-N plot involves sonic, neutron, and density readings to help identify mineral components in complex carbonate lithology (Burke et al., 1969) and modern computer programs such as the proprietary "STATMIN" system can generate reasonably reliable lithological logs by processing digital data from a "basket of logs." The differences between carbonates and siliciclastics are examined further in the following paragraphs where the three categories of rock properties described in Chapter 1—primary (fundamental), secondary (derived), and tertiary—are reviewed to illustrate the unique attributes of carbonate reservoirs.

8.1.1 Fundamental Rock Properties and Depositional Reservoirs

Fundamental rock properties include texture, composition, sedimentary structures, taxonomic diversity, and depositional morphology. The last two properties are not commonly listed as "fundamental rock properties" in most texts but they are important attributes of sedimentary deposits that must be included in thorough reservoir studies. Fundamental rock properties provide the basis for defining lithofacies, or lithogenetic units that make up depositional reservoirs. Diagenetic and fractured reservoirs are simply altered versions of the original depositional version. The most reliable method for identifying these fundamental properties in carbonates is direct observation of cores or cuttings. Cores provide enough sample volume to determine sedimentary textures, grain types, sedimentary structures, and biota. Cuttings usually provide enough volume to determine mineralogy, grain types, and estimates of texture. Logs are not very helpful in identifying fundamental rock properties in carbonates. Facies types can be identified in siliciclastic sandstones by using the shape of the gamma ray and resistivity or, with older logs, the SP–resistivity log traces. When the paired traces outline a bell, a funnel, or a cylinder, the corresponding sandstone facies are assumed to be channel-fill, deltaic, or reworked sheet sands, respectively. Other "type curves" are assumed to be indicators of other of sand–shale depositional successions. The underlying assumption is that the gamma ray, SP, and resistivity logs are sensitive to vertical changes in grain size. In fact, that assumption is false. The logs are not sensitive to grain size. The gamma ray tool measures natural radioactivity that issues from the K, Th, and U found in clay minerals that are commonly incorporated in shales and mudrocks. The tool does not measure grain size. In fact, "hot limes" and "hot dolomites" are commonly found in carbonate reservoirs where particle size has nothing to do with the presence of natural radioactivity. The SP and resistivity tools likewise measure electrical properties of the rock–fluid system and shales tend to have less deflection from the log baseline than coarser grained sections that have bigger fluid-filled pores. Mineralogical composition is used to classify sandstones but not carbonates. Carbonate rock classification is based on grain type and depositional texture. Mineralogy may be strongly correlated with porosity in carbonates but it has much less influence on sandstone porosity. Sedimentary structures and biota can only be determined with complete certainty by observing borehole cores. Sedimentary structures provide clues to the hydrodynamics and directions of flow in ancient environments in both terrigenous sandstones and carbonates. In some cases, image logs and sensitive dipmeters can detect larger sedimentary structures such as large-scale crossbedding in dunes. Fossil content is arguably more important for interpreting depositional environment in carbonates than in terrigenous sandstones probably because most carbonates form in marine environments where fossil assemblages can reveal subtle differences in depositional settings. Diverse assemblages of fossils indicate favorable environment for life. Low diversity indicates a stress environment such as a hyper- or hyposaline lagoon, low oxygen content, or some other limiting factor on life. Low diversity is rarely associated with grain-supported or reef rocks; therefore low diversity can be a negative indicator for depositional porosity in reservoir rocks. There are exceptions. Low diversity but very high abundance of a few tolerant species can result in rocks composed of huge quantities of only one or two fossil species. One example of this is the salt-tolerant bivalve *Fragum hamelini* of Shark Bay, Australia (Logan et al., 1970). It is present in vast numbers in an environment that few other organisms can tolerate.

8.1.2 Reservoir Morphology

Anatomy of the depositional unit—reservoir architecture—is a fundamental, depositional characteristic that is so important that it deserves a separate heading. In depositional reservoirs, it represents the spatial distribution of both depositional facies and their attendant porosity. The 3D morphology, the size, shape, and orientation with respect to depositional dip of siliciclastic reservoirs, can be predicted by using idealized depositional models such as those described by LeBlanc (1972). These standard models range from alluvial fans to deep-sea turbidites and they have been refined in most recent literature to include vertical profiles of the typical textural, compositional, bedform, and petrophysical characteristics for each facies. The method assumes that reservoirs consist exclusively of intergranular, depositional porosity and that depositional architecture can be defined by choosing the best look-alike (analog) from the catalog of known examples. Adjustments for differences between the actual reservoir and the look-alike are usually adjustments in scale of the depositional unit rather than in its morphology. Validation of the look-alike as a model for the real example is commonly done by comparing gamma ray and resistivity log patterns to standard shape templates (generating electrofacies maps), and this method can be quite accurate when used in conjunction with geophysical information about basin architecture and sequence stratigraphy.

The seven standard depositional successions for carbonate ramps and shelves described in Chapter 4 can be used as aids in constructing depositional models for carbonate sequences. Lithologic logs, especially those based on direct observation and description of borehole cores, can reveal which of the seven depositional successions is present at a location. Using the methods outlined in Chapters 4–6, the 3D anatomy of reservoirs with depositional and some types of hybrid porosity can be reconstructed. In practice, because cores are rarely available for reservoir studies, wireline logs are usually the primary source of information for stratigraphic correlation, petrophysical calculations, and lithological logs, including computer-generated, multimineral lithological logs. Seismic data is the primary method to identify structures and large-scale stratigraphic features in the subsurface. When used together, information from logs, seismic records, and direct examination of rocks can enable geoscientists and engineers to construct the best possible reservoir model.

8.1.3 Derived Properties: Porosity and Permeability

Determining porosity–permeability relationships in carbonate reservoirs requires different methods than those used for siliciclastics because porosity in carbonates can be any of three types or some combination of them. Although some rare sandstone reservoirs and aquifers have intragranular porosity in leached feldspars, porosity in siliciclastic sandstones is nearly always intergranular such that semilog plots of sandstone permeability versus porosity are linear with comparatively little point scatter away from the best-fit line. This close correspondence between measured ϕ and k from core analyses is routinely used to estimate siliciclastic reservoir permeability in fields where one or two cores were taken but most wells have log data only. Estimates of permeability are obtained from semilog plots of measured k versus ϕ, which were obtained from routine core analyses. An equation in the form of $y = mx + b$ is obtained from linear regression of the points on the ϕ–k plot. To estimate k in wells without core analyses, log-derived ϕ values are "plugged into" the just-established linear regression equation and estimates of k are "backed-out." This method is also used on carbonate reservoirs, because in many cases it is the only method available to estimate k in the absence of core data. However, caution must be exercised with the estimated values because the method assumes that a semilog linear relationship exists between porosity and permeability. This is not true for carbonate reservoirs that lack interparticle porosity such as those with intraparticle, fenestral, vuggy, moldic, channel, fracture, and dual porosity. It may be valid for some intercrystalline pore systems in pure dolostones with comparatively large crystal sizes.

8.1.4 Tertiary Properties and Petrophysical Characteristics

Tertiary properties are indirectly measured by wireline logs and to an extent by the seismograph. As mentioned before, logging tools do not make direct measurements of primary or secondary rock properties; they measure other properties that serve as proxies for them. Proxy parameters include acoustic transmissivity, electrical properties, magnetic resonance characteristics, and natural radioactivity. Modern logging tools provide relatively good data for interpreting lithology and porosity in many carbonate reservoirs except those with multicomponent mineralogical composition where dolomite, calcite, quartz, clay minerals, and anhydrite may be present in varying percentages. Recent advances in software applications reportedly can compute accurate lithology in rocks with as many as five minerals. Although lithology can be estimated with reasonable accuracy from modern log data, it is not easy to make accurate petrophysical calculations from multimineral carbonate reservoirs, as Holtz and Major (2004) point out in their discussion of Permian dolostone reservoirs in West Texas. They were able to refine their petrophysical calculations by examining cores to determine the combined mineralogical composition and variety of pore types that were prevalent in different reservoir zones, again emphasizing the importance of thorough and accurate rock descriptions. Their work echoes Focke and Munn (1987) by emphasizing that pore geometry has a strong influence on petrophysical properties in carbonate reservoirs. Calculations of permeability and fluid saturation should be made in conjunction with studies of rock samples because carbonate porosity may not correlate with permeability in the expected linear relationship, and calculated S_w values may be far from correct if the Archie m value is not chosen according to the dominant pore types.

Distinguishing between flow units, baffles, and barriers is sometimes called "rock typing" and several methods stand out, including those described in Lucia (1995), Martin et al. (1997), and Gunter et al. (1997). The Lucia rock types are for use on reservoir rocks with only interparticle porosity. In that case, porosity and permeability vary in predictable clusters on semilog plots of particle size and poroperm values. The Martin et al. (1997) and Gunter et al. (1997) rock types are more petrophysical in character in that they are based on relationships between poroperm values and pore throat dimensions determined from capillary pressure measurements. Those methods derive from the "Winland R 35" method. Laboratory measurements such as mercury injection capillary pressure (MICP) and nuclear magnetic resonance (NMR) provide important information about capillarity, saturation, fluid composition, pore characteristics, pore throat sizes, and size distribution within samples. MICP data can be used to compute height of the hydrocarbon column in reservoirs and seal capacity required to prevent hydrocarbon leakage. In addition, median pore throat diameters calculated from MICP data generally correlate well with permeability. MICP measurements can be compared with measurements in thin section of pore geometry and genetic pore category to identify the pore types and sizes that consistently have the highest correspondence with reservoir performance, or flow unit quality.

Research on the relationships between NMR and MICP measurements is in its early stages, at least in academia, but great progress is being made. Current work indicates that genetic pore types in carbonates can be identified by statistical evaluation of NMR relaxation time curves (Genty et al., 2007). Correlations between NMR, MICP, and pore size–shape are powerful tools for predicting reservoir quality and behavior at field scale, especially as new generations of NMR logging tools are developed. Work with laboratory NMR measurements should help develop much more sophisticated interpretations of carbonate reservoir characteristics. The NMR logging tool is used for borehole measurements of total NMR porosity, in situ permeability, fluid saturations, oil viscosity, and bulk volumes of irreducible water. It is particularly useful in discriminating pore sizes and pore characteristics in the subsurface. The principal limitation of this generation of device is its depth of investigation. When results of laboratory studies are incorporated in new generation NMR log interpretations, it may be possible to identify and rank flow units, baffles, and barriers by calibrating NMR data with direct measurements from only one or two cores per field.

The seismograph records shock waves reflected from acoustic interfaces in the subsurface. It does not record texture, grain type, sedimentary structures, or taxonomic diversity and it cannot distinguish directly between depositional, diagenetic, or fracture porosity. Those distinctions have to be inferred by seismic interpreters who, with modern seismic data acquisition and processing techniques, can use reflection amplitude, frequency, phase, polarity, spatial extent, and shear wave characteristics to help identify seismic signatures of reservoirs, particularly gas reservoirs, because gas-filled pores react to seismic pulses much differently than do oil-filled pores (Brown, 1999). Seismic reflectivity depends largely on impedance contrast between the target rock layers and those that enclose it. If a 10-m thick gas reservoir in microporous limestone occurs in a deeply buried carbonate sequence several times as thick as the reservoir zone, that reservoir will be below the limits of detection by the seismograph. There will not be enough impedance contrast and the thickness of the porous zone will be below the limit of separability or one quarter the seismic wavelength, as illustrated by Brown (1999). If the limit of separability is not exceeded and if the impedance contrast between reservoir and enclosing nonreservoir rocks is big enough, the seismograph can be a powerful tool to help identify reservoirs and, in some cases, flow units within reservoirs. Anselmetti and Eberli (1997) studied seismic compressional and shear wave velocities (V_p and V_s) in 295 minicores and found, much as Wang (1997) did in his laboratory study, that different seismic velocities in rocks of equal porosity were the result of different pore types. It is possible, under the right conditions, to estimate carbonate rock and reservoir properties based on their seismic reflection characteristics. But as Lorenz et al. (1997) emphasize, there are caveats in making subsurface interpretations where the scale of the seismic measurement is larger than the scale of the individual reservoir interval.

8.2   DATA REQUIREMENTS

The amount of uncertainty involved with finding and developing carbonate reservoirs depends on how much is already known from previous work in a prospect area. It is relatively easy to diagnose that an oolitic grainstone is a beach–dune deposit on a ramp if it is already known that the depositional succession was deposited on a ramp and that the drilling location is in a depositional dip position that should place it near the ancient strandline, or "pinchout edge" for that formation. It is not easy to determine the depositional model for a rank wildcat in an untested basin where the kind of platform is poorly understood, the depositional dip position in the target formation has not been established, and only a few old, 2D seismic records are available. To make it worse, if cores and full complements of borehole logs are not taken in such frontier exploration ventures, little will be learned from the failures. Without the necessary information about the geology of the first failure, more failures will follow. What kinds of information are necessary to explore for and develop carbonate reservoirs while minimizing risk and uncertainty? The information exists in a hierarchy of scales that extend from regional (platform or basin scale), to field scale (one or more reservoirs), to reservoir scale (one or more depositional or diagenetic facies or fracture zones), to flow unit scale (subdivisions of reservoirs based on genetic pore type), to pore scale (associations of pore types, or pore facies, and their corresponding petrophysical characteristics).

8.2.1 Regional Scale Investigations

Regional studies depend mainly on outcrop studies, syntheses of subsurface geological data, geophysical data such as reflection seismic profiles, seismic attribute analyses, and gravity or magnetic surveys. Outcrop and subsurface geological data are especially valuable if available, because they provide direct measurements and observations on lithology, wireline log character, petrophysical attributes, and nature (size and shape) of the reservoir trap–seal system. Seismic profiles are commonly studied to determine platform geometry, to determine structural anomalies, and to enable interpreters to generate seismic and sequence-stratigraphic models that aid in generating geological concepts for drilling. Seismic attributes can be helpful to identify fractured reservoirs and possibly other pore systems that have distinct acoustic signatures detectible by the seismograph. Gravity and magnetic anomaly maps are used mainly to locate structural anomalies or potential traps. Interpreted regional seismic lines with other geophysical or geological information superimposed can provide basin and platform models for field scale investigations.

8.2.2 Field Scale Studies

Successful drilling that finds commercial quantities of oil or gas leads to more wells being drilled until the boundaries of the productive unit are established. The reservoir or reservoirs within these boundaries comprise a field. Field scale investigations and reservoir characterization involve detailed stratigraphic correlations to outline the anatomy of the depositional, diagenetic, or fracture system that makes up the reservoir. They involve identification of structural or paleostructural characteristics that may have influenced deposition, diagenesis, and fracturing. And they involve identification and characterization of depositional and diagenetic facies or fracture systems that determine reservoir quality. Most of this information must be obtained from direct examination of borehole cores. Lithological logs from cuttings, when used in conjunction with wireline log and seismic data, can sometimes provide enough information for successful field development. However, even the best description of cuttings cannot detect subtle changes in lithology in stratigraphic units less than 10 feet thick because cuttings are collected at 10-foot intervals during drilling. Each sample of millimeter-sized (or smaller) cuttings represents a mixture of every rock type penetrated in the 10-foot interval. The geologist who prepares lithological logs from cuttings must know the local stratigraphic section rather well in order to detect changes in lithology from bag to bag of cuttings without confusing contamination ("cave") from higher in the borehole with in situ rock. In general, lithological logs from cuttings provide useful information about stratigraphic "tops" but they are usually less useful as sources of data on reservoir characteristics. In cases where dolomite crystals, ooids, or fragments of distinctive reef organisms are recognizable in cuttings, they are very useful. Cuttings sometimes contain crystalline cements, exotic minerals, or bits of slickensides (grooves formed by differential slip along polished, gouge-filled fracture surfaces). These features can be helpful in making geological interpretations. At the end of the day, however, it is full-diameter cores that are the best sources of information about carbonate rock and pore characteristics.

Wireline logging tools and the seismograph are limited in what they can detect at reservoir scale. Wireline logging tools are the primary source of data for stratigraphic correlation, subsurface geological mapping, and petrophysical calculations when core analyses are absent. In fields where lithological logs and rock samples are limited or unavailable, even vintage wireline logs can be digitized and processed with powerful new computers and software that can calculate estimates of lithology, porosity, and saturation. Although these old logs provide useful information about reservoirs, they are not reliable substitutes for direct observation of rocks because the logging tools do not make direct measurements on minerals, porosity, or saturation. They measure other petrophysical properties that are used as proxies for mineralogy, porosity, and saturation. Powerful software algorithms are also limited by the number of minerals that can be distinguished in complex, multimineral rocks (typically three in older software and up to five in new-generation versions using modern, digital log data).

Some field scale reservoir zones may be thick enough to register on seismic profiles, but most are not. Some fractured zones can be detected by their seismic attributes, but many fractured horizons are too thin to be detected with reflection seismology. Impedance contrasts are necessary for a reflection of the seismic impulse to occur. In many carbonate reservoirs, impedance contrast is not great enough for the seismograph to detect differences between reservoir and nonreservoir rock. It is necessary in such cases to have as much information as possible about the reservoir rocks. Cores, cuttings, or lithofacies extrapolations from nearby outcrops can provide helpful information. In short, the most accurate results on field scale reservoir characterization studies derive from observation of borehole cores or cuttings worked in concert with log and seismic data. Of course, after drilling, production and well test data are very useful for reservoir definition and flow unit evaluation.

8.2.3 Quality Ranking of Flow Units

Flow units, baffles, and barriers are field scale features. Information required for identifying, describing, and quality ranking flow units includes lithological, petrographic, and petrophysical data. Ideally, these kinds data should be evaluated together to detect and interpret correspondences between them. Lithological data from borehole cores provides detailed descriptions of fundamental rock properties, diagenetic alterations, visible porosity, oil staining, and marker fossils for correlation, and even small interbedded "tight" zones that may be baffles or barriers to flow. Recognizing these small interbeds sometimes aids in *depth shifting* between cores and logs and may provide markers for interwell correlation within fields. Petrographic data from thin section descriptions is used to identify constituent composition, rock texture, pore types, pore geometry, and diagenetic alterations. Petrographic data can be compared with core descriptions and petrophysical data, along with production or flow tests, to identify and rank the quality (combined values of porosity, permeability, and capillary pressure characteristics) of flow units. Wireline log signatures sometimes correspond with rock properties in flow units so that ranked flow units can be mapped from log character, petrophysical, and petrographic data (the method of *rock typing* mentioned earlier). Where measured porosity and permeability values from core analyses are not available, porosity and saturation calculations from wireline logs can be used with a moderate amount of confidence in some carbonate reservoirs except those dominated by vuggy, moldic, and fracture porosity. If the pore system is depositional and intergranular, and if core analyses are available for at least one core, then permeability can be estimated for correlative intervals in wells without cores by using the best-fit equation for the linear relationship between porosity and log of permeability, as described earlier.

8.2.4 Pore Scale Features

Pore scale features include physical characteristics of pores and pore throats such as pore geometry, genetic pore type, pore/pore throat size ratio, and median pore throat size. This information is gathered from thin section petrography, scanning electron microscopy, and MICP and NMR measurements. Pore and pore throat geometries are compared with petrographic or SEM image analyses and MICP data to assess how different pore characteristics relate to capillary pressures, measured porosity, and measured permeability. In this way, pore and pore throat geometries can be compared with their potential to transmit or impede fluid flow, thereby providing a ranking scheme for flow units or a basis for a rock-typing scheme. NMR measurements can provide results to corroborate and extend the usefulness of the flow unit ranking or rock-typing schemes. Once these pore scale characteristics have been determined, it is necessary to scale up to flow unit and reservoir scale. This can usually be done by finding key rock properties to serve as proxies for ranked flow units or rock types. In some cases the proxy rock properties are simply genetic classes of porosity or certain size and shape ranges of pores. These parameters are easily measured with petrographic image analysis (PIA) techniques and can serve as markers for correlating flow units, baffles, and barriers (Layman and Ahr, 2003).

8.3 DEPOSITIONAL RESERVOIRS

Depositional carbonate reservoirs are those in which the dominant pore system is preserved more or less as it looked at the time of deposition. In truth, some diagenetic modification of depositional pores is inevitable. The genetic classification of porosity types was designed to take this into account by including hybrids—diagenetically altered, depositional porosity—as a midpoint along the leg of the triangle between depositional and diagenetic end-member pore types. The genetic classification lists interparticle, intraparticle, fenestral, keystone, shelter, and "reef" porosity as depositional pore categories. Reef porosity may be inter- and intraskeletal porosity in reef organisms along with inter- and intraparticle porosity in detrital material that fills spaces between skeletal masses. Some reef classification schemes were described in Chapter 2, including those by Embry–Klovan (1971) and Riding (2002). The Embry–Klovan scheme is particularly useful for reefs that include large amounts of rigid framework or large skeletal constituents. The Riding classification is more useful for mudstone and cementstone reefs or buildups, as some workers prefer to call them, but it does not include information about porosity or pore types. Each reef reservoir should be investigated individually to determine the amount of skeletal framework, detrital infill, mud, or cementstone components, and how those elements are related to porosity and genetic pore types.

8.3.1 Finding and Interpreting Depositional Reservoirs

There are only about five different pore types in depositional reservoirs. Finding them means identifying locations where one or more of those pore types exist and where pores are connected through a large enough volume to make a commercial hydrocarbon reservoir. Three of the basic pore types exist only in detrital, grain-rich rocks such as grainstones and packstones; they are interparticle, intraparticle, and keystone/shelter pores. The most common pore type in depositional reservoirs is probably interparticle porosity—space between grains. Intraparticle porosity, pores within grains, is more important as a type of reservoir porosity in grain-rich rocks. Keystone pores form on beaches when air is compressed and expelled through grain-rich beach sediment when breaking waves fall on the sediment surface. Keystone pores are rare and are not important in carbonate reservoirs. Shelter pores are formed by accidental preservation of open space beneath some fragment large enough to act as an "umbrella" to block sediment from filling all available spaces beneath it. Shelter pores are insignificant in carbonate reservoirs. Reef pore types include floatstones and rudstones, both of which are grain-rich detrital rocks. Skeletal reefs typically have grain-rich, carbonate sand or gravel infill between the skeletal elements. In short, depositional porosity is virtually limited to rocks with high grain-to-mud ratios. Exceptions are fenestral porosity that forms in mud-rich rocks usually associated with tidal flats and constructional vugs formed during the growth of mudstone and cementstone "reefs" and mounds. Some of these vugs are called stromatactis cavities and they may be especially abundant in deep-water buildups, particularly those of Carboniferous age, where they may contribute most of the hydrocarbon reservoir storage volume (Ahr, 1998, 2000). Most depositional pore types, except stromatactis cavities and constructed vugs in mud or cementstone buildups, occur in grain-rich rocks that required vigorous water motion to concentrate grains and winnow mud. Environments with sustained, vigorous wave and current action occur at shorelines and in shallow subtidal zones where typical waves and currents—those related to average yearly climate—sweep the seafloor almost constantly. Some writers call this the "fair-weather wave base" or the idealized, average depth at which waves influence sediment movement on the seabed. It is an idealized average depth because wave climates vary greatly with weather changes— changes in atmospheric pressure that make waves larger or smaller, depending on wind velocity, duration, and fetch.

Of the seven standard depositional successions, those with the best potential to be reservoirs are the shallow-water, high-energy successions such as beaches with or without dunes, barrier islands, large patch reefs, shelf-edge reefs, and tidal bars. Deep-water deposits such as mudstone—cementstone buildups (mud-mounds), carbonate debrites, and turbidites may also be good depositional reservoir rocks. Stromatactis cavities may store large volumes of hydrocarbons and interparticle porosity in debrites and turbidites may be abundant enough for those deep-water deposits to be classified as reservoirs. Usually hydrocarbon production from stromatactis cavities requires natural fractures to provide permeability. Debris flow and turbidite interparticle pores are commonly enlarged by burial diagenesis, which enhances both connectivity and pore volume. Of course, pores that have been enhanced by diagenesis are hybrids rather than purely depositional pores. Appropriate caution must be used in determining the spatial distribution of hybrid porosity because it may not follow depositional facies boundaries or present structural contours.

Finding most depositional reservoirs in the subsurface involves knowing what type of platform existed in the exploration target area and which sites on the platform are most likely to have grainy or reefy rocks with preserved depositional porosity. If porous facies extend over great distances to form "trends," the trends may be associated with regional structural or stratigraphic features. The Jurassic Smackover of the ancestral Gulf of Mexico is an example of a regional ramp with subregional basement and salt-generated structural anomalies that influenced depositional facies patterns. Many of the Smackover reservoirs are hybrid pore systems; they are not purely depositional. Vast Smackover grainstone deposits blanket the pinchout edge (the ancient Gulf strandplain) for hundreds of miles. If all those rocks were in flow communication, they would create one gigantic reservoir extending from eastern Texas to southeastern Alabama. That does not exist, of course, because the required traps and seals are not equally widely distributed. Some structural or stratigraphic trap has to be present to isolate portions of the regional trend and confine individual reservoir units. For example, structural noses or embayments along the pinchout edge are typically tested to find traps.

Reservoirs are three-dimensional features in which width and length are determined by platform bathymetry and thickness is determined by time and rate of deposition on different platform sectors. The spatial distribution of depositional reservoirs across platforms can be determined with the aid of sequence-stratigraphic methods such as those discussed in Chapter 4. Grain-supported rocks and reefs are usually associated with strandplains and shelf edges. They are also characteristically located at or near the tops of shallowing-upward cycles, or parasequences. Shallowing-upward successions in carbonate reservoirs are common over vast areas on the Permian Central Basin Platform, Northwestern Shelf and Northern Shelf of Texas and New Mexico. They are equally common in Jurassic rocks from the Persian Gulf area, such as the Arab D limestones. Likely locations for grain-supported reservoir rocks may be identified by structural and paleobathymetric mapping (interval isopach mapping) used in conjunction with sequence-stratigraphic analysis to identify the zones on ramps and shelves that are most likely to have grain-rich or porous reef facies by virtue of their paleobathymetric setting and the sea-level phase that existed at the time. For example, beach and barrier island facies should be expected to occur in updip positions on ramps during sea-level highstands. Debrites and coarse, proximal turbidites would be expected at toes of slope on shelves during sea-level lowstands.

Dunes, beaches, barrier islands, and grain-rich patch reefs are common along the shoreline, or pinchout edge, of ramps and out to water depths of about 10 meters except on platforms with extreme climates such as those off present-day Southwest Australia described by James et al. (1992). In storm-swept platforms such as the Lacepede Shelf, sediments may not be deposited in the extremely turbulent waters shallower than 100 meters or more. That situation is extreme compared with most sedimentary basins of the world. Typically, shelf-edge reefs and grainstone tidal bars rarely exist in water deeper than 10 meters. Shallow settings within the fair-weather wave base characterize nearshore zones on ramps and slope breaks on shallow shelves, but similar environments occur on and around bathymetric highs, or shoals, surrounded by deep water. For example, salt domes, horsts, erosional remnants, and preexisting highs such as drowned dunes or reefs may have been shallow enough for waves and currents to sweep across their crests constantly under normal weather conditions, setting the stage for deposition of grainstones or reefs that ordinarily form along shorelines or shelf-edge slope breaks. There are many examples of shallow-water deposits on tops of antecedent highs surrounded by deeper water (Ahr, 1989).

Debrites, turbidites, and deep-water mudstone–cementstone buildups may not be easy to locate even if the platform type and the position of the strandline are known. Abundant subsurface geological data will provide the best clues to the location of debrites and turbidites that would otherwise be difficult to identify on seismic profiles unless the deposits were more than about 50 feet thick (rough estimate of one-quarter wavelength of a seismic impulse) and were surrounded by different rock types with enough seismic contrast to be recorded by the seismograph. Many mud and cementstone buildups have been detected with the seismograph but determining which ones have reservoir-quality porosity and permeability is very difficult until they are tested by the drill or until unique seismic attributes can be extracted to identify potential reservoirs. Many Early Carboniferous mudstone–cementstone buildups are located on the toe of slope in outer ramp environments in basins around Europe and North America (Ahr, 1998). Many of these examples tend to nucleate in transgressive systems tracts (TST) and fully develop during high stands (HST).

Interpreting depositional reservoirs, or confirming the identity of successions penetrated by the drill, is mainly a matter of comparing cored intervals with the seven standard successions discussed in Chapter 5. Sometimes it is possible to identify depositional successions from detailed cuttings logs, but the most reliable and satisfactory work depends on core descriptions. Grainstones and packstones are common on strandplains where dune and beach or barrier island successions are deposited. Those facies are recognizable by their vertical profile of textures, grain types, sedimentary structures, and fossil content, as discussed in Chapter 5. Grain-supported rocks also occur on tops of antecedent highs, around patch reefs on ramps and shelf interiors, and along slope breaks at shelf edges. These deposits lack the upper shoreface and dune segments that occur on beaches and barrier islands because deposits along slope breaks and around paleo-highs are "always wet." That is, there are few, if any, beach and dune deposits around paleo-highs and shelf-edge slope breaks. The anatomy of these deposits is determined by, or limited by, the shape of the bathymetry on which they were deposited, not by the zero isobath of the strandplain. Tidal bars, slope-break deposits such as those on the Great Bahama Bank, are confined to the shelf edge where they occur as a series of lobate sand fingers lying perpendicular to the shelf edge near the drop-off to abyssal depths. Inboard from the sand waves is a shallow, mud-dominated subtidal environment. Similarly, grainstones and packstones on paleo-highs and around patch reefs occur on tops or perimeters of the highs. They also lack beach and dune sedimentary structures, they are not elongate parallel to a pinchout edge, and they may be surrounded by deeper subtidal environments that have distinctive, deep subtidal fossil assemblages.

Published examples of deep-water carbonate reservoirs seem to focus on debrites and turbidites, including chalk turbidites in the North Sea. Some authors probably include rock fall and slump deposits with debrites, and grain-flow deposits may be included with turbidites. It is almost impossible to distinguish subtle differences in these kinds of deposits without borehole cores. The following characteristics are common to debrites and turbidites: (1) they are composed of material derived from upslope or updip in shallower water; (2) they were transported in directions generally at high angles to slope breaks; (3) they usually occur as fan-shaped deposits that may or may not be channelized, although massive debrites may occur as irregular piles of rubble at the base of slope; (4) true turbidites sometimes exhibit fining-upward "Bouma sequences," while debrites lack organized trends in sedimentary structures and textures; and (5) porosity is generally interparticle, some of which is probably inherited from the original deposit. Most writers describe porosity in these deep-water deposits as having been enhanced by burial diagenetic leaching perhaps associated with fluid migration that attends hydrocarbon migration (Mazzullo, 2000). Porosity in the North Sea chalk turbidites is described as interparticle, and porosity preservation to depths of as much as 10,000 feet is attributed to overpressuring and hydrocarbon displacement of interstitial water (Feazel et al., 1990).

In summary, interpreting depositional reservoirs requires interpreting depositional environments and successions because, by definition, diagenesis and fracturing are unimportant. Facies maps become proxies for porosity maps, and reservoir boundaries are defined by facies boundaries, which may parallel present or antecedent structural highs and lows. Permeability corresponds with porosity. High permeability correlates well with high porosity because high porosity occurs in coarser grained deposits with larger pores and pore throats. Identifying zones with high permeability and porosity again depends on examination of cores or cuttings to identify the facies or depositional successions that consistently exhibit coarse and well sorted textures. Remember the checklist for diagnosis and interpretation of depositional reservoirs at the end of Chapter 5.

8.3.2 Selected Examples of Depositional Reservoirs

It is unlikely that purely depositional porosity exists because diagenesis always affects depositional texture, mineralogy, and porosity to some degree. Original mineralogy and microstructure are rarely preserved intact, especially if grain mineralogy was metastable aragonite or Mg-calcite. Instead, altered constituents, textures, fabrics, and porosity are the rule. The field examples discussed in the first part of this section were chosen because (1) they clearly produce from pores in which depositional attributes strongly dominate; (2) porosity is fabric selective; (3) fabrics and pore types alike are facies selective; (4) the depositional successions can be identified among the seven standard successions described in Chapter 5; and (5) the 3D anatomy of the reservoirs is related to platform bathymetry so that it can be predicted with the methods described in Chapter 5. In sum, those five conditions make it possible to create a geological concept for exploration and development. Field examples in the second part of this section produce from depositional porosity that has been significantly changed by diagenesis (hybrid with recognizable depositional attributes) but still retains its relationship with lithofacies boundaries and facies characteristics are still proxies for porosity.

Two fields were chosen for this first section because they offer case histories that include work done by the author and his students. First-hand examination of the rock and reservoir properties in these fields confirmed that these examples are suitable for inclusion in the rather rare category of depositional reservoirs because diagenetic alteration of pore geometry and type has been nominal. The examples are North Haynesville Field in Louisiana (Jurassic Smackover Formation) and Conley Field in Texas (Mississippian Chappel Formation). Production at North Haynesville Field is from slightly modified interparticle porosity in ooid–peloid–rhodolite grainstones (Bishop, 1968; Ahr and Hull, 1983). Production at Conley Field is from slightly altered intraparticle porosity in fragmented, crinoid–bryozoan grainstones and packstones (Ahr and Walters, 1985). In both examples, the depositional facies boundaries closely correspond to the outlines of underlying paleostructural highs. The paleostructural feature at North Haynesville Field is a salt anticline that has changed shape since the time of Smackover deposition such that part of the present structural crest has moved. It no longer conforms exactly to the shape of the grainstone buildup that originally formed on the crest of the salt dome. Skeletal grainstones at Conley Field were deposited on the crest of an anticedent high interpreted to be a horst block that has undergone several episodes of vertical movement accompanied by moderate erosion. In the Conley Field example, present structure closely conforms to the outlines of depositional reservoir bodies indicating that present structure is a reliable guide for determining drilling locations—a first step in the development of a geological concept for exploration and development. In the North Haynesville example, the outlines of the depositional reservoir and present structure are slightly offset but interval isopach maps reveal the shape of the original salt structure that corresponds more closely to depositional facies boundaries.

8.3.2.1 North Haynesville Field

Location and General Information   North Haynesville is one of several Smackover fields in North Louisiana associated with faulted salt anticlines (Figure 8.1). Our study began after this field had been developed so that a large amount of geological data was available. Borehole cores provided us with an excellent opportunity to identify genetic pore types (interparticle in ooid–peloid–rhodolite grainstones and packstones), wireline logs enabled us to correlate stratigraphic intervals and tops with which to map the reservoir body (an elongate grainstone–packstone sand wave, or tidal bar complex, lying on the paleocrest of a faulted salt ridge), and core analyses along with data from wireline logs enabled us to describe reservoir flow units by their porosity and permeability characteristics.

Structural Setting   The present structure at North Haynesville Field (Figure 8.2) was mapped by Bishop (1968) as an elongate, faulted anticline. Our remapping from wireline log data confirmed Bishop's interpretation. Faulted anticlines are typical in the salt basins of the Ancestral Gulf of Mexico. In the East Texas, North Louisiana, and Mississippi salt basins these halokinetic structures produce a variety of structural shapes and sizes depending on the thickness of salt beneath the structure, the distance from the updip edge of the salt, the amount of overburden, and the length of time the structure had been growing. Salt structures are known as "salt ridges, pillows, and domes" depending on the thickness and shape of the structure. Salt

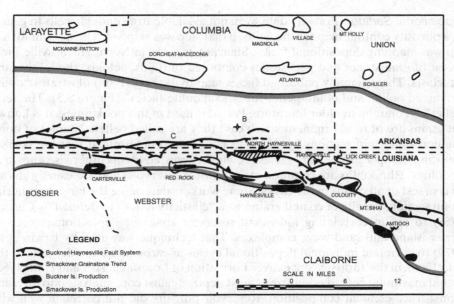

Figure 8.1   Location of North Haynesville Smackover Field, Louisiana. The field is along the boundary between Louisiana and Arkansas near the center of the map area. (Adapted from Bishop (1968).)

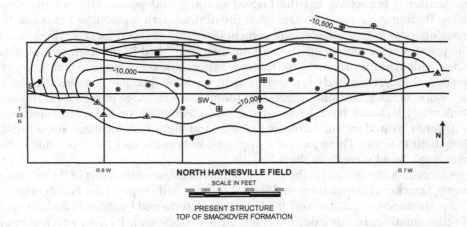

Figure 8.2   Present structure of North Haynesville Field showing the E–W elongate, faulted, salt-generated anticline that formed contemporaneously with Jurassic Smackover deposition. Note that structurally high positions are those nearest the fault zone. Crests of salt-related highs are typically the sites of grainstone accumulations, many of which are oil and gas reservoirs. Adjacent, off-structure rocks are nonreservoir mudstones and wackestones. (Adapted from Bishop (1968).)

ridges are elongate and relatively thin; salt pillows are somewhat like thick, short ridges, and salt domes can be very thick, cylindrical salt bodies. Salt begins to deform with comparatively little overburden, because some low-relief structures were recognized to have formed during or before Smackover deposition (Hughes, 1968).

Depositional Setting   Porosity data were not available from enough wells to generate a porosity contour map for the field, so that step was skipped. The next analytical step was mapping depositional facies. Smackover rocks in North Haynesville Field consist of grainstones and packstones composed of ooids, peloids, rhodoliths, and intraclasts. The originally published facies map by Bishop (1968) illustrates "oolite bar, mixed oolite, and oolite-pellet-superficial oolite facies" (Figure 8.3). The term "pellet" is common in older literature. Peloid is used in this book unless it is known that grains are of fecal origin, in which case they are called pellets. Studies of borehole cores and thin sections by Ahr and Hull (1983) determined that the coarser grains in Bishop's "oolite bar" facies are algal encrusted grains otherwise known as rhodoliths. Rhodoliths are common in Smackover reservoirs and are usually among the coarsest of all grains in the rocks. Some workers determine the largest grain size in thin section, usually of coated grains, as a "clasticity index" as defined by Carozzi (1958) to aid in correlating individual reservoir zones or depositional cycles in barrier island and sand wave complexes. That technique was used by Erwin et al. (1979) to correlate individual depositional cycles in carbonate dune deposits at the Oaks Field, in the Jurassic, Smackover Formation of Louisiana. Ahr and Hull (1983) plotted maximum coated grain size against depth, against core analysis porosity, and against trace element composition. Reservoir porosity did not correlate well with clasticity index, or maximum coated grain size, because rocks with the largest coated grains did not always have good sorting and uniformly large pores. Still, the North Haynesville "oolite bar" facies of Bishop (1968) with its rhodoliths, ooids, peloids, and scattered intraclasts has the highest porosity and permeability of the three facies. Petrographic study revealed that the northeastern segment of the oolite bar facies underwent extensive calcite cementation that reduced its reservoir quality. The primary reservoir occurs in the remaining "bar" and its adjacent oolite–pellet facies. Sedimentary structures and textures in borehole cores indicate that the North Haynesville reservoir body is a marine sand wave, or tidal bar complex, similar to the "slope-break grainstone" succession found along the edge of the Great Bahama Bank near Eleuthra Island. Although the Smackover platform is a ramp, some parts of it were elevated by salt tectonics to form local highs with localized slope breaks along their margins. These localized slope breaks became sites for deposition of the grainstone "sand waves" or "bars."

Antecedent topography—the paleo-high that existed on the crest of the salt ridge during Smackover deposition—influenced the size and shape of the North Haynesville grainstone body. Reservoir thickness was determined by the rate and duration of sedimentation. In this case, the reservoir is thickest near its axis, which is interpreted by Ahr and Hull (1983) to have been directly over the crest of the salt ridge at Smackover time. Deposition kept pace with relative sea-level rise, producing the grainstone–packstone accumulation on the antecedent high where hydrological conditions were most favorable for the formation and accumulation of oolites and rhodolites. As sedimentation and salt movement continued, the oolite sand wave grew thicker and depositional topography was accentuated.

Reservoir Characteristics   Bishop (1968) and Ahr and Hull (1983) did not identify and rank individual flow units in the field. However, petrographic studies by Ahr and Hull revealed that the highest porosity and permeability occur in Bishop's oolite bar facies, except where its northeastern tip has been cemented by calcite.

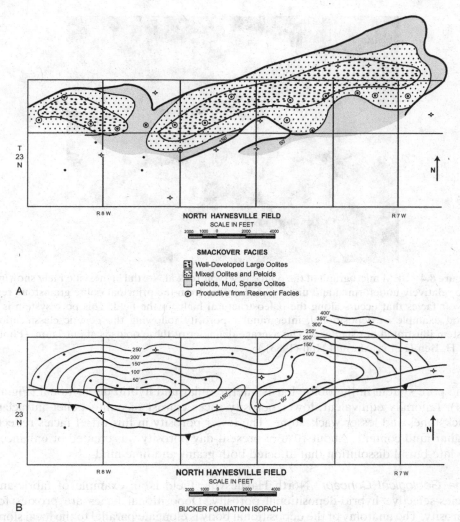

A

NORTH HAYNESVILLE FIELD
SCALE IN FEET
2000 1000 0   2000   4000

SMACKOVER FACIES
- Well-Developed Large Oolites
- Mixed Oolites and Peloids
- Peloids, Mud, Sparse Oolites
- ⊙ Productive from Reservoir Facies

B

NORTH HAYNESVILLE FIELD
SCALE IN FEET
2000 1000 0   2000   4000

BUCKER FORMATION ISOPACH

Figure 8.3 (a) Smackover facies map at North Haynesville without faults. The greatest abundance of large oolites in oolite grainstones is on the crest of the structure that existed at time of deposition, which is somewhat different (more to the north) from the present-day structural shape of these salt structures because the underlying salt continued to deform over time modifying the anticlinal forms. Finding and developing oolite grainstones in this salt basin depends on identifying those structures that were high enough at the time of Smackover deposition to have been within the fair-weather wave base. A typical method of screening prospects is the use of interval isopach maps such as the one in (b), which shows thin zones in the Buckner Formation that immediately overlies the Smackover reservoir facies. Thin intervals indicate sites where the seabed was elevated at the time of Buckner deposition; thick intervals indicate sites where the Jurassic sea was deeper. (Adapted from Bishop (1968).)

Figure 8.4   SEM micrograph of the oolitic reservoir rock at North Haynesville Field showing the relatively undeformed and uncemented oolites from the principal oolite grainstone reservoir facies that occurs along the paleostructural highs in the field. This pore system is a good example of depositional, intergranular porosity following the genetic classification system illustrated in Chapter 2. The average diameter of these ooids is about 1 mm. (Photo by H. Ben Hull in Ahr and Hull (1983).)

The pore system in the oolite bar is intergranular and hybrid depositional (Figure 8.4). Laterally equivalent, lower porosity rocks consist of oolite–pellet–intraclast packstones and lesser wackestone. The lower porosity in this offset facies reflects higher mud content. About 10% of present-day porosity was created or enhanced by late-burial dissolution that affected both grains and lime mud.

The Geological Concept   North Haynesville Field is an example of fabric-and facies-selective hybrid-depositional porosity. Depositional facies are proxies for porosity. The anatomy of the depositional body is elongate parallel to the local slope break along the margin of a salt ridge, and it is thickened over the paleostructural crest. Present structure is slightly offset from the original salt structure that influenced the location of facies boundaries, which in turn correspond generally with reservoir boundaries. The thickest part of the grainstone buildup follows the paleostructural crest and includes the coarsest grainstones with large pores and pore throats—consequently, the highest porosity and permeability. The standard depositional succession is a slope-break grainstone accumulation—or standard depositional succession number 5. The slope break in this case is at the edges of the ancient salt ridge. Much of what is now northern Louisiana was covered by a shallow sea during Smackover deposition; therefore the geological concept for finding this type of reservoir is to search for other paleostructural highs with similar tectonic histories along strike with North Haynesville Field. The paleostructural highs are salt ridges elongate parallel to depositional strike, with or without faulted margins, and that were paleostructurally high at just the right time. By following along strike, the chance of finding a salt ridge with a similar growth history is improved. One evaluates salt ridge growth history by studying interval isopach maps to locate thin zones in the overlying Buckner Formation. These thins indicate that the paleostructural crests were high at the time of Buckner deposition. Elongate paleostructural highs at this dip position on the platform have a strong chance of being blanketed by grainstone deposits. At first, drilling along structural trends seems to be no more sophisticated than drilling a group of structural anomalies. It is more than that because the structural crests had to be high enough at the time of Smackover deposition to have been washed by fair-weather waves and currents that helped form and accumulate coated grains. Salt structures that formed later would not have been in the zone of grainstone formation. Structures that formed too early may have continued to grow through the depth favorable for grainstone deposition to become piercement domes or diapirs on which there was no Smackover deposition. Clearly, just any structural crest is not enough. It must be a paleostructural crest that was in the right water depths of the Smackover sea at the right time for grainstone deposition, for burial before extensive cementation, and for preservation of depositional porosity to exist, only to be enlarged by later burial dissolution.

8.3.2.2 Conley Field

Location and General Information Conley Field is located in the Hardeman Basin of North Texas (Figure 8.5). Early seismic surveys in the 1930s located the structural anomaly that underlies the field, but initial drilling was unsuccessful because examination of well cuttings failed to reveal the presence of oil (Freeman, 1964). After extensive seismic surveys in the 1950s Shell Oil Company drilled the Conley Field discovery well. Production was established in the Ordovician Ellenburger, and the Carboniferous Osage, Palo Pinto, and Chappel Formations. Reservoir porosity in the Mississippian Chappel Formation is primarily intraparticle porosity in bryozoan

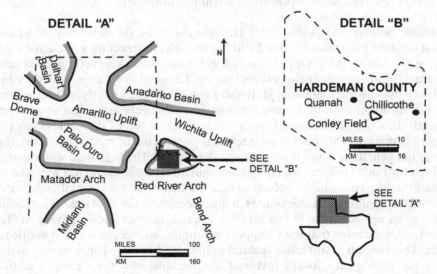

Figure 8.5 Location of Conley (Mississippian) Field in the Hardeman Basin of North Texas.

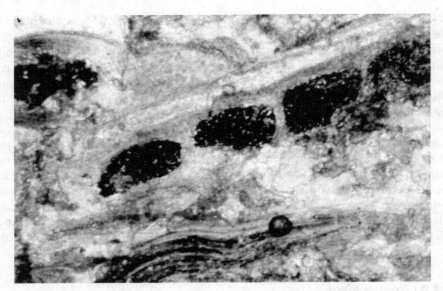

Figure 8.6 Photomicrograph of the principal reservoir rock at Conley Field: intragranular porosity in bioclastic grainstones and packstones composed primarily of fenestrate bryozoan fragments and lesser amounts of crinoidal debris. Most of the reservoir porosity occurs as intraskeletal pores within fenestrate bryozoan fragments that accumulated with criniodal debris on the crest of an antecedent high in the Mississippian Hardeman Basin of North Texas. The nonreservoir zones, although structurally high in part, have diagenetically reduced porosity caused by abundant syntaxial overgrowths on crinoidal grains. The width of the photo is 500 μm.

skeletal grains (Figure 8.6) that has undergone minor diagenetic alteration, making it a hybrid reservoir with depositional attributes dominant.

Structural Setting   As at the North Haynesville Field, the development of a geological concept to explain Conley Field began with constructing a present-structure map on the top of the Chappel Formation (Figure 8.7). A structural cross section (Figure 8.8) through the field reveals that the Chappel Formation is thicker in the field area than away from the field. It also reveals that the underlying Osage Formation (an informal rock unit name) is thinner in the field area than away from it. Beneath the Osage Formation is an unconformity on top of the Ordovician Ellenburger Formation. The thin Osage interval is interpreted to represent a topographic high in the field area at the time of Osage deposition. The high area was apparently too high for thick Osage sediments to accumulate on and around it. Later, the environment on the paleo-high evolved to become favorable for skeletal carbonates to accumulate, resulting in a thickened Chappel section in the field area. This localized thick shows up as a thin in the interval isopach map of the overlying St. Louis Limestone, indicating that the Chappel carbonates had accumulated synoptic relief and had become an in situ loose skeletal grainstone buildup. The next step in developing the geological concept involved determining why the Chappel section is thicker in the field area than away from it. Before cores were made available for study, our initial hypothesis was that the thick section must consist of a localized

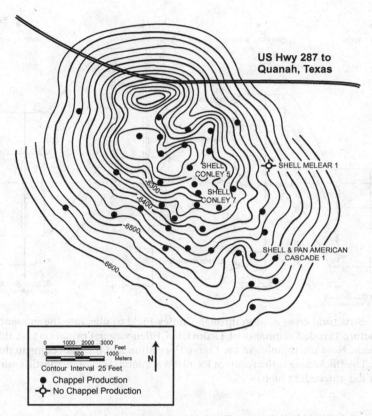

Figure 8.7   Present-structure map on the top of the Chappel (Mississippian) Formation reservoir horizon at Conley Field. Note the position of the present structural highs indicated by the negative numbers along contour lines and compare with the position of an underlying fault block (Figure 8.8) by noting the names and locations of wells.

reef (mudstone or cementstone buildup). After examining borehole cores from several field wells, it became clear that the buildup is composed of bioclastic grainstones instead of a reef or a mudstone–cementstone buildup of the type commonly found on platforms of Mississippian age.

Depositional Succession   After learning that the Chappel buildup at Conley Field is a skeletal grainstone accumulation—not a reef—our next step was to identify which of the seven depositional successions is present. Fragmented and moderately well-sorted skeletal grains along with poorly preserved, low-angle inclined bedding suggested that these bioclastic grain-supported rocks represent a reworked death assemblage composed mainly of crinoids and fenestrate bryozoans. Broken and abraded skeletal remains indicate that the skeletons were vigorously moved by wave and current activity typical of a high-energy shoreline or slope-break environment. Some fragmentation may have been the result of predation, but evidence of chewing or grinding was not readily visible in thin sections, the degree of grain sorting is inconsistent with haphazard predation, and the inclined bedding suggests that the

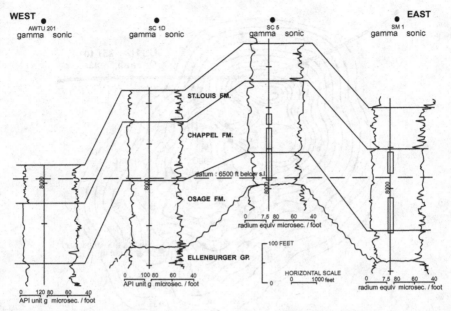

Figure 8.8   Structural cross section through Conley Field to illustrate the influence of antecedent structure (eroded remnants of Ordovician Ellenburger Formation) on thickness of overlying beds. Note the thinning in the Osage Formation and the thickening in the Chappel Formation. The thickening is the result of localized accumulation of bioclastic grainstones on the crest of the antecedent high.

particles were entrained in large sand waves that were washed back and forth across the antecedent high at Conley Field. The depositional succession that most closely fits this deposit is the slope-break succession similar to the tidal bar complex near Eleuthra Island on the Bahamas. Of course, the maximum water depths at Conley Field were far shallower than the abyssal depths in Exuma Sound, where the Bahamian tidal bars now occur. The slope break in this case is the edge of the antecedent high. In effect, the localized high acted as a small, isolated platform with comparatively sharp slope breaks around its margins. Knowing the depositional succession and the paleotopography on which the reservoir rocks were deposited made it possible to predict the anatomy of the depositional unit. It conforms to the crest of the antecedent high. Because the outline of the antecedent high is the same as the outline of present structure, development drilling would follow boundaries of the structural high.

Reservoir Characteristics   Once the geological concept that predicts the location, size, and shape of the Chappel reservoir was completed, it remained to identify and rank individual flow units. At the time our study was completed, no attempts were made to rank flow units on their petrophysical characteristics. However, careful petrographic study combined with poroperm data from core analyses enabled Ahr and Walters (1985) to identify the depositional microfacies with the highest porosity and permeability. Petrographic study of thin sectioned samples from borehole cores revealed that rocks from zones with highest measured porosity and permeability

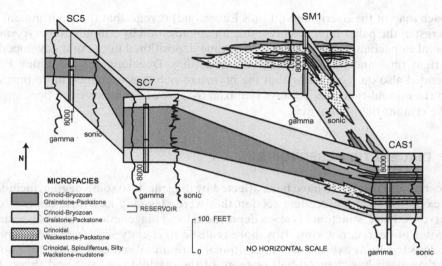

Figure 8.9 A fence diagram showing facies correlation across Conley Field. Reservoir porosity is limited to the crinoid–bryozoan grainstone and packstone facies shown as a darker shade in the panels. Although grainstone facies occur across much of the antecedent high, production is from zones with the greatest abundance of bryozoan fragments in proportion to crinoidal fragments.

were those with highest ratios of fenestrate bryozoan fragments to crinoid fragments. Rocks with high percentages of crinoid fragments were extensively cemented by syntaxial overgrowths that nucleated on crinoids and subsequently bridged pores to plug pores and pore throats. Rocks composed mainly of fenestrate bryozoan fragments retained porosity as intraskeletal pores. The anatomy of bryozoan-rich parts of the sand wave complex was determined by creating a fence diagram of the Chappel section in the field (Figure 8.9). This diagram outlines the productive and non productive zones in the field.

The Geological Concept Conley Field was found because it is easily identified by seismic surveys as a present structural anomaly. In the Hardeman Basin of North Texas, many structural anomalies are present in the Chappel Limestone interval, but only a few of them have reservoir porosity. In many cases the reservoirs produce from fractures and to a lesser extent from matrix porosity. At Conley Field, however, reservoir porosity is confined to bryozoan–crinoid grainstones that accumulated much like the oolites and rhodolites at North Haynesville Field in the previous example. The porosity is intra- and interparticle porosity in bioclastic grainstones and packstones rather than nonskeletal grains that depend largely on in situ chemical precipitation for their origin. Interestingly, the Conley reservoir proves that skeletal grainstone buildups can occur on antecedent highs just as nonskeletal, coated grain buildups do. Finding structural anomalies is relatively easy. Finding anomalies with porous and permeable reservoir rocks is not. The main element in the geological concept that explains the Conley reservoir is that it became a topographic positive, or buildup, during Mississippian time. The thickened Chappel Limestone in the field (indicated on geological cross sections and on the interval isopach map of the overlying St. Louis Limestone) reveals that the environment on the crest of the paleo-high was favorable for colonization by crinoids and bryozoans. Interval isopach mapping is the key to finding depositional thicks that developed at the right time and in the right structural setting. Development of Conley Field depended also on recognizing that the bryozoan-rich rocks retained more porosity than the crinoid-rich ones because syntaxial overgrowth cements filled pore spaces in the crinoid-rich sediments.

8.4   DIAGENETIC RESERVOIRS

All carbonate reservoirs have been affected by diagenesis to some degree, including the examples in the preceding section that were classified as hybrid-depositional reservoirs. The distinction between depositional and diagenetic reservoirs is usually a matter of degree, not kind. It is more realistic to classify slightly altered depositional reservoirs as hybrids with depositional attributes dominant. Once it is known that diagenesis has changed half or more of the original pore size and shape, the reservoir is classified as a hybrid with diagenetic attributes dominant. If the diagenetic changes are so great that none of the original depositional characteristics can be recognized, the reservoir porosity is purely diagenetic. Two good examples of such cases are a dolostone reservoir with intercrystalline porosity that has no visible relationship to depositional texture, fabric, facies, or other depositional characteristics, and a karst or paleocave reservoir in which dissolution diagenesis cuts across all preexisting rock textures, fabrics, and facies to create the massive caves, caverns, and karst features.

8.4.1   Finding and Interpreting Diagenetic Reservoirs

Formulating geological concepts for exploration and development of hybrid reservoirs is not very different from developing concepts for exploration and development in reservoirs with depositional porosity because depositional attributes are the benchmarks that measure how severely diagenesis has modified reservoir porosity. Hybrids with depositional attributes dominant can be treated essentially the same as if they were depositional reservoirs. Hybrids with diagenetic attributes dominant require different strategies and concepts because flow units, baffles, and barriers will not always "line up" with depositional facies boundaries. In these cases, it is necessary to determine how severely diagenesis has altered depositional porosity, what types of diagenesis were involved, in which diagenetic environment the changes took place, and when during burial history the changes took place. Remember that diagenesis includes all the changes that happen to rocks after deposition and before metamorphism. Pore-altering changes that happened early during burial history will have different characteristics than late burial diagenetic changes and the environments in which early diagenesis takes place will be distinctive. For example, early diagenesis must take place in the vadose zone, in one of the phreatic zones, or the shallow subsurface environment. Each environment leaves distinctive traces that can be identified even if later alteration "overprints" the early products. Each environment is usually associated with some geological event that brought the pristine rocks in contact with disequilibrium waters that caused change. These events include such things as uplift and exposure, exposure after sea-level lowering, submergence after sea-level rise, or exposure to mineralized waters that resulted in precipitation of distinctive cements, among other types of change. The characteristics of early and late diagenesis and the environments in which they occur are discussed in Chapter 6.

Interpreting purely diagenetic reservoirs can be tedious if not truly difficult because unraveling the history of diagenesis usually requires determining ancient hydrological history rather than structural or depositional histories. For example, both dolomitization and dissolution are accomplished in nature by rock–water interaction. Dolomitization results when precursor carbonates are exposed to Mg-rich fluids and Mg is exchanged for some of the Ca in the precursor. The paleohydrological question is: When and where did the Mg-rich fluids originate and what ancient hydrological mechanism brought them in contact with the precursor rocks?

Dissolution occurs when undersaturated fluids are brought in contact with carbonates. When and where did those fluids originate and how did they come in contact with the carbonates? An appreciation of these unknowns puts sharp focus on the value of interpreting the time of diagenesis by studying cross-cutting relationships in thin sections. Seemingly insoluble problems can be simplified by establishing the timing of diagenetic changes that had the greatest impact on porosity. Knowing the timing of the diagenetic events relative to the time of deposition and early lithification can enable workers to concentrate on the paleohydrological regime that caused the changes to occur. Dolomitization may have occurred soon after deposition, as in the seepage-reflux model, which postulates that Mg-enriched brines move by density flow outward and downward from coastal lagoons through porous and permeable shoreline carbonates. In this case, it is relatively simple to determine when and where the fluids originated and how they came in contact with the precursor rocks. Cave formation and massive dissolution are common at the top of the meteoric phreatic zone and, in some climates, in the vadose zone (McIlreath and Morrow, 1990). Most of the literature describes porosity associated with collapsed paleocaves and karst that formed in continental environments; however, recent studies (Smart and Whitaker, 2003) indicate that caves formed in mixing-zone environments along marine coastal margins may be better analogs for collapsed paleocave and karst porosity.

Mapping ancient water tables is more difficult than identifying hypersaline lagoons that were adjacent to shorelines at or near the time of deposition because lagoons and shorelines have distinctive lithofacies. Ancient water tables do not. Tracing water tables depends on cement mineralogy, cement crystal habit, trace element content of calcite or dolomite cements, isotopic analyses of cements, and the physical appearance of rocks. Identifying these characteristics requires samples from borehole cores or outcrops. Microscopic study of cements and their links with paleohydrology gave rise to a specialized branch of petrography known as "cement stratigraphy." Meyers (1974) did pioneering work on cement stratigraphy of some Mississippian skeletal grainstones in the Sacramento Mountains of New Mexico to unravel the history of cementation and its relationship to paleohydrology. Later, Grover and Read (1983), among others, used cathodoluminescence to distinguish between burial cements formed in different zones within an ancient aquifer. Cathodoluminescence is sensitive to the trace element composition of carbonate cements, particularly to variations in Mn and Fe. Metallic elements in oxygenated water will generally be in oxidized states and less likely to be incorporated in calcite or dolomite cements that precipitate from those waters. Reducing waters, on the other hand, contain reduced metallic cations that are readily incorporated in calcite and dolomite lattices so that iron-rich, or ferroan, carbonates can be used as indicators of reducing conditions in deeper burial waters. Goldstein's (1988) study of cement stratigraphy in Pennsylvanian carbonates from New Mexico includes an extensive bibliography on cement stratigraphy.

8.4.2 Field Examples of Diagenetic Reservoirs

Nonfractured, diagenetically modified carbonate reservoirs exhibit either hybrid or purely diagenetic porosity and account for so many examples that it is beyond the scope of this book to review all of them. Also note that the literature does not include a category for hybrid pore types. Previously, writers described reservoirs with diagenetic, fractured, or "primary" porosity, if they described the pore system characteristics at all. A sampling of the literature reveals the overwhelming number of fields that produce from either hybrid or purely diagenetic pore systems. Scott et al. (1993) note that of the 509 giant fields having over 500 MMbbls of oil equivalents, 59 produce from Cretaceous carbonate reservoirs, including 8 of the 36 supergiant fields that alone account for 12.9% of the world's estimated ultimate recoverable reserves. Of the 16 Cretaceous carbonate oil and gas fields tabulated by Scott et al. (1993), 13 reservoirs, or 80% of the total number, produce from either moldic or vuggy porosity, or both. Scott's description does not give enough information to determine how many of those fields produce from hybrids as opposed to purely diagenetic porosity. Three of the 16 fields produce at least in part from fracture porosity in addition to diagenetically altered porosity. Assuming that these examples are reasonably representative of carbonate reservoirs in general, a conservative estimate is that 7-in-10 carbonate reservoirs produce from strongly altered porosity.

An extensive compilation of carbonate reservoir case histories by Roehl and Choquette (1985) included a review of the structure, stratigraphy, and reservoir characteristics of 35 carbonate reservoirs of different geological ages and locations. Of the 35 examples, 8 are described as having depositional porosity with little alteration except pore space reduction by cementation (hybrid: depositional attributes dominant). Nine reservoir examples are interpreted to be largely dependent on fracture porosity, and all others exhibit pore systems in which porosity enhancement by diagenesis is in those authors' words "strongly predominant" (hybrid: diagenetic attributes dominant). Karst and paleocave reservoirs are, in general, classified as purely diagenetic and independent of depositional characteristics, although they may have been influenced by fractures and joints. Studies by Whitaker and Smart (1998), Purdy and Waltham (1999), Loucks (1999), Kosa et al. (2003), and Loucks et al. (2004) present extensive discussions on the origins and characteristics of karst and paleocave reservoirs. Detailed illustrations of paleocaves in the Guadalupe Mountains of New Mexico by Kosa et al. (2003) are instructive because they show the scale and complex architecture of paleocaves that developed along syndepositional faults. Along the same line, Whitaker and Smart (1998) describe the hydrological and diagenetic mechanisms that formed fracture-related caves along the east coast of South Andros Island in the Bahamas. The shape and extent of many cave systems is dictated by preexisting fracture systems, and this association may be especially useful for exploring and developing paleocave reservoirs. The caves of northeast Yucatán, South Andros, and in the Guadalupe Mountains are related to fracture systems. It is certainly worth considering that paleocave reservoirs and aquifers may follow preexisting fracture patterns.

Two field examples of hybrid reservoirs with diagenetic attributes dominant were selected for this chapter. One of them illustrates diagenetic microporosity (neomorphic stabilization diagenesis) in a grainstone reservoir at Overton Field, Texas. The other is an example of pore enlargement, new pore formation, and moldic porosity in a Permian grainstone and reef-rubble buildup at Happy Field, Garza County, Texas. These examples were chosen because they represent different styles of diagenesis and pore types.

8.4.2.1 Overton Field

Location and General Background   Overton Field is one of the largest gas fields in the East Texas basin (Figure 8.10). It produces from altered oolite grainstones in the Jurassic Cotton Valley Limestone. Diagenesis affected only the oolitic facies on tops of paleostructural highs while relatively unaltered, nonproductive oolite grainstones are present lower (older) in the stratigraphic section. The relatively unaltered oolite facies are also missing on paleostructural highs in the producing fairway.

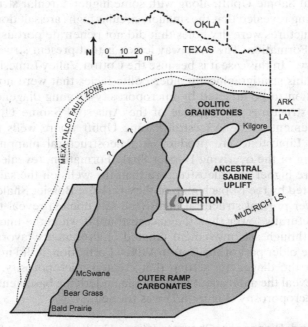

Figure 8.10   Location of Overton Cotton Valley (Jurassic) gas field in East Texas. Note that the field is on the edge of an ancient structural feature known as the Ancestral Sabine Uplift. To the west, off the edge of the ancient uplift, is the East Texas salt basin where salt tectonics has created uplifts, domes, diapirs, and withdrawal basins.

Natural gas is produced from intraparticle microporosity that developed after metastable ooids were altered to microporous, microcrystalline grains that retained the size and shape of the parent ooids but lost their original microstructure and mineralogy. The microcrystalline microporosity is interpreted to be the result of neomorphic stabilization (a type of recrystallization) that transformed metastable, probably Mg-calcite, ooids into spheroidal masses of stable, microrhombic calcite microcrystals with micrometer-scale, intercrystalline porosity. Because microporosity occurs only in oolite grainstones on the crests of paleo-highs, it is facies selective and because only the ooids were altered—other constituents were not similarly altered—the reservoir is also fabric selective. These attributes define the reservoir as a hybrid with diagenetic attributes dominant. But diagenesis only affected oolite facies on paleostructural highs—not all oolite facies in the field. This hybrid reservoir is a special case in which porosity distribution is the result of diagenetic alteration that occurred only on certain paleostructures, meaning that oolite grainstone facies in general are not proxies for porosity. Instead, paleostructure is the key element in the geological concept for finding reservoir porosity. Only the upper Cotton Valley oolite facies on paleostructural highs are productive. Other oolite facies in the formation are not productive and although those oolites may occur higher on present structure than oolite facies in the productive fairway, they were not on highs when pore-forming diagenesis created the reservoir.

Structural Setting   Present structure at Overton Field (Figure 8.11) reveals several NE–SW elongate highs along the western perimeter of a basement feature known as the Ancestral Sabine Uplift, along with some higher, circular structures west of the uplift. Drilling revealed that the smaller, circular highs are salt domes. Wells that tested these structures were dry holes that did not penetrate porous sections of the Cotton Valley Formation. The takeaway lesson is that present structural highs can be "high and dry." In this case it is because the Cotton Valley Limestone on the salt anticlines contains stratigraphically older oolite facies that were not in the paleostructural position to be affected by microporosity-forming diagenesis. The more oval, elongate structures on the edge of the Ancestral Sabine Uplift reflect the underlying basement of the Ancestral Sabine Uplift where wells that tested the Cotton Valley Limestone are productive. Paleostructural mapping, or interval isopach mapping of the overlying Bossier Shale Formation, revealed that the productive wells are higher on paleostructure than the wells on the salt domes to the west, as illustrated by the isopach map of the overlying Bossier Shale (Figure 8.12). Shale thicks over present structural highs (the salt domes) reveal that the domes were paleostructurally lower than the ancestral uplift when the microporosity was formed. Even though salt movement created environments favorable for oolite formation in the older part of the Cotton Valley Formation, it did not put the older oolite facies in the diagenetic setting that created microporosity. Bossier Shale isopach thins reveal the antecedent highs on the underlying basement surface where intraparticle microporosity (Figure 6.7) was formed.

Depositional and Diagenetic Characteristics   Unaltered oolites from dry holes drilled on the salt anticlines west of the Ancestral Sabine Uplift are also older than those in the producing wells on the basement highs. The older oolites were deposited in agitated water as oolite bars on bathymetric highs at a lowstand of Jurassic sea

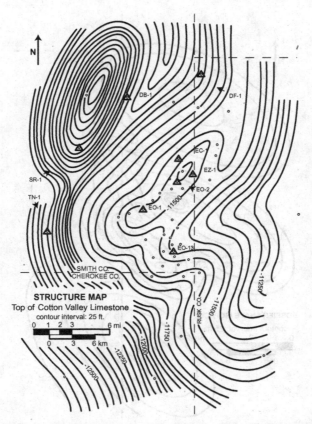

Figure 8.11 Present structure on top of the Cotton Valley Lime Formation at Overton Field. Note that there are two shape trends in the structure: one is generally N–S oriented and the other is NE–SW oriented. The north–south trend reflects the regional margin trend of the underlying basement feature—the Ancestral Sabine Uplift. The NE–SW trends reflect sub-regional paleo-highs on the ancestral uplift. Circular trends in the western part of the map, off the edge of the ancestral uplift, reflect younger salt deformation. There was not enough salt on top of the old basement feature to have been involved in dome formation; therefore salt structures are absent or unimportant on top of the ancestral uplift. Production in this field is from oolite grainstones that were at precisely the right paleostructural elevation at Jurassic time to have been formed as grainstones and subsequently modified by subtle diagenetic change to become microporous. Detecting the precise paleostructural elevation that was necessary for these grainstones to have undergone porosity-enhancing diagenesis requires interval isopach mapping of the overlying Bossier Shale.

level. As relative sea level advanced, the older oolites were cemented and ooid microstructure remained well preserved. In contrast, the younger oolites on basement highs underwent porosity-enhancing diagenesis early in their burial history. Extensive intraparticle microporosity was formed by diagenetic (neomorphic) stabilization of metastable oolites; no evidence of dissolution such as enlarged interparticle pores, molds, or vugs was found. Storage volume and productivity depend instead on the microporosity, but because the micropores are so small, this is a gas reservoir.

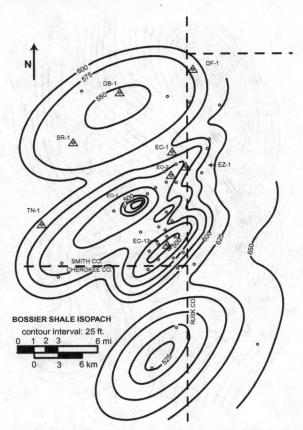

Figure 8.12 Interval isopach of the Bossier Shale Formation that overlies the Cotton Valley Limestone reservoir at Overton Field. Note that thin intervals in the Bossier Shale correspond to highs on the underlying Cotton Valley Limestone surface. The Bossier thins could represent local thickening of the Cotton Valley Limestone or antecedent highs beneath the entire stratigraphic section. In this case, the Bossier thins reflect high areas on the underlying Ancestral Sabine Uplift and the Cotton Valley Limestone facies on these highs are the reservoir rocks. Off the old highs, the Cotton Valley Limestone includes oolite grainstones on some salt domes, but they were not at the right paleostructural elevation at the right time to have been diagenetically altered and they are "tight." An example of the altered ooids and attendant microcrystalline microporosity are illustrated in Figure 6.7.

The younger oolite bodies are interpreted to have formed at a higher stand of relative sea level than the older, non-microporous oolites that were deposited off the Ancestral Sabine Uplift. The combination of shallow water at time of deposition and exposure to undersaturated waters during early burial is interpreted to be the principal cause for microporosity formation, although the origin of this unusual type of porosity is still debated by carbonate sedimentologists. The time of origin of the microporosity at Overton Field has also been debated. Ahr (1989) presented evidence that microporosity formed early in the burial history of the younger oolite buildups, but Dravis (1989) argued for a late burial origin. Fretwell (1994) found evidence that microporosity did form early in the oolite bodies on the basement paleo-highs, and that it was later modified by deep-burial neomorphism and partial cementation.

The depositional succession illustrated in the Cotton Valley oolite buildups at Overton Field is similar in many respects to the tidal bar complexes of the Schooner Cays on the Great Bahama Bank. They are tidal bar or sand wave buildups composed of well-washed grainstones deposited on and along bathymetric slope changes (slope breaks). The Schooner Cay tidal bar belt is associated with the shelf edge of the Bahama Banks and the strong tidal effects that sweep across the shelf margin from the depths of Exuma Sound. Oolite bar buildups at Overton Field were deposited on bathymetric highs over salt anticlines and basement features related to the Ancestral Sabine Uplift and the East Texas Salt Basin. Oolites deposited on the more stable and apparently higher-relief basement structures formed in shallow-water environments that would be the first to undergo exposure to micropore-forming diagenetic waters during relative sea-level retreat and early burial. Oolites deposited on salt-related highs west of the Ancestral Sabine Uplift are interpreted to have remained in the marine phreatic environment without undergoing the early diagenesis that produced the microrhombic microporosity.

Reservoir Characteristics   Porosity at Overton Field is both fabric and facies selective. Pore-forming diagenesis occurred only in the oolite buildups on the tops of basement highs, which are, in turn, reflected as thin zones on interval isopach maps of the Bossier Shale that overlies the Cotton Valley Limestone in the field area. Exploration and development strategies should be based on locating paleo-highs that are reflected in interval isopach (or isotime) thins in strata just younger than the reservoir target zones.

Geological Concept   The depositional succession for the Cotton Valley oolite buildups is interpreted to be a sand wave complex that conformed to slope breaks or slope changes from shallower to deeper water. Sedimentary structures in cored sequences reveal sequences almost identical to the lower part of the beach and barrier island standard succession: that is, slightly muddy lower segments that passed upward into ripple-bedded, mud-free grainstones. The muddier segments commonly contain skeletal fragments and, in some cases, whole coral skeletons up to 10–15 cm in diameter. Paleogeographically, the sand wave complex is located along the slope changes around the Sabine Uplift so that it is assigned to the slope-break standard depositional succession. Tidal bars, or sand waves, are elongate parallel to the slope break in this case, and they may extend inboard from the slope break for a few kilometers. Because diagenesis was fabric and facies selective but affected facies only in certain paleostructural situations, the combination of facies and paleostructural maps provides the key elements for exploration and development at Overton Field. This knowledge gives the explorationist a model—a geological concept—with which to proceed. Understanding the origin and timing of pore-forming diagenesis enables development of the reservoir in a cost-effective manner.

8.4.2.2 Happy Field

Location and General Information   Happy Field is located in the northern Midland Basin about 50 miles southeast of Lubbock, Texas (Figure 8.13). The reservoir is at

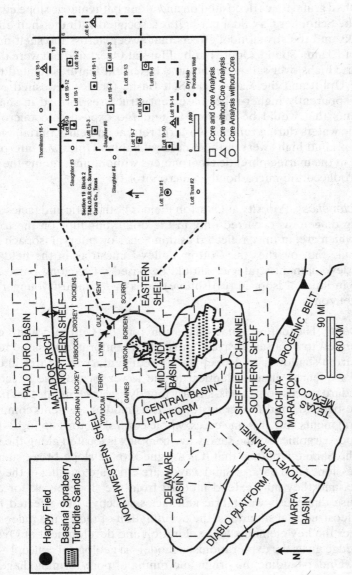

Figure 8.13 Location of Happy Field, Garza County, Texas. Although the field is listed as Happy Spraberry Field, it produces from shallow-water Permian carbonates in the Clearfork Formation. Current work not yet published by University of Texas researchers suggests that these shallow-water deposits are not in their original position but that they were moved downslope as massive slide blocks.

an average depth of 4900 feet and is interpreted to have been deposited originally as a localized buildup that formed on a structural terrace along the eastern margin of the Midland Basin. Work underway at this writing may change the initial interpretation because evidence has been found that the entire carbonate section in Happy Field may have been displaced downslope after having broken away from a shallow, shelf-edge terrace.

Originally known locally as Happy Spraberry Field, it was discovered in 1988 by Bennett Petroleum during recompletion of an existing well. It is referred to here as Happy Field because it produces from shallow-water carbonates interpreted to be in the Permian, lower Clear Fork Formation, not from deep-water Spraberry sandstones that occur in the center of the Midland Basin. After the initial discovery, Bennett Petroleum ran 3D seismic surveys to further define the Happy Field structure. Encouraging results led Bennett and their partners, Torch Energy, to drill six new wells. Torch Operating Company acquired production interests and nine more wells were drilled by 1992 as part of a waterflood that achieved moderate success in the northern part of the field. The small increase in oil and water production after waterflooding was less than the volume of water injected and the reservoir pressures remained virtually unchanged, suggesting that flow units were not in good communication and that potentially significant amounts of oil may have been bypassed. Original oil in place was estimated to be 17.2 MMBO and cumulative production through 1996 was 2.4 MMBO and 4.5 MMCFG.

Structural Setting Happy Field is located on the Eastern Shelf of the Midland Basin. Recent seismic studies by independent operators indicate that the field is a stratigraphic trap located on a paleotopographic terrace that, during lower stands of sea level, was the site of shallow subtidal carbonate sedimentation. However, ongoing work suggests that slope failure may have allowed the entire carbonate section at Happy Field to slump basinward, moving it to a deeper water, downslope environment. The shapes of present structural contours and interval isopach contours of reservoir thickness are strongly similar, indicating that the present-day structure in the field represents depositional relief on the top of the Happy carbonate buildup, not the result of postdepositional tectonic deformation.

Borehole cores reveal that the reservoir consists of two low relief, NE–SW trending pod-like accumulations of oolitic and peloidal grainstones and packstones with lesser amounts of bindstones, rudstones, and floatstones. Minor tilting of the basin margin has not added to the amount of structural closure on the reservoir. The northern pod is the larger of the two; it has structural closure of about 70 feet on its south side. The southern pod has about 50 feet of structural closure to the west. There is no evidence of faulting on wireline log correlation sections. The reservoir pods are separated by an interval isopach thin that may be a submarine channel.

Depositional and Diagenetic Characteristics Happy Field carbonates consist of oolitic and peloidal grainstones to packstones, reef rudstones and floatstones, and, rarely, *Tubiphytes* bindstones, some of which are in growth position. Bindstones and other reef-like facies were originally deposited near the base of a shallowing-upward succession that ended with deposition of crossbedded, oolite grainstones and subordinate packstones. The reservoir is developed mainly in the packstone–rudstone trend where grain-moldic, solution-enhanced intergranular and vuggy porosities dominate. Happy Field rocks underwent a complex diagenetic history beginning with early stabilization, dissolution, cementation, and emplacement of late burial saddle dolomite, along with anhydrite and chalcedony as pore-filling and replacement minerals. The main dissolution porosity is interpreted to have been formed during early diagenesis, where exposure to undersaturated waters created or enhanced porosity in the Happy rocks. Later cementation is interpreted to have relegated the once-porous and permeable grainstones to moderate and poor reservoir quality. Some oolite grainstones show extensive dissolution and widespread grain-moldic porosity, along with vuggy and solution-enhanced intergranular pores, but later burial diagenetic calcite and anhydrite cements filled much of the pores in such irregular distribution patterns that correlation between porosity and permeability in the grainstones is very difficult. Because reservoir flow units are distributed in such complex patterns, it was not possible to identify "straightforward" relationships between depositional or diagenetic attributes that could serve as proxies for porosity and permeability. Instead, a different strategy was developed whereby stratigraphic "slices" 10 feet thick were mapped from top to bottom of the reservoir. The average porosity and average permeability of each 10-foot interval were contoured on translucent paper. Pairs of porosity and permeability maps for each interval were then overlain to identify which 10-foot slices had high porosity coinciding with high permeability. In essence, this is a primitive method of 3D reservoir visualization that could be accomplished with sophisticated, albeit expensive, computer software. More about the method is explained later.

Reservoir Characteristics Petrographic and petrophysical data on (1) depositional texture, (2) pore type and geometry, (3) porosity from core analyses, and (4) permeability from core analyses were studied to define and rank flow units. Spatial distribution of reservoir flow units and their relationship to depositional facies were not obvious from initial study; therefore an alternative strategy was developed to define flow units and test for facies selectivity. The method involved creating "slice maps" of reservoir parameters and depositional facies. The method consisted of calculating average values of porosity and permeability from core analyses for each of the 10-foot stratigraphic slices in the Happy Field carbonate section and then contouring them. Facies types for each 10-foot slice were also mapped. The datum on which the slices were "hung" is a shale marker that is the most reliable and identifiable stratigraphic signature on borehole logs across the field. Examples of porosity and corresponding permeability slices for the intervals 20-to-30 and 60-to-70 feet below the shale marker are shown in Figure 8.14a–d. The averaged values for each parameter were mapped in successive slices until the base of the carbonate section was reached. A reservoir quality ranking scheme was then developed to identify field segments with the greatest, intermediate, and lowest potentials for flow communication and recovery efficiency. Porosity and permeability values were divided into three quality ranges defined simply as high, medium, and low. For example, high porosity values at Happy Field were defined as the highest one-third of the range in porosity values. The range in porosity at Happy Field is 3–30%; therefore the highest one-third of that range is 21–30%. A similar procedure was followed for permeability. The resulting data were put in a 3 × 3 matrix with three porosity and three permeability, or nine "poroperm," reservoir quality rankings to be identified and mapped individually across the field. Each quality zone was outlined by examining porosity maps

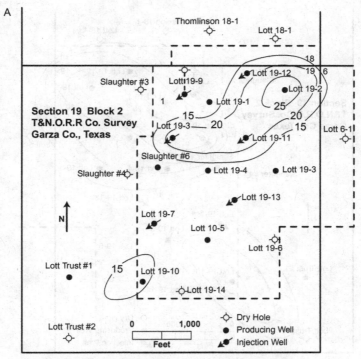

Average Porosity for Interval 20-30 Feet Below Marker

Figure 8.14a   Two example "slice" contour maps of average porosity (8.14a, c) and permeability (8.14b, d) in 10-foot intervals across Happy Field. The two slice intervals are (8.14a, b) 20–30 feet and (8.14c, d) 60–70 feet below a shale horizon that serves as a stratigraphic marker. Note the difference in the spatial distribution of average porosity and permeability at different depths showing that the 3D size and shape of reservoir flow units does not always conform simply to facies boundaries or structural outlines. These differences in flow unit quality reflect variations in genetic pore type and style and degree of diagenesis. (From Ahr and Hammel (1999).)

overlain by permeability maps and assigning a color or black and white pattern to field sectors where the high one-third, intermediate one-third, and low one-third ranges of poroperm values coincide. The composite slice maps made it possible to visualize 10-foot intervals where highest, intermediate, and lowest combined values of porosity and permeability exist in stratigraphic space. Knowing that diagenesis significantly, but irregularly, influenced the amount, type, and spatial distribution of porosity and permeability, depositional facies maps were made for each corresponding 10-foot slice to compare with corresponding poroperm "quality" slice maps. This method made it possible to identify specific zones in each depositional lithofacies where the highest, intermediate, and lowest poroperm values exist within the limits imposed by interpretive contouring. Illustrations of these multiple overlays are unsatisfactory without color; therefore no attempt is made to show them here. The slice mapping method produced better results than other techniques because large

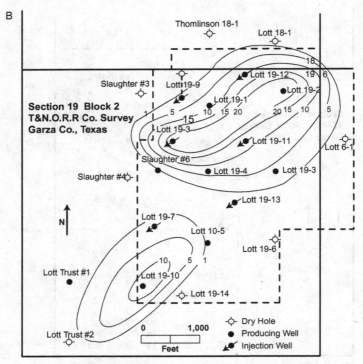

Average Permeability for Interval
20-30 Feet Below Marker

Figure 8.14b (*Continued*).

variations in flow unit quality occur within lithofacies and the variations did not conform to obvious depositional or diagenetic attribute boundaries. The slice maps pinpoint zones within the field where flow unit quality is high, intermediate, or poor. Petrographic examination of samples from those flow units reveals whether diagenesis has enhanced or reduced porosity and permeability, what type of diagenesis caused the alteration, and what concept could be developed to explain reservoir behavior.

Today, the spatial distribution of ranked flow units can be visualized with high-speed computers. For students or those who do not have the expensive visualization software, similar results can be achieved simply by printing color-coded slice maps of poroperm values and depositional lithofacies on transparencies. The entire set of transparencies for all slices can be suspended with monofilament line so that the ranked flow units and depositional microfacies can be seen in 3D space across the entire field or as well control permits. Stratigraphic or structural cross sections and panel diagrams with quality-ranked lithofacies can furnish limited two- and three-dimensional views of the flow unit system, providing the operator with information about flow unit quality variations between borehole locations. Such information is essential in waterflood or enhanced recovery operations.

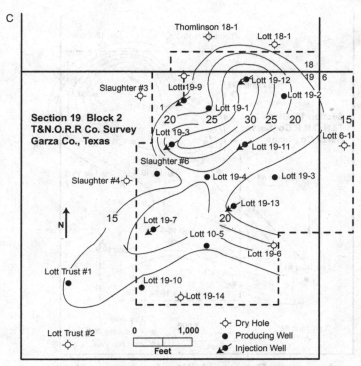

Average Porosity for Interval
60-70 Feet Below Marker

Figure 8.14c *(Continued)*.

Geological Concept Happy Field was discovered during a recompletion process rather than having first been identified as a drilling target. Initial research indicated that the Happy grainstones and reef rubble were formed in place outboard of a slope change on the Eastern Shelf. Ongoing work suggests that the entire carbonate section in the field has been displaced as a huge slump block, or blocks, and that the entire carbonate section has been displaced from its original location. In either case, the oolite grainstone–packstone accumulation along with bindstone reef rubble indicates a shallow-water origin. Because there are slope breaks updip from the present Happy Field location, the slump block (if it is indeed a slump block) probably came from there. It is clear that reservoir porosity at Happy Field was created by several episodes of diagenesis that were mainly facies selective. Most of the porosity at Happy Field is interpreted to have formed early in burial history, probably during a minor drop in relative sea level during or not long after deposition. This timing of diagenetic activity was determined after petrographic study revealed several generations of calcite cements that filled dissolution-enlarged pores. Undersaturated water was probably introduced during a sea-level fall and emersion or exposure to a paleoaquifer. Either situation would have resulted in dissolution of metastable constituents. The great quantity of moldic and solution-enhanced intergranular porosity

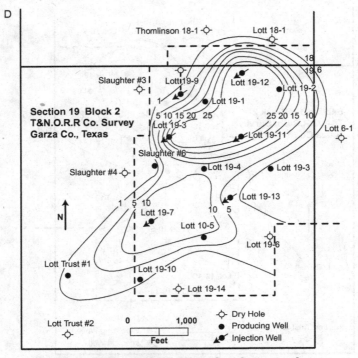

**Average Permeability for Interval
60-70 Feet Below Marker**

Figure 8.14d   (*Continued*).

in upper and middle parts of the buildup suggests that dissolution diagenesis occurred soon after deposition and that it was most effective in the bathymetrically higher part of the accumulation. If dissolution had been caused by deep-burial waters, it should have been more extensive in the lower part of the reservoir, not the upper part. However, there is more saddle dolomite in the lower part of the reservoir than in the upper part, which may obscure the true nature of the predolomite pore geometry. The extensive moldic and enlarged interparticle porosity in the oolitic grainstones was subsequently reduced by later diagenetic cements, including several generations of calcite cement along with late-diagenetic anhydrite, saddle dolomite, and chalcedony. These cements partially plugged the early porosity, but not in a systematic and stratigraphically predictable pattern. The absence of a clear pattern of cement plugging has a negative effect on the correlation between measured porosity and permeability (points on the ϕ–k plot are widely scattered). It was the mud-shrouded packstones and rudstones stratigraphically below and interfingering with the grainstones that became the reservoir rocks, probably because neomorphically stabilized micrite in those beds reduced pore throat sizes and pore-to-pore connectivity that partially shielded the muddy facies from the later cementation. As a consequence, porosity and permeability in the rudstone–packstone facies correlate well enough to allow estimates of permeability from the linear regression equation from a ϕ–k plot. A surprise aspect about this reservoir is that the grainstone section is not the best quality reservoir rock as is usually expected. Instead, the partly muddy rocks ultimately made the best reservoir.

This study offers several take-home lessons: (1) distal segments on carbonate platforms can be the sites of shallow-water deposition during sea-level low stands; (2) significant buildups may occur on insignificant paleostructural features like the original (pre-slump) terrace at Happy Field; (3) multiple episodes of diagenesis may obscure relationships between poroperm values and fundamental rock properties; (4) multiple diagenetic overprints that are difficult to tie to specific events or environments make predicting the spatial distribution and quality rank of flow units very difficult; and (5) stratigraphic slice maps to "take apart" the depositional and diagenetic facies in carbonate reservoirs help unravel complex reservoir attributes to aid in constructing 3D flow unit models.

8.5   FRACTURED RESERVOIRS

Nearly all carbonate reservoirs have been fractured to some extent. That is the essential point: they are fractured to some extent. Whether fractures have a great influence on reservoir characteristics is a separate matter. Nelson's (2001) classification of fractured reservoirs points out that Type IV fractures have a negative influence on reservoir characteristics. Nelson's classification is a prerequisite for working with fractured reservoirs because until the presence of fractures has been confirmed and the reservoir has been tested, it is difficult, if even possible, to evaluate the extent to which fractures contribute to reservoir performance.

8.5.1   Finding and Interpreting Fractured Reservoirs

Exploration for fractured reservoirs usually focuses on tectonic features or regional flexures where fractures are known or suspected to be present. Fractures that generally accompany folds, faults, structural hinge lines, or monoclinal flexures were discussed in Chapter 7. Structural features are relatively easy to identify with modern seismic methods. Knowing the kinds of fractures that commonly occur on specific parts of structures, it is possible to target them. For example, the hanging wall of normal faults usually has the highest fracture intensity, and the types and orientations of fractures that typically occur in folds can be predicted if the orientation of the principal stresses is known. In cases where the quality of seismic data is particularly good, it may be possible to directly identify fractures from seismic reflection character or seismic attributes. Unless fractures are already known to occur in a given area, or they are suspected to be present in buried structures, or some hint of them can be identified in seismic data, finding them in the subsurface is more a matter of chance than of skill.

Once fractured rocks are found, it is necessary to test the reservoir to assess the extent to which fractures influence reservoir performance. It is possible to evaluate the importance of fractures to reservoir performance by well testing, the use of imaging logs, drilling time and caliper logs, various porosity tools, and methodical study of borehole cores or cuttings. Nelson (2001) designed his classification of fractured reservoirs directly with this in mind. Recall his classification from Chapter 7:

Type I. Fractures provide essential reservoir porosity and permeability.

Type II. Fractures provide essential permeability.

Type III. Fractures assist permeability in an already producible reservoir.

Type IV. Fractures provide no additional porosity or permeability but they do impose significant anisotropies such as barriers to flow.

Fracture types I and II are probably the categories that are commonly thought of as true fractured reservoirs. In those cases fractures may provide most of the permeability or porosity—or both—to the reservoir. Types III and IV are also common but do not provide significant amounts of porosity and permeability. Instead, they may offer a moderate assist, as in type III, or even have a negative influence if they act as baffles or barriers to flow, as in type IV reservoirs. Type III and IV fractures include healed fractures such as those with gouge, mineral cements, stylolitized fractures, and discontinuous, small fractures that do not improve permeability. Additional consideration must be given to the relationship between fractures and the in situ stress in the borehole. Rocks may undergo several periods of fracturing such that older fracture sets are cut by younger ones. In those cases, the fracture set that is parallel to the present-day, maximum principal stress is the one most likely to be open and contributing to porosity and/or permeability. Finding this fracture set depends on measuring the in situ stresses within the borehole and identifying the fracture set that is aligned parallel to the principal stress in that regime.

8.5.2 Field Examples of Fractured Reservoirs

Fractured reservoirs make up about one-fifth of the fields listed by Scott et al. (1993) and Roehl and Choquette (1985). North (1985) has an extensive list of examples of fractured reservoirs that include the Asmari Limestone (Tertiary) fields in Iran, Mesozoic carbonates in the Lake Maricaibo area of Venezuela and the Yucatán area of Mexico, Paleozoic carbonates of West Texas, and Mesozoic and Tertiary chalks of the North Sea and North America, among others. Lorenz et al. (1997) describe fractured carbonate reservoirs in mildly deformed, moderately deformed, and severely deformed settings. The two examples of fractured reservoirs selected for inclusion here were chosen because this author worked on both of them and obtained first-hand information about them. These examples are not ideal, however, because in both cases the original geological concept was not built around a search for fractured reservoirs. Also, the degree to which natural fractures influenced reservoir performance was not known until after the discovery wells were completed, but that is the usual case in early exploration drilling. The exploration wells in both of these examples were not targeted at fractured reservoirs, but rather at "mounds" or "reefs" in Carboniferous strata in Texas and the Williston Basin of North Dakota. Such carbonate buildups are usually rather easy to recognize on seismic profiles. Only after studying cores, cuttings, logs, and other data such as well tests was the extent and significance of fracturing fully realized. The number of successful exploration wells that have been targeted—by design—at naturally fractured rocks, as compared with fractured reservoirs discovered by accident, is an interesting unknown. In the examples presented next, both cases exhibit mixtures of depositional, diagenetic, and fracture porosity, but it is fractures that have the dominant influence on reservoir performance.

8.5.2.1 Quanah City Field

Location and General Information Quanah City Field is located in the Hardeman Basin, at the town of Quanah, Texas (Figure 8.5). The one-well field with three offset dry holes is named for the town of Quanah, and was developed by Sun Oil Company during the 1960s. The discovery well, the Sun Carrie Minshew #1, initially produced 338 BOPD. Later, an independent operator drilled a sidetrack to the Sun Oil Company Tabor Oil Unit #1 that produced 2016 BOPD. The target reservoir at Quanah City Field is the Mississippian (Viséan) Chappel Limestone, which also produces in a few wells south of Quanah City Field. Those wells are not included in this discussion.

Structural Setting Both present structure and paleostructure are important in this case. Present structure at Quanah City Field was initially mapped on subsurface geological data as a simple, faulted anticline. The Minshew #1 and the Quanah City #1A wells are deviated holes; consequently, it was difficult to determine true stratal thicknesses and structural attitudes from the vintage logs that were available at the time of the original study. By today's standards, the comparatively primitive 1960s technology for making borehole surveys provided sets of coordinates for constructing azimuth and depth plots of the borehole. There were no dipmeter logs, but by good fortune, a seismic structure map was made available just before research on the field was completed. This map (Figure 8.15) reveals that the bottom-hole locations of the Minshew #1 and Quanah City #1A wells are close to faults that are interpreted to be older, reactivated faults, a common feature in the Hardeman Basin. The faults are also interpreted to have been the primary cause for natural fractures in the Chappel Limestone reservoir.

Paleostructure played an important role in localizing the objective "reef mound" in this field. Although none of the four field wells penetrated the entire Carboniferous section, the location of the Quanah City mound juxtaposed between two large faults suggested to Ahr and Ross (1982) that a fault-related paleostructural high existed on the Ordovician dolomite (the Ellenburger Formation) that underlies the Carboniferous section in this area. The paleo-high was interpreted to have been the site on which the Chappel Limestone mound nucleated and, according to Ross (Ahr and Ross, 1982), field closure was increased by both mound growth and postdepositional fault reactivation.

Depositional and Diagenetic Characteristics The Quanah City mound (Figure 8.16) is interpreted to have grown on an underlying paleo-high in a subtidal, open marine setting. The mound facies consists largely of lime mud interpreted to have been produced in situ, mainly by microbial activity. Fenestrate bryozoans and crinoids are common in the mudstones and wackestones of the mound center. A flanking veneer of crinoidal grainstones and packstones is present. Early diagenetic dolomite is common in the muddy facies and is interpreted to have replaced neomorphic microspar in mud-supported rocks, leaving larger skeletal allochems relatively unaltered. Early in the burial history, spiculiferous zones in the mound facies are interpreted to have been replaced by chert. Subsequent leaching dissolved much of the previously unaltered carbonate allochems, "chalkified" some siliceous crusts, and resulted in solution-collapse brecciation of limestone, dolomite, and "chalky"

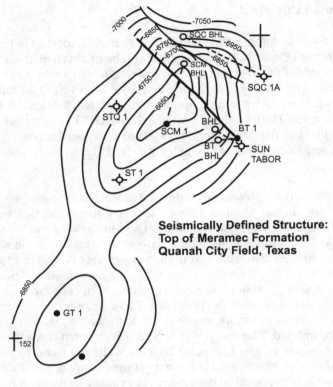

Seismically Defined Structure:
Top of Meramec Formation
Quanah City Field, Texas

Figure 8.15   Present structure map of Quanah City Field based on seismic data. The map reveals that bottom-hole locations of the SQC 1A and SCM 1 wells are close to faults. The location of Quanah City Field is shown on Figure 8.5.

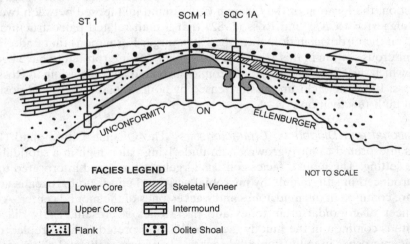

Figure 8.16   A simplified reconstruction of the Mississippian "mud mound reef" at Quanah City Field based on information from petrographic study of borehole core samples and wireline log correlations. Research revealed that reservoir storage capacity is largely in moldic and vuggy pores but flow capacity is primarily through fractures. (From Ahr and Ross (1982).)

chert. Early dissolution (prefracturing) also formed moldic and vuggy porosity in the skeletal "hash" zones in both mound and flank facies.

Further mound growth occurred accompanied by slumping and soft-sediment deformation of grainy flank facies. Finally, a capping facies of brachiopods and bryozoans stabilized the upper surface of the flank deposits. Peloidal, silty lime mud filled intermound areas, elevating the surrounding seabed into fair-weather wave base. Oolite shoals formed over the mound and adjacent flanks. Acicular isopachous rim cement formed on ooids in the marine phreatic environment. A later burial stage of chalcedonic quartz replaced allochems, grains, early cement, and neomorphic microspar matrix. Burial compaction produced tangential, concavo-convex, sutured grain contacts and stylolites. Two episodes of fracturing are evident. The earlier fractures and remaining interparticle pores were filled with coarse, blocky calcite typical of burial cementation. A later generation of fractures was partially filled with saddle dolomite and bitumen. Saddle dolomite crystals are excellent indicators of fractured reservoirs because the crystals are obvious even in rotary cuttings and their occurrences are generally limited to fractures or connected vugs (Ahr, 1982).

Geological Concept   The Quanah City Field reservoir rock is a "mud mound" of the type commonly associated with Mississippian mound facies around the globe. The Quanah City buildup is interpreted to have formed in a shallower setting than the typical Waulsortian mounds of Europe. It is more similar to the Late Mississippian mounds of Derbyshire, in central England. Coincidentally, many of those mounds in the area around Castleton, England are also fractured and are host rocks to sulfide ores similar to MVT ores, along with especially colorful fluorite known as Blue John. Without the extensive diagenesis and fracturing that dramatically altered the Quanah City mound, it is unlikely that enough porosity and permeability would have existed to make a reservoir. Although oolitic grainstones were deposited on top of the mound, they became cemented to the extent that they provided the seal for the reservoir. The extensive leaching that produced vuggy and moldic porosity enhanced the storage capacity of the Quanah City reservoir, but many of the molds and vugs are not touching vugs. They provide a significant amount of porosity but only limited permeability. Porosity and permeability from routine core analyses were plotted against depth and depth-shifted core lithofacies to reveal that high porosity and permeability do not correspond to depositional lithofacies and permeability does not correlate well with porosity. The dominant lithologies are early-replacement dolomite and saddle dolomite, which do not correlate with depositional lithofacies except to the extent that early dolomitization appears to be more abundant in mud-supported rocks. High permeability corresponds to highly fractured zones in cores. Fractures provide essential permeability by connecting the molds, vugs, and other matrix porosity, making this a Type II fracture system. The fractures at Quanah City Field are interpreted to have been formed in conjunction with faulting. Fractures also appear to be oriented parallel to the fault system because productive fractures were penetrated by the Minshew #1 borehole but missed by the Quanah City #1A well. Only the Minshew #1 made a well.

Cores from the Minshew #1 well are replete with moldic and vuggy porosity, natural fractures, and fracture-filling saddle dolomite. The abundant dissolution porosity is interpreted to provide the main storage capacity for this well, but the fractures provide essential permeability, indicating that the reservoir is a Type II fractured reservoir. Another important line of evidence to support the interpretation that the reservoir is fractured is the acoustic log. Although no dipmeter logs, image logs, or pressure tests were available for study, the acoustic log from the Minshew #1 exhibits several distinctive "cycle skips" in the productive zone just above the cored interval. Cycle skips are nearly instantaneous scale changes that must be made to keep up with shifts in acoustic transit time so the tool can continue to record accurately in zones of rapidly changing porosity. Characteristics similar to those in the Minshew #1 well were observed by the author in nearby fields, where in addition to cycle skips in the acoustic log, caliper logs indicated "washouts" in the zone of greatest fracturing, difficulty was encountered in maintaining steady mud circulation because the mud would infiltrate the fractures, drilling time was dramatically decreased as compared to nonfractured zones above and below the fracture horizon, and saddle dolomite crystals were recovered in lagged cuttings known to be from fractured zones. In severely fractured zones it is possible for large pieces of borehole wall to "cave" into the hole, causing the drill bit to jam. Taken together, these lines of evidence are good general indicators that fractures have been encountered during drilling. Direct observations of fractures in borehole core provided confirmation.

The original concept for establishing the drilling location in Quanah City Field was simply to drill a structural anomaly that was detected by seismic surveys. The anomaly turned out to be the Quanah City mound, a buildup that accentuated bathymetric relief by growing on top of a preexisting high—a fault block. After the discovery well was drilled, it was clear that the reservoir was strongly influenced by fractures. The lesson from this experience is that the identification of natural fractures in the subsurface is largely dependent on the ability of geophysicists to find structures that are likely to be fractured or to extract seismic attributes that are known to indicate fractures. In some cases where faults or folds can be mapped on surface outcrops and extrapolated to the nearby subsurface or where structures can be mapped in the subsurface with the seismograph, it may be possible to predict where fractures will occur based on structural geometry, inferred loci of stress concentrations, and knowledge about the brittle characteristics of the host rocks. If fracture orientation can be identified with borehole imaging logs or other directional devices, opportunities may exist to drill horizontal wells to encounter the greatest number of fractures per unit lateral distance in the subsurface.

8.5.2.2 Dickinson Field

Location and General Information   The Dickinson Lodgepole Unit (DLU) is located in the Williston Basin of North Dakota and consists of nine individual fields (Figure 8.17). The discovery well, the Conoco Dickinson State (DLU) #74 well was completed in 1993 in the Carboniferous Lodgepole Formation (Young et al., 1998). The drilling location was chosen to test a structural anomaly identified on seismic lines, not to test for fractured carbonates. The well flowed at an initial stabilized rate of 1150 BOPD and 0 barrels of water per day and produced at rates in excess of 1500 BOPD for six months until curtailment. Two years later Duncan Oil Company drilled a second Lodgepole discovery nearby. These Lodgepole Formation wells are particularly attractive because initial production rates are commonly about

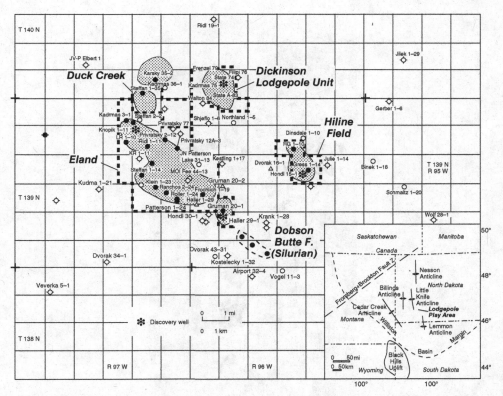

Figure 8.17 Location of the Dickinson area fields, Williston Basin, North Dakota. Note that there are nine individual fields encompassing several large "reef mounds." (From Montgomery (1996).)

1000 BOPD and individual wells may produce over 1 million barrels of oil. Primary recovery has been estimated to be about 28% with original oil in place calculated to be between 70 and 100 million barrels. The Lodgepole mounds produce from moderately low porosity (average 5%), which consists of a complex mixture of depositional, diagenetic, and fracture pore types (Figure 8.18). Depositional porosity consists mainly of stromatactoid vugs in the mounds and grainstones interbedded within the mound complex. Diagenetic porosity includes leached interparticle porosity in grainstones and rather extensive burial dissolution that enlarged vugs and fractures within the mudstone and wackestone mound facies. Vugs and fractures enlarged by late dissolution are also commonly filled or partly filled with saddle dolomite, similar to the type observed in the Hardeman Basin wells discussed earlier, although the stromatactoid vugs in the Dickinson mounds are rare in the Hardeman Basin examples. Fractures are common in whole cores from Dickinson area wells, they are recognized on image logs, and by their influence on reservoir performance pressure transient testing.

Several types of fractures exist in the Dickinson mounds and they formed at different times during mound burial history and subsequent tectonism. Compaction fractures are common in the thin-bedded, lower mound zones where the mound

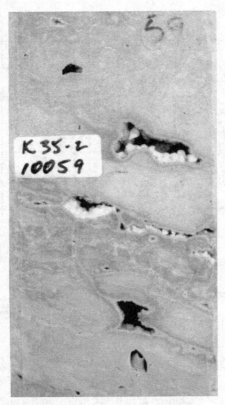

Figure 8.18   Photograph of a core segment from the Kadrmas 35-2 well in the Dickinson area showing depositional (stromatactis), diagenetic (solution-enlarged vugs), and fracture porosity that characterizes the Dickinson Field reservoirs. Note the light colored crystals of fracture-and-vug lining saddle dolomite that were emplaced as late burial diagenetic features. The core is approximately 4 inches in diameter.

facies consists of a series of platy layers a few centimeters to 15 cm thick. The upper mound zones are typically massive mudstone and cementstone beds up to a meter or more in thickness and lacking the well-defined bedding that is associated with compaction fractures in basal mound strata. Compaction fractures are interpreted to have formed by overburden loading as the thin, platy layers were continually fractured by overburden load from the massive, upper buildup accumulation. Compaction fractures formed a network of small, brittle fractures that occur only in the platy, lower portions of the buildups. Larger (multicentimeter- and meter-scale) fractures are interpreted to have formed by tectonism. A third, even larger scale set of fractures at multimeter scale are interpreted to have been formed during regional or subregional faulting. This interpretation was confirmed by monitoring pressure communication across the field. Reservoir permeability determined by pressure transient tests varies from 200 to 2000 md and wells 2–3 miles apart in a single field showed pressure responses to tests in minutes to no longer than a few hours (Young et al., 1998). This pressure communication exists between widely separated wells that are not connected by porous or permeable depositional facies. Moreover, permeability greater than 1000 md is not consistent with average unfractured matrix rock porosity of only 5%. This reservoir is interpreted to be a Type II fractured reservoir because fractures provide essential permeability. There may be parts of the lower, more intensely fractured mounds that behave as Type I fracture systems.

Structural Setting   The Dickinson Unit lies in a portion of the Williston Basin where major north–south tectonic features overlie zones of vertical, basement faulting. Major displacement along these zones took place during the Carboniferous Ancestral Rocky Mountain Orogeny and again during the subsequent Laramide Orogeny (Montgomery, 1996). Mapping on a Precambrian datum indicates that a major tectonic boundary exists in this region of the Williston Basin, an interpretation supported by estimates of paleoheat flow determined from maturity measurements in the Bakken Shale (Montgomery, 1996). DLU reservoirs are in a multimound complex about one township in size and each field within the unit produces from one or more of the mounds in the complex, although subregional and regional faults with attendant fracture permeability may provide flow communication between mounds. Mounds were initially identified from seismic data as structural anomalies, not as Waulsortian mounds. Information from the early discoveries identified the anomalies as microbial mudstone and cementstone mounds and revealed that the underlying Devonian–Mississippian Bakken Shale is thicker beneath and in the near vicinity of the mounds than farther away from them. There is disagreement about whether the thickened Bakken Shale served as a nucleation site on which the mounds developed. The thickened Bakken Shale is interpreted to be the result of dissolution of underlying salt combined with faulting (Young et al., 1998). Continued or rejuvenated tectonism probably formed the large-scale fractures.

Depositional and Diagenetic Characteristics   The Dickinson Lodgepole mounds closely resemble the Early Carboniferous (Tournaisian to Viséan) Waulsortian mounds worldwide. The Dickinson mounds, like all typical Waulsortian mounds, began to develop in an outer ramp environment near the toe of slope transition between ramp and basin floor. Typically, Waulsortian mounds begin growth in the early stages of transgressions—in the TST segments of stratigraphic sequences. The presence of heterozoans and the absence of photozoans in the Dickinson area buildups, along with other deep-water characteristics in the bedded, off-mound strata, corroborate the distal ramp, deep-water environmental interpretation. The DLU mounds range in size from about mile in diameter to 4 miles in longest axial length. Mounds grew in complexes rather than as single domes and the complexes vary in vertical thickness from 250 to 350 feet. The mounds consist primarily of mudstones and wackestones with a macrofauna composed of crinoids, bryozoans, brachiopods, bivalves, sponges, and scattered solitary corals. The lower portions of the mounds are commonly bedded, while the upper parts are more massive. A microbiota similar to other Waulsortian mounds around the globe is also present. Taxonomic diversity in both macro- and microbiota tends to increase toward the tops of mounds. Although the mounds are composed of mud-supported rocks, some grainstones and packstones are present as pockets or pods of loose, skeletal allochems that filled low areas between and around mound growth centers. Because the detrital pockets are not laterally continuous over wide areas, the grainy rocks do not significantly affect reservoir volume or performance. Locally, however, leached grainstones and packstones are among the most porous mound facies. *Stromatactis*-like cavities are common and volumetrically abundant in the lower and middle parts of the mounds. These vuggy pores are commonly formed around shelter cavities beneath fenestrate bryozoan fronds or where sponge bodies existed. Radiaxial calcite usually lines the cavities. Although variations in microfacies character are present in the Dickinson mounds, the microfacies are not organized in systematic patterns that would simplify exploration and development in the fields. Instead, the Dickinson mounds make reservoirs because of the three main types of fractures that connect diagenetically enhanced depositional porosity—particularly the solution-enlarged, stromatactoid vugs.

The DLU mounds were deposited as microbially mediated or microbially generated peloidal mud, skeletal meshworks (especially fenestrate bryozoan meshworks), and submarine cementstones. Early marine diagenetic radiaxial calcite cements around fenestrate bryozoans, along with syntaxial overgrowth cements on crinoids, created a bindstone to cementstone fabric. Radiaxial calcite commonly coats fenestrate bryozoan sheets and polyform surfaces to form ornate, lined cavities similar to classical *Stromatactis* vugs (Figure 8.18). Many of these vugs were enlarged by early burial dissolution that resulted in enhanced storage capacity for DLU hydrocarbons. After the early dissolution, at least two generations of calcite cements were deposited: a blocky calcite pore-filling spar and a dogtooth (scalenohedral) calcite pore-lining druse. Some of these burial cements were partly to completely dissolved by a second invasion of undersaturated diagenetic waters. Both vuggy porosity and fracture porosity were enhanced by this late stage leaching. Etched vug and fracture surfaces were overlain or partly replaced by saddle dolomite (Figure 8.18), some of which exhibits post-saddle-dolomite leaching and flooding by migrating hydrocarbons still present as oil stains. Some saddle dolomites are partly replaced and overlain by late-diagenetic anhydrite and some saddle dolomites that partially fill fractures are overlain by small, equant, subhedral dolomite crystals that appear to have sifted into fractures as internal sediment.

Geological Concept   The discovery well at Dickinson Field was drilled on a seismically defined prospect in the deeper, Ordovician–Silurian Interlake Formation. Disappointing results led to testing in the lower Mississippian Lodgepole Formation. This decision was based on gas shows in the mud log and oil stains in cuttings, log analyses of the zone, and problems with lost mud circulation at that stratigraphic level (Young et al., 1998). Prior to the discovery at Dickinson Field, according to Montgomery (1996), the Lodgepole was known as a low-volume, sporadic reservoir along the Nesson anticline, where fracturing was evident. The formation tested tight with only occasional oil shows elsewhere in the Williston Basin. After the Lodgepole mounds were recognized as the reservoir at Dickinson Field, 3D seismic surveys were acquired. The new seismic data along with existing well data enabled the operators to define the general outlines of the mounds that comprise the Dickinson complex, but the resolution of the data was not sufficient to outline separate microfacies or fractured zones. It was after pressure transient testing and study of whole cores and image logs that the significance of fractures on reservoir performance was confirmed (Young et al., 1998). As in many other cases, this fractured reservoir was discovered by accident. It was not recognized as an exploration target and was tested only after the failure of the first well to produce from a deeper objective zone. There are lessons to be learned from this experience. Note that some of the reasons for testing the Lodgepole zone included problems with lost circulation. If close study had been made on drilling time, mud circulation, caliper and acoustic log character, and especially on the presence of saddle dolomite in cuttings, the mound interval would probably have been identified immediately as a fractured carbonate reservoir. Whole core and pressure tests confirmed that the reservoir is fractured. Image logs could have been much more useful had they been studied more extensively in conjunction with whole cores to identify fracture density, spacing, and spatial orientation.

8.6 CONCLUSIONS

We have focused on fundamentals of geology related to carbonate reservoirs and on the application of those fundamentals in exploration and development. Much of the information in this book comes from the author's more than forty years of experience with carbonate rocks and reservoirs in both industry and academia. The hope is that this book will be useful to geoscientists and engineers who work with carbonate reservoirs and aquifers, and especially that it will open new vistas for university students. A few main points about carbonate reservoirs bear repeating as we conclude. Carbonate reservoirs are rock bodies but they do not necessarily conform to stratigraphic boundaries because reservoirs are defined by porosity and permeability. In strong contrast to sandstone reservoirs, porosity and permeability in carbonates can be independent of depositional facies or formation boundaries, as exemplified by diagenetic and fracture porosity that cut across depositional facies boundaries. Many carbonate reservoirs have pore systems that formed long after sedimentation. Removal or replacement of original rock texture and fabric can create pore characteristics that did not exist at the time of deposition. In siliciclastic sandstone reservoirs, diagenesis such as cementation or authigenic clay formation may reduce porosity, but rarely does diagenesis increase porosity. If the number of papers about fractured reservoirs in sandstones as compared to carbonates is any indication, fractured reservoirs are less common in sandstones than in carbonates. Because porosity in carbonate reservoirs and aquifers is not always the result of a single geological process and because porosity in carbonates can undergo repeated episodes of change during burial, we developed a simplified genetic classification of porosity to focus on the three end-member pore types; namely, depositional, diagenetic, and fracture pores along with their respective hybrids.

Geoscientists and engineers must recognize pore types by origin so that they can design exploration and development projects around geological concepts that take pore origin into account as one of the methods for correlating at reservoir scale. It is far too risky to plan a strategy simply around depositional facies, stratigraphic units, or present-day structural anomalies. Recognition of genetic pore types enables one to develop concepts to exploit diagenetic reservoirs where porosity and permeability are related to ancient water tables or episodes of rock–water interaction during later burial. Similarly, one can exploit fractured reservoirs that conform to the geometry of tectonic faults and folds or in situ stresses rather than to stratal geometries. Recognizing carbonate pore types by origin requires direct observation of the reservoir rocks as cores or cuttings. As of now, there are no indirect methods of determining pore origin in carbonates, although NMR logs do show promise. If pore origins cannot be determined, then reservoir size, shape, and connectivity cannot be predicted with enough certainty to provide realistic economic assessments of drilling and development prospects. If porosity can be classified by direct observation of rock samples, then reservoir characterization, economic predictions, and reservoir simulation can be done with much greater accuracy. The next step in reservoir analysis, after recognizing genetic pore types, is the identification of flow units, baffles, and barriers.

We define flow units as high-connectivity zones within reservoirs. High connectivity means high porosity and permeability and low capillary resistance to fluid flow. Baffles are zones with low connectivity but they are limited in lateral and vertical extent so that fluids can still flow at low rates around, over, through, or under them. Baffles can be compared to islands in a stream. Barriers are zones that do not allow fluids to flow at reasonable rates and barriers may be laterally or vertically extensive, or both. Barriers can be compared to dams on a stream. Some barriers may even be seals that prevent fluid escape from reservoirs. We examined ways in which to test seal capacity based on capillary pressure differences between the seal rock and the reservoir rock. Rocks with high seal capacities are barriers. Another way of defining seal capacity is the maximum height of hydrocarbon column that can be supported by a given reservoir rock, or h_{max}. We search for reservoirs with high connectivity but that are closed by tight seals. Almost all stratigraphic and most structural settings have some internal baffles, but the baffles only interfere with flow to various degrees. If the baffles are comparatively small in size and allow some fluid flow through them, then they are more a nuisance than a real problem. When baffles are large and have much higher capillary resistance to flow than the reservoir rock, they become problems because they may create reservoir compartments that can be difficult to exploit. Commonly encountered baffles are strata with irregular distribution of muddy depositional textures mixed with porous and permeable grain-supported textures. Other types include rocks with irregular distribution of diagenetic cements or compaction-reduced pores and pore throats.

We can estimate the degree of connectivity within and between flow units by constructing a data matrix of porosity and permeability values, preferably measured porosity and permeability from core analyses. This ranking method is one way to assess the relative quality of reservoir flow units and identify the ones that are best, intermediate, or poor in quality. In order to identify the best, intermediate, and poor connectivity zones from the matrix of porosity and permeability data, a system of slices or layers of arbitrary thickness can be constructed to trace the quality-ranked segments from the top to the bottom of the reservoir interval. This is especially useful in hybrid reservoirs where the distribution of diagenetic effects is difficult to map or predict. The spatial distribution of ranked flow units can be identified and visualized at field scale with slice maps. In practice, stratigraphic slices about 10 feet thick are the limit in practical applications because that is about the lower limit of resolution on most wireline logs. Extreme highs and lows of porosity and permeability in the reservoir can be excluded to simplify the results, providing that the extremes also have low frequency of occurrence. The same is done with permeability; that is, the most representative range of values is chosen.

In a hypothetical example, let the most representative range of average porosity in the entire reservoir sample be 5–35% and the average permeability range be 15–105 md. The average porosity values for the reservoir can be divided arbitrarily into three categories of 10% each, or 5–15%, 16–25%, and 26–35%. Permeability values can be divided into three categories of 15–45, 46–75, and 76–105 md. Then, constructing a symmetrical matrix of 3 × 3 (or 9) possible values for porosity and permeability pairs simplifies the selection process for best, intermediate, and poor ranking.

After the ranking matrix is constructed from poroperm data, the portions of each 10-foot slice that are best, intermediate, and poor in rank can be identified within the field. One method is to overlay contour maps of average porosity and average permeability for each 10-foot interval of vertical thickness and locate the areas where contours of high average porosity overlie areas of high average permeability, and so on (Ahr and Hammel, 1999). Around boreholes where cores were taken and poroperm values were measured directly, accuracy is not a problem. Where no cores were taken, porosity must be estimated from log calculations. If no pressure tests are available to determine permeability in those wells lacking cores, then permeability must be estimated by comparing poroperm data from core analyses with noncored well wireline log data. Linear regressions comparing porosity and permeability from core analyses, and then from wireline log data, can provide estimates of permeability for noncored wells. Map overlays of porosity, permeability, and facies character enable one to identify best, intermediate, and poor flow units, baffles, and barriers—the latter two being rocks that have porosity and permeability values below the limits used in the quality ranking matrix.

If reservoir flow units do not correspond to depositional facies characteristics then rock samples must be examined to determine the degree and kind of diagenesis that has altered them. Has diagenesis increased or decreased original depositional porosity? By what processes? Finally, after determining how much and what kind of diagenetic change has taken place, it only remains to determine the chronological history of diagenetic events that led to present-day reservoir quality. Did replacement or dissolution influence flow unit quality, for example? Was dissolution early or late in burial history? Is the imprint of dissolution cross-cut by a later form of diagenesis? What is the impact of each of the diagenetic events on the final outcome—today's reservoir flow unit quality?

Fractured reservoirs are special cases that require special methods of analysis. Fractures may contribute essential porosity and permeability or not. They may influence reservoir performance in a positive or negative manner. They may be caused by a variety of mechanical processes, but the fact that they are formed mechanically as deformation due to stress is incontrovertible. They are not depositional or diagenetic. Determining reservoir quality in fractured systems is at least partly a process of determining whether the reservoir is Type I, Type II, Type III, or Type IV fractured system according to Nelson (2001). The size and shape of fractured rock bodies in the subsurface can probably be estimated by evaluating pressure transient data with methods such as Horner plots. There is no reason to expect fractured zones to correspond with facies maps or with types of diagenesis, although it is possible to some extent.

Now we have a common language about carbonates and reservoirs. We learned how and where carbonate rocks form. We learned about fundamental, derived, and tertiary rock properties that must be studied in order to get reproducible results in our research on carbonates and to communicate those results in understandable terms. Some of those terms include classification terminology for both rocks and porosity. In particular, we learned that a genetic classification of carbonate porosity is a necessary and powerful tool to help construct geological concepts to explain the depositional, diagenetic, and tectonic histories that created reservoirs and aquifers.

We found that many fundamental rock properties are related to reservoir properties. Porosity and permeability are derived properties that correlate directly with depositional sedimentary texture, with sedimentary structures, and to a degree with carbonate grain types. We focused attention on the rock properties that correspond most closely with reservoir characteristics that, in turn, influence reservoir quality. There we learned about saturation, wettability, capillarity, capillary pressure, and how those reservoir characteristics are related to fundamental, derived, or tertiary rock properties. After learning how rock and reservoir properties are related, we learned how carbonate sediments are formed, how sediment accumulations become stacked in stratigraphic arrays, and how the stratigraphic units can be correlated, mapped, and exploited as reservoir units. We found that seven basic depositional successions may be deposited in lateral arrays that characterize two end-member depositional platform types—ramps and shelves. The variety of environments that exist on these platforms produces that array of depositional successions, each of which represents specific locations on the platforms. The lateral array of these successions—from beaches and dunes to basinal rhythmites—is predictable for ramps and shelves. Given the location of an ideal depositional succession, we can predict the locations of other successions in updip or downdip directions. Each of the seven ideal depositional successions on ramps and shelves has characteristic textures, grain types, sedimentary structures, and ranges of taxonomic diversity that provide key information to use in constructing depositional reservoir models for ramps and shelves.

Knowing the array of ideal depositional successions across ramps or shelves enables us to construct two-dimensional facies models for the platforms. Time is the third dimension. Time and the accompanying sedimentation, erosion, or nondeposition comprise the 3D depositional model—the stratigraphic succession. We made a conceptual jump from rock and reservoir properties and from 2D sedimentary blankets that encompass only small increments of time to 3D stratigraphic units that may include thousands or millions of years' worth of deposition. The mental jump led to principles of stratigraphy, correlation, geological time, and sequence stratigraphy, or chronostratigraphy. Part of dealing with stratigraphy is that sedimentary successions are created and influenced by sediment source and supply (the carbonate factory), by platform subsidence or uplift, and by relative sea-level fluctuations. These large-scale phenomena usually occur in cyclical fashion so that the stratigraphic record is replete with cyclical repetitions of depositional successions. Tracing only the successions that have the greatest reservoir potential requires understanding of the total process–response system.

The basic information about sedimentology and stratigraphy led to the heart of this book—the three end-member types of carbonate porosity and how profoundly they influence reservoir characteristics. Depositional reservoirs reflect depositional processes. Sedimentology and stratigraphy are the primary tools needed to distinguish facies and diagnose reservoir characteristics.

After the depositional successions and stratigraphic relationships are worked out and after identifying facies that are predicted to have good reservoir potential, it is sometimes unfortunately discovered that the predicted reservoir performance was not achieved. Depositional characteristics may not always govern reservoir characteristics; rather, the postdepositional changes that enhanced, created, or reduced porosity and permeability instead govern reservoir quality. Diagenetic porosity and accompanying permeability are commonly the key factors that determine carbonate reservoir quality. Fractured reservoirs are not only distinctive in their mode of origin, but fracture porosity and permeability are computed in different ways than for depositional and diagenetic reservoirs.

Many variations are possible for these three end-member types of reservoirs and there is no single model that satisfies all conditions for all genetic reservoir types. In fact, some reservoirs may have geological characteristics that are so specialized that their case histories are applicable only to those unique situations. This leads to a philosophical point: the most important thing in working with carbonate reservoirs is to keep an open mind, a fertile imagination, and a basic understanding of the main processes that create end-member pore types in carbonates. Avoid the temptation to choose an "analog" or a look-alike. Analyze and diagnose the reservoir on the basis of its fundamental rock properties, how those rock properties relate to reservoir (petrophysical) characteristics, how the reservoir units are distributed in stratigraphic space on different platform geometries, and whether diagenesis or fracturing has influenced reservoir behavior. This requires not only a keen mind, good observational powers, and analytical thinking but also direct observation of rock samples in cuttings or full-diameter cores.

Studying carbonates can be greatly rewarding because they contain about half of the hydrocarbons on the planet, they are aquifers in many parts of the world, they are hosts to sulfide ores on several continents, they are used for construction material and road metal, they are an essential ingredient in cement, some forms of calcite are used in optics (the original nicol polarizing prism was cut from Iceland Spar), and dolomite is sold as a health-food supplement. Reservoirs and aquifers in carbonate rocks are intriguing. They can pose challenges to the best geoscientist. They can be complex but not so complicated that they are difficult to understand. A closing reminder is that probably 60% of the already-discovered oil around the world still remains in place and at least half that is in carbonate reservoirs. The knowledge of how to extract some of the remaining oil at prices competitive with exploration is within our grasp. The key is in understanding the spatial distribution of porosity and permeability, the combined amounts of which define flow units, baffles, and barriers. More accurate maps of poroperm systems can be developed to improve efficiency of sweeps made by injected fluids as they move from injection to extraction wells. It seems certain that the economics of the petroleum industry will increasingly focus on optimizing development of new discoveries and especially on improved recovery in established fields. Improved knowledge of carbonate pore systems will also aid in exploring for and management of groundwater resources in carbonate aquifers. It will aid in developing better ways to predict contaminant dispersal in groundwater aquifers. It is just waiting to be done.

REVIEW QUESTIONS

8.1. Which wireline logs are most useful for determining facies characteristics in carbonates? Explain.

8.2. Are lithological logs based on examination of cuttings just as useful as "lith logs" based on cores? Explain.

8.3. How much and what kind of information is necessary to identify any one of the seven standard depositional successions?

8.4. Porosity and permeability commonly exhibit a linear relationship on semilog plots. Is that always true in carbonate reservoirs? Explain.

8.5. What did Focke and Munn (1987) discover about the relationship between carbonate pore types and petrophysical characteristics?

8.6. Would regional, field, or reservoir scale data be most helpful for developing sequence-stratigraphic models of carbonate depositional systems?

8.7. Which of the seven standard depositional successions are most likely to be depositional reservoirs?

8.8. Assume that a porous and permeable depositional reservoir trend extends along a depositional strike for many miles. Would you expect the entire trend to be productive? If not, how would you choose exploration targets within the trend?

8.9. A productive oolite grainstone facies with characteristics like those of the slope-break standard succession was found in an outer-ramp environment otherwise dominated by mudstones and wackestones. How do you explain the presence of the grainstone succession?

8.10. Porosity in the oolite grainstone from Question 8.9 was found to be a solution-enhanced intergranular hybrid (Type IA). What possible depositional and burial history could lead to the formation of this type of porosity in an outer-ramp environment?

8.11. What conditions are necessary to create and preserve high values of interparticle porosity in chalks?

8.12. What clues in carbonate reservoir rocks might indicate that diagenesis took place in or near a paleo water table?

8.13. What is a "slice map" and what purpose does it serve?

8.14. When more than one episode of fracturing has occurred and several sets of fractures can be identified by image logs, which are the fractures most likely to influence production and why?

REFERENCES

Abegg, F. E., Loope, D. B., and Harris, P. M. (2001). Deposition and diagenesis of carbonate eolianites. In: *Modern and Ancient Carbonate Eolianites: Sedimentology, Sequence Stratigraphy, and Diagenesis*. SEPM Special Publication No. 71, pp. 17–32.

Ahlbrandt, T. S., Charpentier, R. R., Klett, T. R., Schmoker, J. W., Schenk, C. J., and Ulmishek, G. F. (2005). *Global Resource Estimates from Total Petroleum Systems*. AAPG Memoir No. 86, Tulsa, OK., 324 pp.

Ahr, W. M. (1971). Paleoenvironment, fossil algae and algal structures in the upper Cambrian of Texas. *J. Sediment. Petrol.* 41:205–216.

Ahr, W. M. (1973). The carbonate ramp, an alternative to the shelf model. *Trans. Gulf Coast Assoc. Geol. Soc.* 23:221–225.

Ahr, W. M. (1982). Saddle crystal dolomites as fractured reservoir indicators, Mississippian biohermal facies, Hardeman Basin, Texas. *Trans. Gulf Coast Assoc. Geol. Soc.* 32: 195–198.

Ahr, W. M. (1985). Limestone depositional sequences on ramps and shelves: modern and ancient. *Geol. Today* 1:84–89.

Ahr, W. M. (1989). Early diagenetic microporosity in the Cotton Valley Limestone of East Texas. *Sediment. Geol.* 63:275–292.

Ahr, W. M. (1998). Mississippian mounds as reservoir rocks. *SW Section AAPG Trans.*, pp 43–47.

Ahr, W. M. (2000). *Frameless Reef Reservoirs*. West Texas Geological Society Publication 00-109, pp. 1–8.

Ahr, W. M. and Ross, S. (1982). Depositional and diagenetic history of Chappel (Mississippian) biohermal reservoir, Quanah City Field, Hardeman County, Texas. *Trans. Gulf Coast Assoc. Geol. Soc.* 32:185–194.

Geology of Carbonate Reservoirs: The Identification, Description, and Characterization of Hydrocarbon Reservoirs in Carbonate Rocks
By Wayne M. Ahr   Copyright © 2008 John Wiley & Sons, Inc.

Ahr, W. M. and Hull, H. Ben (1983). Burial diagenesis and reservoir development in the North Haynesville (Smackover) Field, Louisiana. *Trans. Gulf Coast Assoc. Geol. Soc.* 33:1–9.

Ahr, W. M. and Walters, J. K. (1985). Conley Field, Hardeman County, Texas: Chappel (Mississippian) production from facies-selective porosity in carbonate sand buildups. *Trans. Gulf Coast Assoc. Geol. Soc.* 35:1–10.

Ahr, W. M. and Hammel, B. (1999). Identification and mapping of flow units in carbonate reservoirs: an example from Happy Spraberry (Permian) Field, Garza County, Texas USA. *Energy Exploration and Exploitation* 17:311–334.

Ahr, W. M., Allen, D., Boyd, A., Bachman, H. N., Smithson, T., Clerke, E. A., Gzara, K. B. M., Hassall, J. K., Murty, C. R. K., Zubari, H., and Ramamoorthy, R. (2005). Confronting the carbonate conundrum. *Schlumberger Oilfield Rev.* Spring 2005, 18–29.

Alberty, M. W. (1992). Basic open and cased-hole tools. In: *Development Geology Manual.* AAPG Methods in Exploration Series No. 10, Tulsa, OK, pp. 144–149; 151–153.

Allen, J. R. L. (1985). *Principles of Physical Sedimentology.* Allen & Unwin, London, 272 pp.

Amare, K. (1996). Mapping flow units in a heterogeneous carbonate reservoir: Reeves Field, Yoakum County, Texas. Unpublished Ph.D. thesis, Texas A&M University, 144 pp.

The American Heritage® Dictionary of the English Language, 3rd edition (1992). Houghton Mifflin, Boston. Electronic version licensed from INSO Corporation.

Amthor, J. E., Mountjoy, E. W, and Machel, H. G. (1994). Regional-scale porosity and permeability variations in Upper Devonian LeDuc buildups: implications for reservoir development and prediction in carbonates. *AAPG Bull.* 78:1541–1559.

Amyx, J. W., Bass, D. M. Jr., and Whiting, R. L. (1960). *Petroleum Reservoir Engineering.* McGraw-Hill, New York.

Anselmetti, F. S. and Eberli, G. P. (1997). Sonic velocity in carbonate sediments and rocks. In: *Carbonate Seismology*, I. Palaz and K. J. Marfurt (Eds.), Geophysical Developments Series No. 6, SEG, Tulsa, OK, pp. 53–74.

Anselmetti, F. S. and Eberli, G. P. (1999). The velocity-deviation log: a tool to predict pore type and permeability trends in carbonate drill holes from sonic and porosity or density logs. *AAPG Bull.* 83:450–466.

Archie, G. E. (1942). The electrical resistivity log as an aid in determining some reservoir characteristics. *Trans. AIME* 146:54–67.

Archie, G. E. (1952). Classification of carbonate reservoir rocks and petrophysical considerations. *AAPG Bull.* 36:278–298.

Asquith, G. B. (1985). *Fracture Porosity and the Cementation Exponent (m).* AAPG Methods in Exploration No. 5, Tulsa, OK, Chap. VIII, p. 43.

Asquith, G. B. and Gibson, C. R. (1982). *Basic Well Log Analysis for Geologists.* AAPG Methods in Exploration No. 3, Tulsa, OK, 216 pp.

Asquith, G. B. and Jacka, A. D. (1992). Petrophysics of bimodal porosity: lower Cretaceous Rodessa Limestone, Running Duke Field, Houston County, Texas. *Trans. Gulf Coast Assoc. Geol. Soc.* 42:1–12.

Asquith, G. B. and Krygowski, D. (2004). *Basic Well Log Analysis*, 2nd ed. AAPG Methods in Exploration No. 16, Tulsa, OK, 244 pp.

Bagrintseva, K. I. (1977). *Carbonate Rocks, Oil and Gas Reservoirs.* Izdated'stvo Nedra, Moscow, 231 pp.

Barnaby, R. J. and Read, J. F. (1992). Dolomitization of a carbonate platform during late burial: Lower to Middle Cambrian Shady Dolomite, Virginia Appalachians. *J. Sediment. Petrol.* 62:1023–1043.

Bathurst, R. G. C. (1975). *Carbonate Sediments and Their Diagenesis*, 2nd ed. Developments in Sedimentology Volume 12, Elsevier, Amsterdam, 660 pp.

Beard, D. C. and Weyl, P. K. (1973). Influence of texture on porosity and permeability of unconsolidated sand. *AAPG Bull.* 57:349–369.

Berg, R. R. (1970). Method for determining permeability from reservoir rock properties. *Gulf Coast Assoc. Geol. Soc. Trans.* 20:303–317.

Bishop, W. F. (1968). Petrology of Smackover Limestone in North Haynesville Field, Claiborne Parish, Louisiana. *AAPG Bull.* 52:92–128.

Bliefnick, D. M., Kaldi, J. G., Bissmeyer, S. K., and Dang, T. T. (1990). Multidisciplinary reservoir description, Walker Creek Field, Columbia and Lafayette Counties, Arkansas. In: *The Integration of Geology, Geophysics, Petrophysics, and Petroleum Engineering in Reservoir Delineation, Description, and Management.* Proceedings of the 1st Archie Conference, Houston, Texas. AAPG Publication, Tulsa, OK, 441 pp.

Boggs, S. Jr. (2001). *Principles of Sedimentology and Stratigraphy*, 3rd ed. Prentice Hall, Englewood Cliffs, 726 pp.

Boreen, T., James, N. P., Wilson, C., and Heggie, D. (1993). Surficial cool-water carbonate sediments on the Otway continental margin, southeastern Australia. *Mar. Geol.* 112:35–56.

Brasher, J. E. and Vagle, K. R. (1996). Influence of lithofacies and diagenesis on Norwegian North Sea chalk reservoirs. *AAPG Bull.* 80:746–768.

Brown, A. R. (1999). *Interpretation of Three-Dimensional Seismic Data*, 5th ed. AAPG Memoir No. 42, Tulsa, OK, 514 pp.

Brown, A. R. (2004). *Interpretation of Three-Dimensional Seismic Data*, 6th ed. AAPG Memoir No. 42, Tulsa, OK, 560 pp.

Brown, L. F. and Fisher, W. L. (1977). Seismic-stratigraphic interpretation of depositional systems: examples from Brazilian rift and pull-apart basins: application of seismic reflection configuration to stratigraphic interpretation. In: *Seismic Stratigraphy—Applications to Hydrocarbon Exploration.* AAPG Memoir No. 26, Tulsa, OK, pp. 213–248.

Bubb, J. N. and Perry, D. (1968). Porosity in some synthetic dolomites: notes. *J. Sediment. Petrol.* 38:247–249.

Budd, D. A. (2001). Permeability loss with depth in the Cenozoic carbonate platform of west-central Florida. *AAPG Bull.* 85:1253–1272.

Burchette, T. P. and Wright, V. P. (1992). Carbonate ramp depositional systems. *Sediment. Geol.* 79:3–57.

Burke, J. A., Schmidt, A. W., and Campbell, R. L. Jr. (1969). The lithoporosity cross plot. *The Log Analyst* 10:25–43.

Carozzi, A. V. (1958). Micro-mechanisms of sedimentation in the epicontinental environment. *J. Sediment. Petrol.* 28:133–150.

Chave, K. (1967). Recent carbonate sediments: an unconventional view: *Geol. Educ.* 5:200–204.

Chilingar, G. V., Dominguez, G. C., Samaniego, V. F., Rieke, H. H., Mazzullo, S. J., and Chilingarian, G. V. (Eds.) (1992). *Carbonate Reservoir Characterization: A Geologic-Engineering Analysis.* Elsevier Developments in Petroleum Science, Vols. 30, 44. 1054 pp.

Choquette, P. W. and Pray, L. C. (1970). Geological nomenclature and classification of porosity in sedimentary carbonates. *AAPG Bull.* 54:207–250.

Coniglio, M. and Dix, G. R. (1992). Carbonate slopes. In: *Facies Models: Response to Sea Level Change*, R. G. Walker and N. P. James (Eds.). Geological Association of Canada, pp. 349–373.

Corbett, K. P., Friedman, M., Wiltschko, D. V., and Hung, J. H. (1991). *Controls on Fracture Development, Spacing, and Geometry in the Austin Chalk Formation, Central Texas: Considerations for Exploration and Production.* Dallas Geological Society, Field Trip #4, 49 pp.

Cossé, R. (1993). *Basics of Reservoir Engineering* (English Translation). Editions Technip, Paris, 342 pp.

Craze, R. C. (1950). Performance of limestone reservoirs. *Trans. AIME* 189:287–294.

Darcy, H. (1856). *Les fontaines publiques de la ville de Dijon.* Dalmont, Paris, 647 pp.

Davies, D. J., McInalley, A., and Barclay, F. (2003). Lithology and fluid prediction from amplitude versus offset (AVO) seismic data. *Geofluids* 3:219–232.

Dawson, W. C. (1988). Stylolite porosity in carbonate reservoirs. *AAPG Bull.* 72:176 (Abst.).

Dickson, J. A. D. and Saller, A. H. (1995). Identification of subaereal exposure surfaces and porosity preservation in Pennsylvanian and Lower Permian shelf limestones, eastern Central Basin Platform, Texas. In: *Unconformities and Porosity in Carbonate Strata,* D. A. Budd, A. H. Saller, and P. M. Harris (Eds.). AAPG Memoir No. 63, Tulsa, OK, pp. 239–257.

Dickson, J. A. D., Montanez, I. P., and Saller, A. H. (2001). Hypersaline burial diagenesis delineated by component isotopic analysis, Late Paleozoic limestones, West Texas. *J. Sediment. Res.* 71:372–379.

Dravis, J. J. (1989). Deep-burial microporosity in Upper Jurassic Haynesville oolitic grainstones. *East Texas Sediment. Geol.* 63:325–341.

Dunham, R. J. (1962). Classification of carbonate rocks according to depositional texture. In: *Classification of Carbonate Rocks,* W. E. Ham (Ed.). AAPG Memoir No. 1, Tulsa, OK, pp. 108–121.

Dunham, R. J. (1970). Stratigraphic reefs versus ecologic reefs. *AAPG Bull.* 54:1931–1932.

Ebanks, W. J. Jr. (1987). Flow unit concept—integrated approach to reservoir description for engineering projects. *AAPG Bull.* 71:551–552 (Abst.).

Ebanks, W. J. Jr., Scheihing, M. H., and Atkinson, C. D. (1992). Flow units for reservoir characterization. In: *Development Geology Manual.* AAPG Methods in Exploration Series No. 10, Tulsa, OK, pp. 282–285.

Ehlers, E. G. and Blatt, H. (1982). *Petrology: Igneous, Sedimentary, and Metamorphic.* W. H. Freeman, San Francisco, 732 pp.

Embry, A. F. and Klovan, J. E. (1971). A Late Devonian reef tract on northeastern Banks Island, NWT. *Bull. Can. Petroleum Geol.* 19:730–781.

Emery, D. and Meyers, K. (Eds.). (1996) *Sequence Stratigraphy.* Blackwell Science, London, 304 pp.

Enos, P. and Perkins, R. D. (1977). *Quaternary Sedimentation in South Florida.* Geological Society of America Memoir No. 147, 198 pp.

Erwin, C. R., Eby, D. E., and Whitesides, V. S. (1979). Clasticity index: a key to correlating depositional and diagenetic environments of Smackover reservoirs, Oaks Field, Claiborne Parish, Louisiana. *Trans. Gulf Coast Assoc. Geol. Soc.* 29:52–62.

Feazel, C. T., Knight, I. A., and Pekot, L. J. (1990). Ekofisk Field—Norway, Central Graben, North Sea. In: *AAPG Treatise of Petroleum Geology, Structural Traps IV: Tectonic and Non-tectonic Fold Traps,* AAPG, Tulsa, OK, pp. 1–25.

Flugel, E. (1982). *Microfacies Analysis of Limestones,* Springer-Verlag, Berlin, 633 pp.

Focke, J. W. and Munn, D. (1987). Cementation exponents in Middle Eastern carbonate reservoirs. *SPE Formation Evaluation* 2:155–167.

Folk, R. L. (1959). Practical petrographic classification of limestones. *AAPG Bull.* 43: 1–38.

Folk, R. L. (1962). Spectral subdivision of limestone types. In: *Classification of Carbonate Rocks*, W. E. Ham (Ed.). AAPG Memoir No. 1, Tulsa, OK, pp. 62–84.

Folk, R. L. (1965). Some aspects of recrystallization in ancient limestones. In: *Dolomitization and Limestone Diagenesis*, L. C. Pray and R. C. Murray (Eds.). SEPM Special Publication No. 13, pp. 14–48.

Folk, R. L. (1974). The natural history of crystalline calcium carbonate: effect of magnesium content and salinity. *J. Sediment. Petrol.* 44:40–53.

Fornos, J. J. and Ahr, W. M. (1997). Temperate carbonates on a low-energy, isolated ramp: the Balearic Platform, Spain. *J. Sediment. Res.* 67:364–373.

Fornos, J. J. and Ahr, W. M. (2006). Present-day temperate carbonate sedimentation on the Balearic Platform, Western Mediterranean: compositional variation along a low-energy isolated ramp. In: *Cool-Water Carbonates: Depositional Systems and Palaeoenvoironmental Controls*, H. M. Pedley and G. Carannante (Eds.). Geological Society, London, Special Publication No. 255, pp. 71–84.

Fouret, K. L. (1996). Depositional and diagenetic environments of the Mississippian Leadville Limestone at Lisbon Field, Utah. In: *Geology and Resources of the Paradox Basin*, A. C. Huffman, W. R. Lund, and L. H. Godwin (Eds.). Utah Geological Association Guidebook 25, pp. 129–138.

Freeman, J. C. (1964). The Conley field, Hademan County, Texas. *Tulsa (OK) Geol. Soc. Digest*, 32:126–130.

Fretwell, P. N. (1994). Some aspects of intra-ooid microporosity in limestones. Unpublished Ph.D. thesis, Department of Earth Sciences, University of Cambridge, England, 205 pp.

Friedman, M. and Wiltschko, D. V. (1992). An approach to exploration for naturally fractured reservoirs, with examples from the Austin Chalk. In: *Geological Studies Relevant to Horizontal Drilling; Examples from Western North America*, J. W. Schmocker, E. B. Coalson, and C. A. Brown (Eds.). Rocky Mountain Association of Geologists, pp. 143–153.

Garber, M. G. (2001). Borehole-image applications in Silurian, Devonian, and Mississippian Midcontinent reservoirs. *Oklahoma Geol. Surv. Circular* 105:151–161.

Genty, C., Jensen, J. L., and Ahr, W. M. (2007). Distinguishing carbonate reservoir pore facies with nuclear magnetic resonance measurements. *Natural Resources Res.* 16:45–54.

Ginsburg, R. N. (1956). Environmental relationships of grain size and constituent particles in some South Florida carbonate sediments. *AAPG Bull.* 40:2384–2427.

Ginsburg, R. N. and James, N. P. (1974). Holocene carbonate sediments of continental margins. In: *The Geology of Continental Margins*, C. A. Burke and C. L. Drake (Eds.), Springer-Verlag, Berlin, pp. 137–155.

Goldstein, R. H. (1988). Cement stratigraphy of the Pennsylvanian Holder Formation, Sacramento Mountains, New Mexico. *AAPG Bull.* 72:425–438.

Grabau, A. W. (1960). *Principles of Stratigraphy*. Dover Publications, New York, in two volumes, 1185 pp. Reprinted from 1924 edition.

Grace, L. M. and Pirie, R. G. (1986). Stratigraphic interpretation using formation imaging and dipmeter analysis. *Soc. Petroleum Eng.* Paper No. 15611, pp. 1–4.

Grover, G. and Read, J. F. (1983). Paleoaquifer and deep burial related cement defined by regional cathodoluminescent patterns, Middle Ordovician carbonates, Virginia. *AAPG Bull.* 67:1275–1303.

Gunter, G. W., Finneran, J. M., Hartmann, D. J., and Miller, J. D. (1997). Early determination of reservoir flow units using an integrated petrophysical method. *Soc. Petroleum Eng.* Paper No. 38679, pp. 373–380.

Ham, W. E. and Pray, L. C. (1962). Modern concepts and classifications of carbonate rocks. In: *Classification of Carbonate Rocks*. AAPG Memoir No. 1, Tulsa, OK, pp. 2–19.

Handford, C. R. and Loucks, R. G. (1993). Carbonate depositional sequences and systems tracts; responses of carbonate platforms to relative sea-level changes. In: *Carbonate Sequence Stratigraphy: Recent Developments and Applications*, AAPG Memoir No. 57, Tulsa, OK, pp. 3–41.

Hardie, L. A. (1967). The gypsum–anhydrite equilibrium at one atmosphere pressure. *Am. Mineralogist* 52:171–200.

Hashmy, K. and Alberty, M. (1992). Difficult lithologies. In: *Development Geology Manual*. AAPG Methods in Exploration Series No. 10, Tulsa, OK, pp. 186–191.

Helgeson, H. C. (1968). Geologic and thermodynamic characteristics of the Salton Sea geothermal system. *Am. J. Sci.* 266:129–166.

Herrington, P. M., Pederstad, K., and Dickson, J. A. D. (1991). Sedimentology and diagenesis of resedimented and rhythmically bedded chalks from the Eldfisk Field, North Sea Central Graben. *AAPG Bull.* 75:1661–1674.

Hodgkins, M. A. and Howard, J. J. (1999). Application of NMR logging to reservoir characterization of low-resistivity sands in the Gulf of Mexico. *AAPG Bull.* 83:114–127.

Hoefs, J. (2004). *Stable Isotope Geochemistry*. Springer, New York, 244 pp.

Holtz, M. H. and Major, R. P. (2004). Integrated geological and petrophysical characterization of shallow-water dolostone. *SPE Reservoir Evaluation Eng.* 7:47–58.

Hubbert, M. K. (1940). The theory of ground-water motion. *J. Geol.* 48:785–944.

Hughes, D. J. (1968). Salt tectonics as related to several Smackover fields along the northeast rim of the Gulf of Mexico basin. *Trans. Gulf Coast Assoc. Geol. Soc.* 18:320–330.

Hulen, J. B., Kaspereit, D., Norton, D. L., Osborn, W., and Pulka, F. S. (2003). Refined conceptual modeling and a new resource estimate for the Salton Sea geothermal field, Imperial Valley, California. www.saltonsea.ca.gov/ltnav/library_content/PhysicalGeography/geothermalresourcesatss.pdf.

Hurlbut, C. and Klein, C. (1977). *Manual of Mineralogy*. 21st ed. John Wiley & Sons, New York, 681 pp.

James, N. P. (1983). The reef environment. In: *Carbonate Depositional Environments*. AAPG Memoir No. 33, Tulsa, OK, pp. 416–440.

James, N. P. and Mountjoy, E. W. (1983). *Shelf-Slope Break in Fossil Carbonate Platforms: An Overview*. SEPM Special Publication No. 33, pp. 189–206.

James, N. P., Bone, Y., Von Der Borch, C. C., and Gostin, V. A. (1992). Modern carbonate and terrigenous clastic sediments on a cool-water, high-energy, mid-latitude shelf: Lacepede, southern Australia. *Sedimentology* 39:877–903.

James, N. P. (1997). The cool water depositional realm. In: *Cool Water Carbonates*. SEPM Special Publication No. 56, pp. 1–20.

James, N. P. and Clarke, J. (1997). *Cool Water Carbonates*. SEPM Special Publication No. 56, 440 pp.

Kerans, C. and Loucks, R. G. (2002). Stratigraphic setting and controls on occurrence of high-energy carbonate beach deposits: Lower Cretaceous of the Gulf of Mexico. *Trans. Gulf Coast Assoc. Geol. Soc.* 52:517–526.

Kohout, F. A. (1967). Ground-water flow and the geothermal regime of the Floridian Plateau. *Trans. Gulf Coast Assoc. Geol. Soc.* 17:339–354.

Kopaska-Merkel, D. C. (1991). *Capillary-Pressure Characteristics of Smackover Reservoirs in Alabama*. Geological Survey of Alabama Circular No. 170, 38 pp.

Kopaska-Merkel, D. C. and Mann, S. D. (1993). Upward shoaling cycles in Smackover carbonates of southwest Alabama. *Gulf Coast Assoc. Geol. Soc. Trans.* 43:173–181.

Kosa, E., Hunt, D., Fitchen, W. M., Bockel-Rebelle, M. O., and Roberts, G. (2003). The heterogeneity of paleocavern systems developed along syndepositional fault zones: the Upper Permian Capitan Platform, Guadalupe Mountains, U.S.A. In: *Permo-Carboniferous Carbonate Platforms and Reefs*, W. M. Ahr, P. M. Harris, W. A. Morgan, and I. Somerville (Eds.). SEPM Special Publication No. 78 and AAPG Memoir No. 83, Tulsa, OK, pp. 291–322.

Layman, J. and Ahr, W. M. (2003). Porosity characterization utilizing petrographic image analysis: implications for identifying and ranking reservoir flow units, Happy Spraberry field, Garza County, Texas. In: *The Permian Basin: Preserving Our Past and Securing Our Future*, T. Hunt and P. Lufholm (Eds.). West Texas Geological Society Symposium Proceedings, Publication No. 02-111, pp. 107–113.

Laubach, S. E. (1988). Subsurface fractures and their relationship to stress history in east Texas basin sandstone. *Tectonophysics* 156:37–49.

Laubach, S. E. (1997). A method to detect fracture strike in sandstone. *AAPG Bull.* 81:604–623.

Laubach, S. E., Reed, R. M., Gale, J. F. W., Ortega, O. J., and Doherty, E. H. (2002). Fracture characterization based on microfracture surrogates, Pottsville Sandstone, Black Warrior Basin, Alabama. *Trans. Gulf Coast Assoc. Geol. Soc.* 52:585–596.

LeBlanc, R. J. (1972). The geometry of sandstone reservoir bodies. In: *Underground Waste Management and Environmental Implications*. AAPG Memoir No. 18, Tulsa, OK, pp. 133–189.

Lee, W. J. (1992). Pressure transient testing: Part 9. Production engineering methods. In: *Development Geology Reference Manual*. AAPG Methods in Exploration No. 10, Tulsa, OK, pp. 477–481.

Lees, A., and Buller, A. T. (1972). Modern temperate-water and warm-water shelf carbonate sediments contrasted. *Mar. Geol.* 13:67–73.

Lees, A. (1975). Possible influence of salinity and temperature on modern shelf carbonate sedimentation. *Mar. Geol.* 19:159–198.

Lochman-Balk, C. and Wilson, J. L. (1958). Cambrian biostratigraphy in North America. *J. Paleontol.* 32:312–350.

Logan, B. W., Harding, J. L., Ahr, W. M., Williams, J. D., and Snead, R. G. (1969). *Carbonate Sediments and Reefs, Yucatan Shelf, Mexico*. AAPG Memoir No. 11, Tulsa, OK, pp. 1–198.

Logan, B. W., Read, J. F., and Davies, G. R. (1970). *Carbonate Sedimentation and Environments, Shark Bay, Western Australia*. AAPG Memoir No. 13, Tulsa, OK, 223 pp.

Longman, M. W. (1980). Carbonate diagenetic textures from nearsurface diagenetic environments. *AAPG Bull.* 64:461–487.

Lonoy, A. (2006). Making sense of carbonate pore systems. *AAPG Bull.* 90:1381–1405.

Lorenz, J. C., Teufel, L. W., and Warpinski, N. R. (1991). Regional fractures, I: a mechanism for the formation of regional fractures at depth in flat-lying reservoirs. *AAPG Bull.* 75:1714–1737.

Lorenz, J. C., Farrell, H. E., Hanks, C. L., Rizer, W. D., and Sonnenfeld, M. D. (1997). Characteristics of natural fractures in carbonate strata. In: *Carbonate Seismology*, I. Palaz and K. J. Marfurt (Eds.). Geophysical Developments Series No. 6, SEG, Tulsa, OK, pp. 179–201.

Loucks, R. G. and Sarg, J. R. (Eds.) (1993). *Carbonate Sequence Stratigraphy*. AAPG Memoir No. 57, Tulsa, OK, 545 pp.

Loucks, R. G. (1999). Paleocave carbonate reservoirs: origins, burial-depth modifications, spatial complexity, and reservoir implications. *AAPG Bull.* 83:1795–1834.

Loucks, R. G., Mescher, P. K., and McMechan, G. A. (2004). Three-dimensional architecture of a coalesced, collapsed-paleocave system in the Lower Ordovician Ellenburger Group, central Texas. *AAPG Bull.* 88:545–564.

Lucia, F. J. (1983). Petrophysical parameters estimated from visual descriptions of carbonate rocks: a field classification of carbonate pore space. *J. Petroleum Technol.* pp. 629–637.

Lucia, F. J. (1995). Rock-fabric/petrophysical classification of carbonate pore space for reservoir characterization. *AAPG Bull.* 79:1275–1300.

Lucia, F. J. (1999). *Carbonate Reservoir Characterization.* Springer-Verlag, Berlin, 226 pp.

Lucia, F. J. (2000). Dolomitization: a porosity-destructive process (Abst.). *AAPG Bull.* 84:1879.

Lucia, F. J. and Major, R. P. (1994). Porosity evolution through hypersaline reflux dolomitization. In: *Dolomites, A Volume in Honor of Dolomieu*, B. H. Purser, M. E. Tucker, and D. H. Zenger (Eds.) International Associational Sedimentologists, Special Publication No. 21, pp. 325–341.

Machel, H. G. (1987a). Cathodoluminescence activation and zonation in carbonate rocks; an experimental approach. *Geol. Mijnbouw* 66:67–74.

Machel, H. G. (1987b). Saddle dolomite as a by-product of chemical compaction and thermochemical sulfate reduction. *Geology* 15:936–940.

Machel, H. G. (2005). Investigations of burial diagenesis in carbonate hydrocarbon reservoir rocks: *Geosci. Can.* 32:103–128.

Magoon, L. B. and Dow, W. G. (1994). The petroleum system: Chapter 1, Introduction. In: *The Petroleum System—From Source to Trap.* AAPG Memoir No. 60, pp. 3–24.

Manfrino, C. and Ginsburg, R. N. (2001). Pliocene to Pleistocene depositional history of the upper platform margin. In: *Subsurface Geology of a Prograding Carbonate Platform Margin, Great Bahama Bank; Results of the Bahamas Drilling Project*, R. N. Ginsburg (Ed.). SEPM Special Publication No. 70, pp. 17–39.

Major, R. P. and Holtz, M. H. (1997). Predicting reservoir quality at the development scale: methods for quantifying remaining hydrocarbon resource in diagenetically complex carbonate reservoirs. In: *Reservoir Quality Prediction in Sandstones and Carbonates*, J. A. Kupecz, J. Gluyas, and S. Bloch (Eds.). AAPG Memoir No. 69, Tulsa, OK, pp. 231–248.

Martin, A. J., Solomon, S. T., and Hartmann, D. J. (1997). Characterization of petrophysical flow units in carbonate reservoirs. *AAPG Bull.* 81:734–759.

Mazzullo, S. J. (1994). Lithification and porosity evolution in Permian periplatform limestones, Midland Basin, Texas. *Carbonates Evaporites* 9:151–171.

Mazzullo, S. J. (2000). Organogenic dolomitization in peritidal to deep-sea sediments: *J. Sediment. Res.* 70:10–23.

Mazzullo, S. J. and Harris, P. M. (1992). Mesogenetic dissolution: its role in porosity development in carbonate reservoirs. *AAPG Bull.* 76:607–620.

McIlreath, I. A. and Morrow, D. W. (1990). *Diagenesis.* Geoscience Canada Reprint Series No. 4, 338 pp.

McKee, E. D. (1966). *Structures of Dunes at White Sands National Monument, New Mexico (and a Comparison with Structures of Dunes from Other Selected Areas).* Blackwell Science, Cambridge, MA.

McKee, E. D. and Gutschick, R. E. (1969). *History of the Redwall Limestone of Northern Arizona.* Geological Society of America Memoir No. 114, 726 pp.

McKee, E. D. and Ward, W. C. (1983). Eolian environment. In: *Carbonate Depositional Environments.* AAPG Memoir No. 33, Tulsa, OK, pp. 131–170.

Meyers, W. J. (1974). Carbonate cement stratigraphy of the Lake Valley Formation (Mississippian) Sacramento Mountains, New Mexico. *J. Sediment. Petrol.* 44:837–861.

Meyers, W. J. (1980). Compaction in Mississippian skeletal limestones, southwestern New Mexico. *J. Sediment. Petrol.* 50:457–474.

Milliman, J. D. (1974). *Recent Sedimentary Carbonates*. Springer-Verlag, Berlin, 375 pp.

Milsom, J. (2003). *Field Geophysics*, 3rd ed. The Geological Field Guide Series, John Wiley & Sons, Chichester, UK, 204 pp.

Monicard, R. P. (1980). *Properties of Reservoir Rocks; Core Analysis*. Gulf Publishing, Houston, 165 pp.

Montañez, I. P. and Read, J. F. (1992). Fluid–rock interaction history during stabilization of early dolomites, upper Knox Group (Lower Ordovician), U. S. Appalachians. *J. Sediment. Petrol.* 62:753–778.

Montgomery, S. L. (1996). Mississippian Lodgepole Play, Williston Basin: a review. *AAPG Bull.* 80:795–810.

Moore, C. H. (2001). *Carbonate Reservoirs: Porosity Evolution and Diagenesis in a Sequence Stratigraphic Framework*. Elsevier, Amsterdam, 444 pp.

Moore, C. H. and Druckman, Y. (1981). Burial diagenesis and porosity evolution, Upper Jurassic Smackover, Arkansas and Louisiana. *AAPG Bull.* 65:597–628.

Moore, H. B. (1939). Fecal pellets in relation to marine deposits. In: *Recent Marine Sediments*, P. D. Trask (Eds.). AAPG, Tulsa, OK, pp. 516–524.

Morrow, D. W. (1990). Dolomite—Part 1: the chemistry of dolomite and dolomite precipitation. In: *Diagenesis*, I. A. McIlreath and D. W. Morrow (Eds.). Geoscience Canada Reprint Series No. 4, pp. 113–123.

Morrow, D. W. (2001). Distribution of porosity and permeability in platform dolomites: insight from the Permian of West Texas: discussion. *AAPG Bull.* 85:525–532.

Morse, J. W. and Mackenzie, F. T. (1990). *Geochemistry of Sedimentary Carbonates*. Developments in Sedimentology Series No. 48, Elsevier, Amsterdam, 707 pp.

Morton-Thompson, D. and Woods, A. M. (1992). *AAPG Development Geology Reference Manual*. AAPG Methods in Exploration Series No. 10, Tulsa, OK, 550 pp.

Moshier, S. O. (1989). Microporosity in micritic limestones; a review. *Sediment. Geol.* 63: 191–213.

Mowers, T. T. and Budd, D. A. (1996). The quantification of porosity and permeability reduction due to calcite cementation using computer-assisted petrographic image analysis techniques. *AAPG Bull.* 80:309–322.

Murray, R. C. (1960). The origin of porosity in carbonate rocks. *J. Sediment. Petrol.* 30:59–84.

Nelson, R. A. (1981). Significance of fracture sets associated with stylolite zones. *AAPG Bull.* 65:2417–2425.

Nelson, C. S., Hancock, G. E., and Kamp, P. J. J. (1982). Shelf to basin, temperate skeletal carbonate sediments, Three Kings Plateau, New Zealand. *J. Sediment. Petrol.* 52:717–732.

Nelson, R. A. (1985). *Geologic Analysis of Naturally Fractured Reservoirs*. Gulf Publishing, Houston, 320 pp.

Nelson, R. A. (2001). *Geologic Analysis of Naturally Fractured Reservoirs*, 2nd ed. Gulf Publishing, Houston, 332 pp.

Neugebauer, J. (1973). Fossil-diagenese, No. 3; the diagenetic problem of chalks; the role of pressure solution and pore fluid. *Neues Jahrb. Geol. Palaeontol. Abh.* 143:223–245.

Neugebauer, J. (1975). Fossil-diagenese in der Schreibkreide: coccolithen: Neues Jahrb. *Geol. Palaeontol. Monatsch.*, 489–502.

North, F. K. (1985). *Petroleum Geology*. Allen & Unwin, London, 607 pp.

Newbury, D., Kchlin, P., Joy, D., Lyman, C., Lifshin, E., Sawyer, L., Michael, J., and Staniforth, M. (2003). *Scanning Electron Microscopy and X-Ray Microanalysis*, 3rd ed., J. Goldstein (Ed.). Kluwer Academic/Plenum Publishers, New York.

Nugent, W. H. (1984). Letters to the Editor. *The Log Analyst* 25:2–3.

Olson, T. M., Babcock, J. A., Prasad, K. V. K., Boughton, S. D., Wagner, P. D., Franklin, M. H., and Thompson, K. A. (1997). Reservoir characterization of the giant hugoton gas field, Kansas. *AAPG Bull.* 81:1785–1803.

Palaz, I. and Marfurt, K. J. (Eds.) (1997). *Carbonate Seismology*. Geophysical Developments Series No. 6, SEG, Tulsa, OK, 443 pp.

Parsons, R. W. (1966). Permeability of idealized fractured rock. *Soc. Petroleum Eng.* Paper No. 1289, pp. 126–136.

Pettijohn, F. J. (1975). *Sedimentary Rocks*, 3rd ed. Harper & Row, New York, 628 pp.

Pickell, J. J., Swanson, B. F., and Hickman, W. B. (1966). Application of air–mercury and oil–air capillary pressure data in the study of pore structure and fluid distribution. *Soc. Petroleum Eng.* Paper No. 1227, pp. 55–61.

Pittman, E. D. (1992). Relationship of porosity and permeability to various parameters derived from mercury-injection capillary pressure curves fro sandstone. *AAPG Bull.* 76:191–198.

Pomar, L. (1993). High-resolution sequence stratigraphy in prograding Miocene carbonates; application to seismic interpretation. In: *Carbonate Sequence Stratigraphy: Recent Developments and Applications*. R. G. Loucks and J. F. Sarg (Eds.). AAPG Memoir No. 57, Tulsa, OK, pp. 389–407.

Pomar, L. and Ward, W. C. (1999). Reservoir-scale heterogeneity in depositional packages and diagenetic patterns on a reef-rimmed platform, upper Miocene, Mallorca, Spain. *AAPG Bull.* 83:1759–1773.

Posamentier, H. W. and James, D. P. (1993). An overview of sequence stratigraphic concepts: uses and abuses. In: *Sequence Stratigraphy and Facies Associations*, H. W. Posamentier, C. P. Summerhayes, B. U. Haq, and G. P. Allen (Eds.). International Association of Sedimentologists Special Publication No. 18, pp. 3–18.

Potter, P. E. (1963). *Late Paleozoic Sandstones of the Illinois Basin*. Report of Investigations No. 217, Illinois Geological Survey, 92 pp.

Purcell, W. R. (1949). Capillary pressures—their measurement using mercury and the calculation of permeability therefrom. *Petroleum Trans. AIME*, Feb. Issue, pp. 39–48.

Purdy, E. G. (1963a). Recent calcium carbonate facies of the Great Bahama Bank: Part 1, petrography and reaction groups. *J. Geol.* 71:334–355.

Purdy, E. G. (1963b). Recent calcium carbonate facies of the Great Bahama Bank: Part 2, sedimentary facies. *J. Geol.* 71:472–497.

Purdy, E. G. (1974). Karst-determined facies patterns in British Honduras: Holocene carbonate dedimentation model. *AAPG Bull.* 58:825–855.

Purdy, E. G. and Waltham, D. (1999). Reservoir implications of modern karst topography. *AAPG Bull.* 83:1774–1794.

Purser, B. H. (1980). *Sédimentation et diagenèse des carbonates néritiques récents*, in 2 vols. Editions Technip, Paris.

Radke, B. M. and Mathis, R. L. (1980). On the formation and occurrence of saddle dolomite. *J. Sediment. Petrol.* 50:1149–1168.

Rasmus, J. C. (1983). A variable cementation exponent, m for fractured carbonates. *The Log Analyst* 24:13–23.

Read, J. F. (1982). Carbonate platforms of passive (extensional) continental margins; types, characteristics and evolution. *Tectonophysics* 81:195–212.

Read, J. F. (1985). Carbonate platform facies models. *AAPG Bull.* 66:860–878.

Reading, H. G. (Ed.) (1996). *Sedimentary Environments: Processes, Facies, and Stratigraphy*, 3rd ed. Blackwell Science, Cambridge, MA, 688 pp.

Reeder, R. J. (Ed.) (1983). *Carbonates: Mineralogy and Chemistry*. Reviews in Mineralogy, Vol. 11, Mineralogical Society of America, 394 pp.

Reineck, H. E. and Singh, I. B. (1973). *Depositional Sedimentary Environments*. Springer-Verlag, New York, 439 pp.

Rhett, D. (2001). Pore pressure controls on the origin of regional fractures: experimental verification of a model. *AAPG Bull.* 85:1–7:(Abst.).

Rider, M. (1996). *Geological Interpretation of Well Logs*, 2ṇd ed. Whittles Publishing, Latheronhouse, UK, 280 pp.

Riding, R. (2002). Structure and composition of organic reefs and carbonate mud mounds; concepts and categories. *Earth-Sci. Rev.* 58:163–231.

Roedder, E. (1984). Fluid inclusions: an introduction to studies of all types of fluid inclusions, gas, liquid, or melt, trapped in materials from earth and space, and their application to the understanding of geologic processes. In: *Reviews in Mineralogy*, Vol. 12. Mineralogical Society of America, 644 pp.

Roehl, P. O. (1967). Stony Mountain (Ordovician) and Interlake (Silurian) facies analogs of recent low-energy marine and subaerial carbonates, Bahamas. *AAPG Bull.* 51:1979–2032.

Roehl, P. O. and Choquette, P. W. (1985). *Carbonate Petroleum Reservoirs*. Springer-Verlag, New York, 622 pp.

Ruppel, S. C. (1992). Controls of platform development in the Leonard Series (Middle Permian) of West Texas; significance of multifrequency cyclicity and paleotopography. *AAPG Annu. Mtg. Absts.*, p. 112.

Saller, A. H., Dickson, J. A. D., and Matsuda, F. (1999). Evolution and distribution of porosity associated with subaerial exposure in Upper Paleozoic Platform Limestones, West Texas. *AAPG Bull.* 83:1835–1854.

Saller, A. H. and Henderson, N. (1998). Distribution of porosity and permeability in platform dolomites: insight from the Permian of West Texas. *AAPG Bull.* 82:1528–1550.

Saller, A., Barton, J., and Barton, R. E. (1989). Slope sedimentation associated with a vertically building shelf, Bone Spring Formation, Mescalero Escarpe Field, southeastern New Mexico. In: *Controls on Carbonate Platform and Basin Development*, P. D. Crevello, J. L. Wilson, J. F. Sarg, and J. F. Read (Eds.). SEPM Special Publication No. 44, pp. 275–288.

Sarg, J. F. (1988). Carbonate sequence stratigraphy. In: *Sea-Level Changes: An Integrated Approach*. C. K. Wilgus, B. S. Hastings, C. A. Ross, H. Posamentier, and J. Van-Wagoner (Eds.). SEPM Special Publication No. 42, pp. 155–181.

Schenk, H. G. and Muller, S. W. (1941). Stratigraphic terminology. *Geol. Soc. Am. Bull.* 52:1419–1426.

Schlager, W. (1981). The paradox of drowned reefs and carbonate platforms. *Geol. Soc. Am. Bull.* 92:197–211.

Schlager, W. and Ginsburg, R. N. (1981). Bahama carbonate platforms—the deep and the past. *Marine Geol.* 44:1–24.

Schmoker, J. W. and Halley, R. B. (1982). Carbonate porosity versus depth; a predictable relation for South Florida. *AAPG Bull.* 66:2561–2570.

Scholle, P. A. (1977). Chalk diagenesis and its relation to petroleum exploration: oil from chalks, a modern miracle? *AAPG Bull.* 61:982–1009.

Scholle, P. A. (1978). 35 mm slides to accompany *A Color Illustrated Guide to Carbonate Rock Constituents, Textures, Cements, and Porosities*. AAPG Memoir No. 27, AAPG Publications, Tulsa, OK, 241 pp.

Scholle, P. A., Bebout, D. G., and Moore, C. H. (Eds.) (1983). *Carbonate Depositional Environments*. AAPG Memoir No. 33, Tulsa, OK, 708 pp.

Scholle, P. A. and Ulmer-Scholle, D. (2004). *A Color Guide to the Petrography of Carbonate Rocks: Grains, Textures, Porosity, Diagenesis.* AAPG Memoir No. 77, AAPG Publications, Tulsa, OK, 474 pp.

Scoffin, T. P. (1987). *An Introduction to Carbonate Sediments and Rocks.* Blackie, 274 pp.

Scott, R. W., Simo, J. A., and Masse, J. P. (1993). Economic resources in Cretaceous carbonate platforms: an overview. In: *Cretaceous Carbonate Platforms*, J. A. Simo, R. W. Scott, and J. P. Masse (Eds.). AAPG Memoir No. 56, Tulsa, OK, pp. 15–23.

Selley, R. C. (1985). *Elements of Petroleum Geology.* W. H. Freeman and Co., New York, 449 pp.

Sheriff, R. E. (2002). *Encyclopedic Dictionary of Applied Geophysics*, 4th ed. Society Exploration Geophysicists, Tulsa, OK, 429 pp.

Shinn, E. A. (1963). Spur and groove formation on the Florida reef tract. *J. Sediment. Petrol.* 33:291–303.

Shinn, E. A. (1969). Submarine lithification of Holocene, carbonate sediments in the Persian Gulf. *Sedimentology* 12:109–144.

Shinn, E. A., Ginsburg, R. N., and Lloyd, R. M. (1969). Recent supratidal dolomite from Andros Island, Bahamas. In: *Dolomitization and Limestone Diagenesis—A Symposium*, L. C. Pray and R. C. Murray (Eds.). SEPM Special Publication No. 13, pp. 112–123.

Shively, M. L. (1991). Analysis of mercury porosimetry for the evaluation of pore shape and intrusion–extrusion hysteresis. *J. Pharm. Sci.* 80:376–379.

Siemers, W. T. and Ahr, W. M. (1990). Reservoir facies, pore characteristics, and flow units; Lower Permian Chase Group, Guymon–Hugoton Field, Oklahoma. *Soc. Petroleum Eng.* Paper No. 20757, pp. 417–428.

Sloss, L. L. (1963). Sequences in the cratonic interior of North America. *Geol. Soc. Am. Bull.* 74:93–113.

Smart, P. L. and Whitaker, F. (2003). Are models of collapsed paleocave breccias based on continental caves adequate? 12th Annual Conference of Carbonate Sedimentologists (Bathurst Mtg.), Durham, England, Abst. p. 102.

Sneider, R. M. (1988). Practical petrophysics for exploration and development. Unpublished short course notes, Robert M. Sneider Exploration, Houston.

Stearns, D. W. (1968). Certain aspects of fracture in naturally deformed rocks. In: *NSF Advanced Science Seminar In Rock Mechanics*, R. E. Rieker (Eds.). Special Report, Air Force Cambridge Research Laboratories, Bedford, MA, pp. 97–118.

Stearns, D. W. and Friedman, M. (1972). Reservoirs in fractured rocks. In: *Stratigraphic Oil and Gas Fields—Classification, Exploration Methods, and Case Histories*, R. E. King (Ed.). AAPG Memoir No. 16, Tulsa, OK, pp. 82–106.

Stricklin, F. L. (1973). Environmental reconstruction of a carbonate beach complex—Cow Creek (Lower Cretaceous) Formation, Central Texas. *Geol. Soc. Am.* 84:1349–1367.

Sun, S. Q. (1995). Dolomite reservoirs; porosity evolution and reservoir characteristics. *AAPG Bull.* 79:186–204.

Telford, W. M., Geldart, L. P., and Sheriff, R. E. (1991). *Applied Geophysics*, 2nd ed. Cambridge University Press, Cambridge, UK, 790 pp.

Tiab, D. and Donaldson, E. C. (2004). *Petrophysics*, 2nd ed. Gulf Professional Publishing Houston, 889 pp.

Tinker, S. W. (1996). Building the 3-D jigsaw puzzle: applications of sequence stratigraphy to 3-D reservoir characterization, Permian Basin. *AAPG Bull.* 80:460–485.

Tucker, M. (Ed.) (1988). *Techniques in Sedimentology.* Blackwell Scientific, Oxford, 394 pp.

Tucker, M. and Wright, V. P. (1990). *Carbonate Sedimentology.* Blackwell Scientific, Oxford, 482 pp.

Ulmer-Scholle, D. S. and Scholle, P. A. (1994). Replacement of evaporites within the Permian Park City Formation, Bighorn Basin, Wyoming, USA. *Sedimentology* 41:1203–1222.

Ulmer-Scholle, D. S., Scholle, P. A., and Brady, P. V. (1993). Silicification of evaporites in Permian (Guadalupian) back-reef carbonates of the Delaware Basin, West Texas and New Mexico. *J. Sediment. Petrol.* 63:955–965.

Vail, P. R., Mitchum, R. M., and Thompson, S. III (1977a). Seismic stratigraphy and global changes of sea level: Part 4, global cycles of relative changes of sea level. In: *Seismic Stratigraphy—Applications to Hydrocarbon Exploration*, C. E. Payton (Ed.). AAPG Memoir No. 26, Tulsa, OK, pp. 63–81.

Vail, P. R., Todd, R. G., and Sangree, J. B. (1977b). Seismic stratigraphy and global changes of sea level: Part 5, chronostratigraphic significance of seismic reflections. In: *Seismic Stratigraphy—Applications to Hydrocarbon Exploration*, C. E. Payton (Ed.). AAPG Memoir No. 26, Tulsa, OK, pp. 99–116.

Van Wagoner, J. C., Posamentier, H. W., Mitchum, R. M., Vail, P. R., Sarg, J. F., Loutit, T. S., and Hardenbol, J. (1988). An overview of the fundamentals of sequence stratigraphy and key definitions. In: *Sea-Level Changes: An Integrated Approach*, C. K. Wilgus, B. S. Hastings, C. G. St. C. Kendall, H. W. Posamentier, C. A. Ross, and J. C. Van Wagoner (Eds.). SEPM Special Publication 42, pp. 39–45.

Van Wagoner, J. C., Mitchum, R. M., Campion, K. M., and Rahmanian, V. D. (1990). *Siliciclastic Sequence Stratigraphy in Well Logs, Cores, and Outcrops: Concepts for High-Resolution Correlation of Time and Facies*. AAPG Methods in Exploration Series No. 7, Tulsa, OK, 55 pp.

Vavra, C. L., Kaldi, J. G., and Sneider, R. M. (1992). Capillary pressure: In: *Development Geology Manual*. AAPG Methods in Exploration Series No. 10, Tulsa, OK, pp. 221–225.

Walker, R. G. and James, N. P. (1992). *Facies Models: Response to Sea Level Change*. Geological Association of Canada, St. Johns, 409 pp.

Walter, L. M. (1985). Relative reactivity of skeletal carbonates during dissolution; implications for diagenesis. In: *Carbonate Cements*, N. Schneidermann and P. M. Harris (Eds.). SEPM Special Publication No. 36, pp. 3–16.

Walther, J. (1894). *Einleitung in die Geologie als historische Wissenschaft, Bd. 3, Lithogenesis der Gegenwart*. G. Fischer, Jena, pp. 535–1055.

Wang, Z. (1997). Seismic properties of carbonate rocks. In: *Carbonate Seismology*, I. Palaz and K. J. Marfurt (Eds.), Geophysical Developments Series No. 6, SEG, Tulsa, OK, pp. 29–52.

Ward, W. C. (1975). Petrology and diagenesis of carbonate eolianites of northeastern Yucatan Peninsula, Mexico. In: *Belize Shelf: Carbonate Sediments, Clastic Sediments, Ecology*. AAPG Studies in Geology No. 2, Tulsa, OK, pp. 500–571.

Ward, W. C. (1997). Geology of coastal islands, northeastern Yucatán peninsula. In: *Geology and Hydrogeology of Carbonate Islands*, H. L. Vacher and T. Quinn (Eds.). Developments in Sedimentology No. 54, Elsevier Science, New York, pp. 275–298.

Wardlaw, N. C. (1976). Pore geometry of carbonate rocks as revealed by pore casts and capillary pressure. *AAPG Bull.* 60:245–257.

Wardlaw, N. C. (1979). Pore systems in carbonate rocks and their influence on hydrocarbon recovery efficiency. In: *Geology of Carbonate Porosity*, D. Bebout, G. Davies, C. H. Moore, and P. S. Scholle (Eds.). AAPG course notes 11, pp. 1–24.

Wardlaw, N. C. and Taylor, R. P. (1976). Mercury capillary pressure curves and the interpretation of pore structure and capillary behaviour in reservoir rocks. *Bull. Can. Petrol. Geol.* 24:225–262.

Wardlaw, N. C. and Cassan, J. P. (1978). Estimation of recovery efficiency by visual observation of pore systems in reservoir rocks. *Bull. Can. Petroleum Geol.* 26:572–585.

Wentworth, C. K. (1922). A scale of grade and class terms for clastic sediments. *J. Geol.* 30:377–392.

Weyl, P. K. (1960). Porosity through dolomitization—conservation-of-mass requirements. *J. Sediment. Petrol.* 30:85–90.

Whitaker, F. F. and Smart, P. L. (1998). Hydrology, geochemistry and diagenesis of fracture blue holes, South Andros, Bahamas. *Cave Karst Sci.* 25:75–82.

Wilson, J. L. (1970). Depositional facies across carbonate shelf margins. *Trans. Gulf Coast Assoc. Geol. Soc.* 20:229–233.

Wilson, J. L. (1974). Characteristics of carbonate-platform margins. *AAPG Bull.* 58: 810–824.

Wilson, J. L. (1975). *Carbonate Facies in Geologic History*. Springer-Verlag, New York, 471 pp.

Witkowski, F. W., Blundell, D. J., Gutteridge, P., Horbury, A. D., Oxtoby, N. H. and Qing, H. (2000). Video cathodoluminescence microscopy of diagenetic cements and its applications. *Mar. Petroleum Geol.* 17:1085–1093.

Wright, V. P. (1992). A revised classification of limestones. *Sediment. Geol.* 76:177–185.

Young, S. W., Camano, E., Jackson, D. B., Morgan, W. A., Sheedlo, M. K., and Ahr, W. M. (1998). North Dakota's Lodgepole play: a look at the reservoir and producing characteristics. *Soc. Petroleum Eng.* Paper No. 39804, pp. 441–456.

Yu, L. and Wardlaw, N. C. (1986a). The influence of wettability and critical pore-throat size ratio on snap-off. *J. Colloid Interface Sci.* 109:461–472.

Yu, L. and Wardlaw, N. C. (1986b). Mechanisms of nonwetting phase trapping during imbibition at slow rates. *J. Colloid Interface Sci.* 109:473–486.

INDEX

Accessibility, 34, 66, 72. *See also* Pore-to-pore throat (size) ratio
Accommodation, 79
Agha Jari field, Iran, 191
Amal field, Libya, 190
Analogs ("look-alikes"), 7, 156
Analysis, complete core, 58
Anatomy, depositional units, 89–90
Andros Island, Bahamas
  fracture-related caves on, 226–227
  tidal flats on, 117
Angle, contact (wettability), 62
Anhydrite, pore filling and replacement, 166, 248
Anhydrite-gypsum, stability relationships of, 166
Anoxia, 98
Anoxic, 130
Antecedent topography (bathymetry), types of, 82
Aquifers, groundwater, 4
API units (gamma ray log), 161
Arab D Formation, 61, 211
Archie
  cementation exponent (m), 10, 59
  equation, 59
  formation factor (F), 59
  resistivity index (I), 61
  resistivity ratio (R_t/R_o), 61
  saturation exponent (n), 62
  tortuosity exponent (a), 59
Architecture
  basin, 76
  reservoir, 77, 86, 203
  sequence, 102
Asmari field, 184
Asmari Limestone, Kirkuk field, 191, 240
Attached
  beach-dune succession, 96, 111
  shoreline, 82
Australia, 124. *See also* "Roaring 40s"
  Lacepede shelf, 125, 211
  Shark Bay, 203

Baffles, to reservoir flow, 1, 41, 107, 145
Balearic Islands, 115
Barriers, to reservoir flow, 1, 41, 107, 145, 250
Berm, storm, 112
Bioturbation, 15
Black Sea, 129
Bossier Shale Formation, isopach of, Overton field, Texas, 228
Boundaries
  facies, 77
  time, 77
Breccias, karst related, crackle and mosaic, 162
Brittle domain, 177. *See also* Material, behavior under stress

Geology of Carbonate Reservoirs: The Identification, Description, and Characterization of Hydrocarbon Reservoirs in Carbonate Rocks
By Wayne M. Ahr   Copyright © 2008 John Wiley & Sons, Inc.

Buildups, carbonate
  chemogenic, 82
  mounds, microbialite (Cambrian) Texas, 181
  "mud mounds", 148, 181
  mudstone-cementstone, Early Carboniferous,
      212, 247
Bunter Sandstone, Triassic (Germany), 87
Buttress and chute structures, 29, 125

Calcisiltite, 16
Caliche, 115
Cambro-Ordovician, North America, 117.
      See also Transcontinental Arch
Capillarity, 56
Capillary
  attraction, 63–64
  drainage curve, 71
  imbibition curve, 71
  injection curve, 71
Capillary pressure, 6–8, 17, 65, 67
  curves, 7, 64
  defined, 64–66
  mercury, measurements of (MICP), 7, 107,
      145, 205, 209
Carbonate(s)
  defined, 2
  eolianites, 111
  factory, 81, 102, 129
  lacustrine, nonmarine, 109
  marine. *See also* Carbonate, factory
  minerals, natural occurrences of, 4
  particles, *see* Constituents, carbonate
  precipitation, inorganic, 81
  production
      biogenic and chemogenic, 81
      principal zone of, 97, 102. *See also*
          Carbonate, factory
  rocks, classification of, 20–21, 25–30
Carbonate compensation depth (CCD), 98,
      129–130
  aragonite and calcite, 129
  factors determining depth of, 129
Carlsbad Caverns, New Mexico, 153
Capping facies, cycle, 28
Caves
  coastal zone, 162
  continental, 162
Cathode luminescence (CL), *see* Luminescence,
      cathode
CCD, *see* Carbonate compensation depth
Cements
  botryoidal, 167
  ferroan calcite, 169
  isopachous, 167–168
  meniscus, 168
  poikilotopic, 169
  pore-lining, 168

Central Basin Platform, Texas, 152, 211
Chalk
  Austin (Cretaceous) Texas, 132–133
      fracture patterns in, 185
  constituents of, 16, 131
  Ekofisk field, North Sea, reservoirs in, 185
  Europe, Middle East, and North America,
      examples in, 124, 131
  North Sea
      classification of, 132–133
      fractured reservoirs in, 240
      porosity and burial depth, 132
      typical age and depositional setting, 124
      turbidites, 213
Chalkification (degradational diagenesis), 150
Chert and chalcedony, 131, 238, 243. *See also*
      Facies, basinal
"Chicken wire" fabric, *see* Environments, tidal
      flat and lagoon
China, mainland, 109
Clay minerals, K, Th, and U in, 202
Coccolithophorids (coccoliths), 16, 131. *See also*
      Chalk
Condensed interval, 100
Conformities
  correlative, 85
  stratigraphic, 101
Conglomerates, flat pebble, *see* Environments,
      tidal flat and lagoon
Conley field (Mississippian) Texas, 124
  Chappel Formation in, 219–221, 223
  depositional reservoir, as example of, 214,
      219–224
  Ellenburger Formation in, 219
  Palo Pinto Formation in, 219
Constituents, carbonate. *See also* Minerals,
      metastable
  biological, 9, 16
  chemical, 9
  depositional sedimentary, 108
  grain types, 15
  mineralogical, 15
  nonskeletal, 2
  skeletal, 20
Contact inhibition, 152
Converting MICP data to oil-water equivalents,
      69–70
Coordination number, 66, 72–73
Coriolis force, 127
Correlations
  geochronological, 84
  layer-cake, 99–100
  stratigraphic methods for, 86–87
Cotton Valley Formation (Jurassic) Texas
  "chalky" porosity in, 151
  neomorphic microporosity in, 159–160, 228
  salt domes and sedimentation of, 136

Cow Creek Formation (Lower Cretaceous)
  Texas, 111
Crossplots
  density–neutron, 202
  porosity–permeability, 194
  Schlumberger *M–N*, 202
Cross-cutting relationships
  in rock properties, 2, 156
  in thin sections, 156
Cross sections, structural and stratigraphic, 87
Crystal boundaries, compromise, 151
Crystal forms
  aragonite cements, 148, 167
  calcite cements, 147–148, 167
  dolomite, saddle, 148
  Mg-calcite, 167
Crystal systems of common carbonates, 3
Currents
  contour, 98, 127
  density, 98, 122, 127, 130
  geostrophic, 122, 127, 130
  longshore, 111
  rip, 111
  thermohaline (density), 128
  turbidity, 98, 122, 127
Cycle skipping, acoustic log, 161, 244
Cycles, stratigraphic
  "greenhouse and icehouse climates", influence
    on, 101
  Milankovich, 101
  origins of, 101
  order of, 101
  shallowing-upward, 28, 102

Darcy (laminar) flow, 44, 187
Density, bulk, 31
Depth shifting, core-to-log, 208
Depositional
  bodies, typical shapes of, 88
  dip, 88
  strike, 88
  successions, ideal (standard), 92–93, 96–98,
    106, 203. *See also* Ideal depositional
    successions and environments
Detached
  beach–dune succession, 96, 111
  shoreline (barrier island), 82, 111
Diachroneity, 80
Diagenesis, 9
  bioerosion, 146
  cementation, 146, 164, 170
  compaction, mechanical, 146, 164, 170–171
  deep burial, 146
  definition of, 145
  dissolution, 146, 150
    and cave formation, 225
    mesogenetic, 157, 160

fresh water, 116
  inversion, mineralogical, 159. *See also*
    Neomorphism
  marine phreatic, 116
  mechanisms of, 146
  mixing-zone, 116
  neomorphism, 148, 159, 165
  recrystallization, 146, 158–159, 170
  replacement, 146, 170
  stabilization, neomorphic, 150, 159
  vadose, 116
Diagenetic environments, 9, 224–225
  classification, basis for, 3, 154
  fresh-water (meteoric) phreatic, 153, 156
  marine phreatic, 153
  mixing zone, 153
  subsurface (burial), 153, 167
  vadose, 153, 156, 167
Diagenetic facies, mapping of, 155
Dickinson field (Mississippian) North Dakota,
    244–249. *See also* Williston Basin
  Lodgepole Formation (Carboniferous) in,
    244
  Lodgepole mounds in, 245
Discoasters, 131. *See also* Chalk
Disconformities, stratigraphic, 101
Diversity, taxonomic, 98, 115
Dolomite, "hydrothermal", 148
Dolomicrites, 26
Dolomitization
  and reservoir porosity, 151–153
  "excess", porosity reduction by, 152
Drill cuttings, microscopic examination of, 9
Drilling breaks, 161, 193. *See also* Fractures,
    presence in borehole, indirect evidence of
Dukhan field, Qatar, 191
Dunes, coastal, 109
Dysoxia, 98

El Abra Formation (Mexico), 46
Enterolithic structures, *see* Environments, tidal
    flat and lagoon
Environments. *See also* Ideal depositional
    successions and environments
  abyssal, 98
  "always wet", 112
  aphotic, 98
  basinal, 129–133
  bathyal, 98
  beach–dune–barrier island, 110–117
  diagenetic, *see* Diagenetic environments
  shallow subtidal (neritic), 121–124
  slope and slope-toe, 126–129
  slope-break, 124–126
  temperate, 81, 112–113, 115
  tidal flat and lagoon, 117–121
  tropical, 113

Epicontinental seas, 129
Events, climatic, storms, tropical and
    "northers", 111

Fabric, 15
    biogenic, 18
    depositional, 18
    diagenetic, 18
Facies
    basinal, 129
    biological, 91
    defined, 91–92
    depositional, 8, 81
    diachronous, 83
    electro, 10, 49, 52
    eolian, 114
    high energy, 78
    micro, 91
    pore, 69, 156
    standard micro, 92–93
    time-transgressive (diachronous), 84, 94
Factor analysis, 91
Faults
    listric normal, 185
    graben-in-graben, 185
Fields, giant and supergiant, 226
Fizz test, to distinguish calcite from dolomite, 2
Flooding surface
    marine, 101
    maximum, 101
Florida
    Key Largo, 93
    keys, 80
    shelf, 93
    White Bank, 93, 125–126
Flow units, 1, 26, 41, 107, 145, 250
    mapping of, 173
Fluid
    flow, parallel plate theory of in fractures, 187
    nonwetting, 46
    recovery factor, in fracture systems, 182
    saturations, in fractures, 182
    wetting, 46, 57
Folds
    anticlinal, fractures on, 185
    monoclinal flexures, fractures on, 185
Fractures
    classification of, genetic, 178–181
    conjugate shear, 178
    in Cretaceous carbonates, Lake Maricaibo
        area, 240
    differential compaction and, 181, 245
    extension and tension, 178
    four types, Nelson's, 190–191, 251
    induced and natural, 188
    intensity, 192
    morphological types of, 182

presence in borehole
    direct confirmation of, 192
    indirect evidence of, 192–194
    slickensided, 182
    spacing, 186, 188
    spacing and intensity, factors that influence,
        195
    surface-related, 181
    on tectonic structures, orientation of, 180
    trends in natural, 9
    types I- IV, reviewed, 239–240
    width, 186
Fragum hamelini, 203

Gahwar field (Jurassic) Saudi Arabia, 103
Gas, as nonwetting fluid, 63
Golden Lane trend (Mexico), 46
Great Bahama Banks, 80, 134
    Eleuthra Island, 222
    Exuma Sound, 222, 231
    Schooner Cays, 231
Great Salt Lake, Utah, 109
Grain-to-mud ratio, 27
Grain size (texture)
    beach-dune deposits, 112
    categories in Grabau's rock classification, 15
    categories on Wentworth scale, 15
    measurement techniques, 15
Gravity, measurements of in exploration, 8
Green River Formation (Eocene), 109
Guadalupe Mountains, New Mexico, 226
Guymon-Hugoton field (Permian), 102
Gypsum
    presence of and log calculations, 51
    dewatering (transformation) of, 166

Haft Kel field, Iran, 191
Halokinetic (salt tectonic) structures, 214
Happy field (Permian, Clearfork Formation),
    Texas, 231
Hardgrounds, 155
Hassi Messaoud field, Algeria, 191
Heterozoan biota, 123, 247
HFS, see High-frequency depositional
    sequences
High-frequency depositional sequences (HFS),
    99
High-stand systems tract (HST), 212
Horner plot, 194, 251 see also Fractures,
    borehole, indirect evidence of
"Hot" lime and dolomite, 202
HST, see High-stand systems tract

Ideal depositional successions and environments
    basinal, 129–133
    beach-dune, 110–117
    illustrations of all, 133–139

shallow subtidal (neritic), 121–124
slope and slope toe, 126–129
slope-break, 124–126
tidal flat and lagoon, 117–121
Image analysis, petrographic (PIA), 209
Impedance contrast, 8, 53, 206
Isooctane, in wettability experiments, 63

James Limestone Formation (Cretaceous)
    Texas, 124

Karst, 147
    caves, 162
    caverns, 147, 162
    paleocaves, 226
        as reservoirs, 162–163
    pinnacles, 162
    Puckett field (Lower Ordovician) Texas, 162
    sinkholes, 155, 162
    towers, 162
    Yates field (Permian) Texas, 162
Kerogen, 145
Keuper, evaporites, Triassic (Germany), 87
Kirkuk field, Iran, 191
Kohout circulation, 155

Lattice, crystal, deformation, types of, 201
Limestones, bituminous, 109
Lisbon field, Utah, 161
"Lith logs", 8
    computer processed interpretation (cpi) log,
        52
    creating synthetic (Schlumberger *M–N* plot),
        51
    from well cuttings, precautions in using,
        207–208
    software applications to compute synthetic, 51
Lithofacies, synthetic, 52
    "STATMIN", computer program for
        generating, 202
Lithogenetic units, 83. *See also* Facies
Logs, wireline
    acoustic, 8, 107, 192
    caliper, 193
    cased hole, 47
    characteristics, as proxies for fundamental
        rock properties, 204
    density, 8, 107
    dipmeter, 18, 107
    FMI®, FMS®, UBI® and fractures, 182, 193
    FMI® output illustrated, 183
    gamma ray, 8,10
    imaging, 9,18,49, 107, 249
    neutron, 8, 107
    NMR, 49, 107, 205, 209
    open hole, 47
    photoelectric effect, 8

resistivity, 8,10
    shapes, as indicators of depositional
        environments in siliciclastics, 49, 202
    "signatures" as indicators of rock and pore
        types, 202
    sonic amplitude, 192
    spontaneous potential (SP), 59
    velocity deviation, 50
Luminescence, cathode (CL), 168, 225

Magetism, earth, in exploration, 8
Maps
    depositional facies, 8
    electrofacies, 10, 49, 107
        statistical (computer processed), 52
    interval isopach (paleostructure), 8, 219
    pore facies, 69
    subsurface, 84
Mariana Trench, 129
Material, behavior of under stress, 44, 177
Megafossils, benthic, 13
Metamorphism
    contrasted with diagenesis, 145
    organic, 145
Micrite, 15
Micritization, 29
Microbes, calcified, 28
Microcalcite, microrhombic, 150–151. *See also*
        Porosity, "chalky"
Microporosity, diagenetic, 91
Microscopy, scanning electron (SEM), 209
Microstructures, internal, 28
Midland Basin, Texas, 231, 233
    Eastern Shelf of, 233, 237
Minerals
    accessory, 201
    metastable
        aragonite, 26, 126, 131, 148, 153, 166, 213
        calcite, magnesian (Mg), 26, 148, 153, 166,
            213
Mineralization
    exotic in reservoirs, 161, 169
    metamorphic, 146
    Mississippi Valley Type (MVT), 146
Muschelkalk, limestone, Triassic (Germany), 87

Naphthenic acid, in wettability experiments, 63
Neomorphism. *See also* Diagenesis
    aggradational, 148
    defined, 159
    degradational, 148
    of limestones, 151
Neritic (shallow subtidal) environment, 97
Nesson Anticline, 248
North Haynesville (Smackover) field, Louisiana,
        223
    depositional reservoir, example of, 214–219

Nuclear magnetic resonance (NMR), 8, 107, 145.
See also Logs, wireline
T2 relaxation time, 53

Oaks field (Smackover Formation), Jurassic,
North Louisiana, 117, 216
barrier island sequence in, 117
Oil
column, calculating height above free water
level, 70
shales, 109
"window", 145
Olenellus, trilobite, 84
OOIP, see Original oil in place
Oozes, siliciclastic, 98
Organic compounds, polar, 63. See also
Wettability
Organic matter, sapropelic, 18
Organisms
porosity in, 108–109
reef-building, 81, 108
rim-forming, 125
Original oil in place (OOIP), calculation of, 61
Orogeny, Ancestral Rocky Mountain
(Carboniferous), 247
Overcompaction and stylolites, 154, 165, 243
Overton field, Jurassic, Texas, 90, 159. 165, 227–
231. See also Porosity, "chalky"

Paleosols, 155
Pathway, diagenetic, 164. See also Cross-cutting
relationships
Periplatform talus, 128
Permeability
absolute, 45
capacity to transmit fluids, 34
Darcy–Ritter equation for (Darcy's law), 31,
186–187
flow test, 194
fracture, 182, 186
intrinsic, 186
matrix, 187
relative, 46
specific, 45
statistical relationship with porosity, 7
Permian Basin, Texas, 128
Persian Gulf, 211
water depth, 129
tidal flats (sebkhas), 117
Petrographic image analysis (PIA), see Image
analysis, petrographic
Petroleum system, elements in, 76
Petrophysical calculations from wireline logs
density, 49
lithology, 49
porosity, 49
resistivity, formation water (R_w), 49

saturation, water (S_w), 49
Photozoan biota, 123
Platform(s)
antecedent, 103
carbonate, defined, 77
carbonate and siliciclastic, slopes on, 127
environmental "cells" or subdivisions on, 81
isolated, 79
margins, bypass and depositional, 127
modern carbonate, examples of, 110
paleotopography of, 135
slope failure, types of, 127
West Florida, 88
Polymorphs, $CaCO_3$, 3
Pore(s), see Porosity
cavernous, 151, 156, 176
channel, 160, 176
facies, 69
fracture, 44, 176–177
geometry, 107, 205
interbreccia (karst), 162
intercrystalline, 51, 144, 151
interparticle, 10
enlarged, 160
intraskeletal, 223
moldic, 10, 147, 151, 156, 160, 162, 226
roughness factor (a), in withdrawal efficiency,
73
vuggy, 10, 147, 151, 156, 160, 162, 176, 226
vuggy and fracture, petrophysical behavior of,
190
vugs, stromatactis, 210, 245, 248. See also
Buildups, carbonate, "mud mounds"
Pore categories (types)
depositional, 14
diagenetic, 14
in detrital rocks, 108
fracture-related, 14
Pore throats, 7
dimensions of, 17
effective radius of, 66
median diameters of, 107
sheet-like, 66
size-sorting of, 66
Pore-to-pore throat (size) ratio, 34
Pore volume, minimum unsaturated, 66
Porosity
Ahr genetic classification of, 26, 42–44
Archie classification of, 35–36
bimodal, 30, 50, 160
"bird's eye", 108
calculated from neutron logs, 194
capacity to store fluids, 34
core (measured), 194
"chalky", 151. See also Cotton Valley
Formation
Choquette and Pray classification of, 36–39

dependence on rock properties, 31–33
diagenetic, 30, 144, 150
effective, 7
equation defining, 31
estimates of reservoir quality based on,
    33–34
fracture, 182, 186
    calculating saturation (S_w) in, 189
    scale-dependency of, 188
Lonoy classification of, 40–41
Lucia classification of, 39–40, 108
proxies for, 26
reduced during burial, 33
sandstone, 9
secondary index (Schlumberger SPI), 50
separate vug, influence on Archie m
    exponent, 60
total, 107
Poza Rica trend (Mexico), 46
Pressure
    buildup tests, 193, 251. *See also* Fractures,
        presence in borehole, indirect evidence of
    communication, 246
    confining, subsurface, 177, 187
    displacement, capillary, 66–67
    entry, capillary, 66
    threshold, capillary, 66
    transient test, 8, 246
Properties, rock
    capillary, 57. *See also* Capillary pressure
    dependent (derived), 14, 30, 204
    fundamental (intrinsic), 14, 116, 202
    primary, 13, 106
    secondary, 13–14
    tertiary (latent), 14, 47

Quanah City field, Texas, 241–244
    Chappel Formation (Mississippian) in, 241
    Ellenburger Formation (Ordovician) in, 241
    Spiculiferous zones in, 241

Ramp(s)
    distally-steepened, 78, 82
    environmental subdivisions on, 121
    homoclinal, 78, 82
    inner, 121
    middle, 121
    outer, 121
    slope angles on, 78, 82
Reef(s)
    conditions favorable for growth of, 82
    framestone, 125
    framework/detritus ratio, 30
    patch, 82, 124
    Stuart City trend (Cretaceous) Texas, 126
    trends, continuous, 82
Reflux, brine and dolomitization, 152, 225

Reservoir(s)
    characterization, 5
    compartmentalized, 170
    depositional, checklist for identifying and
        exploiting, 136–140
    description, 5
    diagenetic, checklist for identifying and
        exploiting, 172–173
    engineering, 5
    facies-selective, 2
    fabric-selective, 2, 36–37
    fracture, checklist for identifying and
        exploiting, 195
    fractured, defined, 177
    geology, 5
    hybrid
        depositional-diagenetic (type I), 106
        diagenetic-fracture (type II) 176
    net pay calculations in, 87
    net sand calculations in, 87
    oil-wet, 63
        values of saturation exponent in, 62
    quality (rankings), 17
        slice-map method, 234–236, 250
    recovery efficiency, 71–72, 73
        in karst reservoirs, 162
    rocks, multicomponent, 201
    stratabound, 2
    visualization of, 3D, 234
    water-wet, 63
Resistivity
    flushed zone (R_{xo}), 58
    formation at 100% water saturation (R_o),
        58
    formation water (R_w), 59
    invaded zone (R_i), 58
    true formation (R_t), 58
Rhizocretions, 115
"Roaring 40s" latitude (environment in), 124
Rock typing, 34–35, 107, 205
    Winland "R 35", 205
Rock units
    hierarchical classification of, 83
    time-transgressive, 88
Rudstone, 15

Sabine Uplift, ancestral, 228, 231
Sacramento Mountains, New Mexico, 225
Saddle dolomite
    late, in fractures, 148, 166, 248
    late diagenetic, 238, 243
    thermochemical sulfate reduction (TSR) and,
        148, 166
Saint Louis Limestone Formation, 224. *See also*
    Conley field
Salinity, hyper and hypo, 123
Salt Basin, East Texas, 231

San Andres Formation (Permian), Texas and
New Mexico, peritidal deposits in, 117
Sand waves, Great Bahama Banks, 126
Sands, tight gas, 46
Saturation, 56
equilibrium and diagenesis, 147
oil, 57
water, 57
effective, 62, 160
total, 62, 160
Scaling-up, pore-to-reservoir, 206–207
Scanning electron microscopy (SEM), *see*
Microscopy, scanning electron
Sealing capacity (h_{max}), calculating, 70–71,
250
Seals, 5, 69
anhydrite plugging, 166
Sediment production vs retention, 81
Seepage reflux, *see* Reflux, brine and
dolomitization
Seismic
amplitude versus offset (AVO) analysis,
191
attributes, 8, 191, 244
data, 3D, 8
reflections, 53, 206
traces, as correlation aids, 88
velocities, 54
wave characteristics, 53
Seismology
reflection, 8
3D, 53
Separability, limit of, 54, 206
Shelves
environmental subdivisions on, 81
Guadalupian (Permian),West Texas–New
Mexico, 126
open, defined, 81
rimmed, 79–80, 93
rims, absence of, 81
South Florida, 125
Shoals, grainstone, 124
Siliciclastic sandstones
diagenesis in, 9
ideal depositional successions (models) in, 95,
203
Slickensides, 182
Slope breaks, deep water, 82
Smackover Formation (Jurassic)
Alabama
cyclicity in, 170
pore facies in, 69
Arkansas, capillary pressure curves from, 67
Gulf Coast, 61, 211
Louisiana, 161
salt tectonics and depositional patterns in,
136

Snap-off, 72
Source rocks, 5
Spits, barrier, 116. *See also* detached shoreline
Spraberry Sandstone trend, Texas, 191, 233.
See also Happy field
Spur and groove structures, 29, 125
Stratigraphic
Code of Stratigraphic Nomenclature, 83
correlation, 85
International Guide To Stratigraphic
Classification, 83
mechanical, influence on fracturing, 185
stacking patterns, sequence, 102
Stratigraphy
allo, 100
cement, 168, 171, 225
chrono, time and time-rock units in, 83
defined, 77
genetic, 100
litho, rock units in, 83
parasequences, 101–102
seismic, 53, 99
sequence, 53, 80, 84
defined, 99
Strain, defined, 177
Stress
compression, 178
defined, 177
extension, 178
principal, intermediate (σ_2), maximum (σ_1),
minimum (σ_3), 178
shear, 178
Structures, sedimentary, 15, 20
in beach-dune deposits, 113–114
in tidal flat and lagoon deposits, 119–121

Tension
adhesion, 62, 65
interfacial, 62, 66
surface, 63
Texture. *See also* Grain size
depositional, 9, 15
grain packing, 17
grain shape, 17
sorting, 17
Thamama Group (Cretaceous) Middle East,
reefs and grainstones in, 126
Tidal prism, 119
Time, geological
absolute, 84–85
relative, 84–85
ways to measure, 84–85
Time-rock units, classification of, 86
Trace fossils, in basinal environments, 130
Transcontinental Arch of North America,
Cambro-Ordovician tidal flat deposits on,
120

Transgressive systems tract (TST), 212, 247
Traps, 5
TST, *see* Transgressive systems tract
Turbidites
   channelized, proximal, 128
   distal, 131
"Turtle" structure, 124

Unconformities, 84, 90, 101, 155

Vents and seeps, seafloor, 82

Walther's rule, 93
Washouts, wellbore, 161, 244
Water
   bearing zone in reservoir, 57
   conate, 57
   formation, resistivity of, 59
   interstitial, chemical composition of, and
      cement mineralogy, 167
   subsurface (burial diagenetic), composition
      of, 154
Waulsortian (Mississippian) mounds, 243, 247.
     See also Buildups, carbonate
Wave(s)
   Airy, 111
   base, fair-weather, 78
   breaking, 111

climate, factors determining, 111
   period, 111
   shoaling transformation, 111
   solitary, 111
Wells, horizontal in fractured reservoirs,
   244
Wettability, 56, 62
Williston Basin
   fractured Mississippian "mud mounds" in the,
      181, 185, 240, 244
   Paleozoic tidal flat deposits in the, 119
Winnowing, 18
Withdrawal. *See also* Reservoir, recovery
   efficiency
   curve, capillary pressure, 71
   efficiency, mercury, 71

Yucatán
   Campeche Bank, 89
   fractured reservoirs in, 240
   Isla Cancun, 89
   Isla Mujeres, 89
   oolite grainstones on NE coast of, 89, 115

Zone, reservoir
   productive, 57
   transition, 57
   water-bearing, 57

Printed in the United States